Sustainable Feed Ingredients in Freshwater Aquaculture

Sustainable Feed Ingredients in Freshwater Aquaculture

Guest Editors

Zsuzsanna Sandor
Csaba Hancz

MDPI • Basel • Beijing • Wuhan • Barcelona • Belgrade • Novi Sad • Cluj • Manchester

Guest Editors
Zsuzsanna Sandor
Institute of Aquaculture and Environmental Safety
Hungarian University of Agriculture and Life Sciences (MATE)
Szarvas
Hungary

Csaba Hancz
Institute of Aquaculture and Environmental Safety
Hungarian University of Agriculture and Life Sciences (MATE)
Kaposvár
Hungary

Editorial Office
MDPI AG
Grosspeteranlage 5
4052 Basel, Switzerland

This is a reprint of the Special Issue, published open access by the journal *Animals* (ISSN 2076-2615), freely accessible at: https://www.mdpi.com/journal/animals/special_issues/YZ0NJ4AF8R.

For citation purposes, cite each article independently as indicated on the article page online and as indicated below:

Lastname, A.A.; Lastname, B.B. Article Title. *Journal Name* **Year**, *Volume Number*, Page Range.

ISBN 978-3-7258-6284-9 (Hbk)
ISBN 978-3-7258-6285-6 (PDF)
https://doi.org/10.3390/books978-3-7258-6285-6

Contents

About the Editors

Zsuzsanna Sandor

Zsuzsanna Sandor is a senior researcher at the Research Centre for Fisheries and Aquaculture, part of the Hungarian University of Agricultural and Life Sciences (MATE). Her work involves assessing alternative protein and oil sources for aquafeeds and their effects on fish metabolism and physiology. She is performing in vivo digestibility studies and conducting nutritional trials to determine the optimal mixing ratio of these ingredients in the diets for freshwater fish. This process involves the evaluation of various factors, including production and feed utilization parameters, the immune status and the function of the antioxidant system, fatty acid metabolism, and the chemical composition of fish flesh. In addition, she focuses on novel feed additives, such as immunostimulants, that play a role in protecting fish health. Throughout her research on fish nutrition, chemical, biochemical, and molecular analyses are conducted.

Csaba Hancz

Csaba Hancz is currently Professor Emeritus at the Kaposvár Campus of the Hungarian University of Agricultural and Life Sciences. He began his research and teaching career at Kaposvár Agricultural College in 1977, following a six-month internship at the Bikal State Farm fishery. Between 1986 and 1990, he worked in Brazil as a fish culture specialist with the AGROBER-AGROINVEST team under Professor Elek Woynarovich. After returning to Hungary, he resumed teaching and research at Kaposvár, which later became an independent university. He retired as professor emeritus in 2015, with a primary research focus on fish nutrition, and continues to mentor PhD students in aquaculture.

animals

MDPI

Review

Fish Responses to Alternative Feeding Ingredients under Abiotic Chronic Stress

Julieta Sánchez-Velázquez [1], Guillermo Abraham Peña-Herrejón [2] and Humberto Aguirre-Becerra [3,*]

1 Facultad de Ingeniería, Campus Amazcala, Universidad Autónoma de Querétaro, El Marqués 76265, Querétaro, Mexico; julieta.sanchez@uaq.mx
2 Centro de Investigación y Desarrollo Tecnológico en Materia Agrícola Pecuaria Acuícola y Forestal (CIDAF), Facultad de Ingeniería, Universidad Autónoma de Querétaro, Campus Concá, Arroyo Seco 76410, Querétaro, Mexico; guillermo.pena@uaq.mx
3 Cuerpo Académico de Bioingeniería Básica y Aplicada, Facultad de Ingeniería, Campus Amazcala, Universidad Autónoma de Querétaro, El Marqués 76265, Querétaro, Mexico
* Correspondence: humberto.aguirreb@uaq.mx

Simple Summary: Environmental factors are considered abiotic stressors when they negatively affect the integrity and welfare of fish. In aquaculture, many stressors occur under intensive fish farming (e.g., poor water quality, handling, crowding, and transportation). Abiotic chronic stress results in higher mortality and disease incidence, lowering fish growth and affecting aquaculture production systems' performance. Feeding strategies can alleviate abiotic chronic stress in fish. Novel ingredients possess nutrients that confer antioxidant potential, improve the immune system, and reduce cortisol in blood plasma, the primary hormone in stress conditions. This management of chronic stress by alternative feeding is a sustainable form of obtaining food for human consumption, considering the animal welfare in intensive production systems. Alternative ingredients such as animal by-products, bacteria, fungi, and insect meal are described and discussed in this review as an alternative to mitigate fish chronic stress in aquaculture.

Abstract: Aquaculture has become one of the most attractive food production activities as it provides high-quality protein for the growing human population. However, the abiotic chronic stress of fish in intensive fish farming leads to a detrimental condition that affects their health and somatic growth, comprising productive performance. This work aims to comprehensively review the impact of alternative and novel dietary protein sources on fish somatic growth, metabolism, and antioxidative capacity under environmental/abiotic stressors. The documental research indicates that ingredients from rendered animal by-products, insects, bacteria as single-cell proteins, and fungal organisms (e.g., yeast, filamentous fungus, and mushrooms) benefit fish health and performance. A set of responses allows fish growth, health, and survival to remain unaffected by feeding with alternative ingredients during chronic environmental stress. Those ingredients stimulate the production of enzymes such as catalase, glutathione peroxidase, and selenoproteins that counteract ROS effects. In addition, the humoral immune system promotes immunoglobulin production (IgM) and cortisol plasmatic reduction. Further investigation must be carried out to establish the specific effect by species. Additionally, the mixture and the pre-treatment of ingredients such as hydrolysates, solid fermentations, and metabolite extraction potentialize the beneficial effects of diets in chronically stressed fish.

Keywords: alternative feeding; chronic stress; cortisol; environmental stressors; animal welfare

1. Introduction

Abiotic stressors are environmental factors that affect fish growth, health, and survival, resulting in decreased production, profit, and product quality. Abiotic stress requires special attention in aquaculture [1–3]. Fish cells, tissues, and organs are negatively affected when any abiotic factor, such as oxygen concentration, temperature, or salinity, is far from optimal [4,5]. A stress response occurs when the homeostatic equilibrium of an organism is altered due to intrinsic or extrinsic stimuli that trigger the General Adaptation Syndrome (GAS). The GAS is a three-phase cascade of physiological responses: (1) Activation of the adrenergic system through sensory organs to increase plasma adrenaline and noradrenaline (catecholamines) [6]. (2) Hypothalamic–Pituitary–Interrenal (HPI) axis activation for plasma cortisol liberation, which, together with catecholamines, promote physiological modifications that include metabolic, hydromineral, hematological, immunological, and structural changes to mobilize energy resources [7]. In this phase, the stress responses are reversible, and the fish may return to a healthy pre-stress condition [8]. (3) The exhaustion phase results from continuous stress [2], where the energy and body resources meet the energy demand associated with the stress instead of being used in vital life processes, such as reproduction and growth, with detrimental consequences [9].

The stress response depends on fish adaptability, species, and stressor persistence or duration, which distinguishes acute from chronic stress [3,10–12]. Acute stress is a low-stress situation that lasts for a short time (minutes to hours) and may have stimulating, beneficial (eustress), or compensatory responses. In contrast, chronic stress lasts for days, weeks, or months and overstimulates coping responses that can negatively affect organisms (distress) [13,14]. Fish adaptability relates to regulatory mechanisms that compensate for the stressor effects [15]. Chronic stress must be considered a severe problem in aquaculture, which can be mitigated through sustainable management strategies (e.g., feeding and water quality control) that improve fish performance and welfare [16].

Dietary interventions have only been developed to manage acute stress in fish rather than chronic stress [17–20]. Feeding strategies to mitigate the effects of stress in aquaculture are explained by Kumar et al. (2015) and classified into nonchemical (biological) and chemical. Nonchemical methods involve environmental management (i.e., temperature, dissolved oxygen, ammonia, stocking density), while chemical methods include dietary supplementation [21]. Ciji and Akhtar (2021) classified feeding additives into nutritional (protein, amino acids, fatty acids, phospholipids, vitamins, minerals, carotenoids, nucleotides) and non-nutritional (biological products: prebiotics, probiotics, bacterial derivates, animal and plant extracts, yeast and its derivates, lactoferrins, micro/macroalgae, and polysaccharides; synthetic chemicals: propylene glycol, organic acids, clay additives, levamisole) [22].

Aquaculture nutritionists are developing diets with alternative ingredients to replace fishmeal with animal by-products, microalgae, bacteria, fungi, and insects. These inputs contain the required protein level for fish and nutrients such as antioxidants or immunostimulants, allowing the formulation of specific diets to manage chronic stress. However, there are few studies on the effect of alternative ingredients in aquaculture diets on fish's health, growth, performance, and welfare as a possible strategy for alleviating chronic stress. Therefore, this article aims to review the use of dietary intervention in chronic stress under abiotic factors. First, the results from scientific evidence about the physiological, metabolic, and growth performance under abiotic chronic stress and dietary interventions with additives are summarized. Then, alternative ingredients such as animal by-products, bacteria, fungi, and insect meal are described and discussed to help cope with the environment or some abiotic stressors in fish under experimental conditions. Finally, a perspective on the feeding strategies to mitigate chronic stress in aquaculture is discussed.

2. Abiotic Chronic Stress Response and Feeding Intervention

The stress response of fish is a set of changes in their organism that neutralize the stressor to regain homeostatic equilibrium [23]. Many authors have pointed out several abiotic stressors and their harmful effects on fish., e.g., high-salinity and high-temperature stress on African catfish (*Clarias gariepinus*) [24], netting, confinement, flashlight, salinity and hypoxia challenges on European sea bass (*Dicentrarchus labrax*) [25], salinity or seawater culture on milkfish (*Chanos chanos*) [16], and chronic hypoxia stress on Nile tilapia

(*Oreochromis niloticus*) [26]. However, authors do not use the term "abiotic stressor"; they usually use "environmental stressors", "husbandry stressors", and "water quality that includes physical or chemical factors" (Table 1). This review uses the term "abiotic stressor" to refer to any environmental factor negatively impacting fish.

Table 1. Terms used in literature to refer to abiotic stressors.

Stressors	Terminology	Reference
Netting, confinement, and flashlight	Chronic stress protocol	[25]
Temperature, pH, turbidity, toxicants, pathogens, predators, handing	Environmental factors	[14]
Temperature	Water quality variables	[27]
Cold current, high temperature, low precipitation, ammonia, harmful algal blooms, warming	Abiotic factor	[28]
Chemicals (insecticides, pesticides), dissolved oxygen, pH, temperature	Abiotic factor	[11]
Temperature, chemical contamination, photoperiod, salinity	Abiotic factor	[29]

The negative impact of chronic stress caused by abiotic factors includes (1) somatic growth, (2) metabolism, and (3) oxidative stress. A search of scientific studies on chronic stress in fish due to abiotic factors was performed to understand how dietary interventions modulate this condition. Two groups of keywords were used: (1) "chronic stress", "feed", "diet", "aquaculture", and (2) "chronic abiotic stress", "feed", and aquaculture between 2017 and 2023. A total of 15,060 and 2240 articles were found, respectively. Articles in which nutritive additives (e.g., amino acids, fatty acids, phospholipids, vitamins, and minerals) were used in diets with positive results in mitigating chronic stress on fish under abiotic stress were also considered (Table 2).

2.1. Somatic Growth

In aquaculture, faster fish growth is one of the most critical selection traits [30]. Stress is one of the most important physiological factors affecting growth. Several biotic and abiotic stressors (e.g., routine hatchery practices such as handling and sorting, poor water quality, and overpopulation) suppress growth [30]. Fish growth is the increase in length and weight, where muscle contributes over half of the body mass; thus, changes in muscle size are of primary importance to growth. Few studies have addressed the stress effect of cortisol stimulation on fish muscle function. Most have focused on the liver as the metabolic target for stress adaptation and cortisol action. Growth hormones (GHs) directly control somatic growth through receptors on the sarcolemma muscle and indirectly by initiating the production and release of insulin-like growth factor (IGF) in the liver and peripheral tissues [31]. IGF is a somatomedin produced in the liver that regulates the growth and metabolism of fish. The IGF system comprises IGF-I, IGF-II factors, several receptors IGF-IR and IGF-IIR, and six binding proteins (IGFPs). The IGF system is one of the central pathways that regulate protein synthesis in skeletal muscle [31]. The IGF-I and IGF-II factors bind to IGF-IRs (IGF receptors), which activate the PI3K-Akt-TOR pathway. This action results in translation and protein synthesis to form muscle by stimulating cell proliferation, differentiation, hypertrophy, and inhibiting muscle atrophy [32]. However, cortisol inhibits protein synthesis by decreasing the myogenic gene expression and reducing the action of the GH–IGF axis [30]. Therefore, blocking cortisol in stressed fish mitigates adverse effects on muscles.

Tryptophan (Trp) is an essential amino acid metabolized into serotonin, melatonin, and kynuenine [33]. Serotonin directly blocks cortisol [34], while kynurenine provides energy substrates, such as Acetyl CoA and pyruvate, that enter the Krebs cycle to respond to the energy demand under stress [35]. In this sense, *Oncorhynchus mykiss* fed with the Trp diet significantly reduced stress-induced plasma cortisol [36]. *Labeo rohita* fed with a 1.4% Trp-enriched diet and exposed to temperature and salinity stress significantly increased the weight gain to 21.46 ± 1.00% compared to fish fed without Trp under both stress conditions with 12.72% [37]. Dietary Trp serves not only for synthesizing neurotransmitters but also for body protein deposition. Trp conference resistance and promote fish growth despite stress. Herrera et al. (2020) observed the maximum values for final weight when meager *Argyrosomus regius* was fed with a Trp-supplemented diet and subjected to chronic stress (Table 2) [38]. Additionally, Trp supplementation counteracts the stress-induced elevations of plasma cortisol in rainbow trout (*Oncorhynchus mykiss*) under high stocking density [36].

Phenylalanine (Phe) is an essential amino acid metabolized through two metabolic pathways: Tyrosine (Tyr) oxidation and transamination to phenylpyruvate. Tyr is a direct precursor of the catecholamine hormone, a compound related to stress response, and the hydroxyphenyl pyruvate as part of the energy metabolism [39]. Phe and Tyr provide the energy demand when fish are under acute or chronic stress, but they also boost fish growth under chronic stress and reduce stress markers such as cortisol [40]. Under chronic stress, Phe enriched diet in Gilthead seabream (*Sparus aurata*) decreases cortisol levels but decreases fish growth. Authors suggest an unbalanced diet (amino acid excess) results in malnutrition and reduced growth [41]. Including amino acids guarantees positive effects on fish growth during chronic stress.

2.2. Metabolic and Cellular Markers

The stress response involves changes in metabolic pathways [42]. Cortisol modulates energetic metabolism. Under stress conditions, the increase in plasma cortisol promotes protein, lipids, and glucose mobilization in the skeletal muscle, which is the principal target of cortisol [43]. A critical metabolic role for cortisol during stress is elevating circulating glucose levels by increasing hepatic gluconeogenesis to fuel the increased energy demand, especially during environmental fluctuation and stress response [44]. The preceding is accompanied by enhanced muscle protein degradation, which increases the availability of amino acids as substrates for gluconeogenesis in the liver [45]. In addition, this provides energy to organs, such as the brain, to overcome the stressor. Cortisol exerts a critical control over skeletal muscle glucose metabolism, inhibiting glucose uptake and utilization and glycogen synthesis [43].

Cortisol in the blood provides information on the stress response. The average resting levels of plasma cortisol in fish can be as low as 13.8 nM, while fish with chronically activated stress response can have a resting level greater than 27.5 nM [46]. However, cortisol cannot be just related to negative stressors; not all cortisol fluctuations indicate distress. Thus, it is essential to differentiate the baseline from a standard range of stress-related concentrations [29]. Cortisol is a hyperglycemic hormone as it causes the elevation of plasma glucose, providing fish with energy to escape or cope with an adverse situation [47]. Elevation in plasma cortisol stimulates glycogenolysis, i.e., the conversion of glycogen stored in the tissue to glucose released into the blood. Increased plasma glucose in the blood is a relatively slow response to a stressor, and pre-stress plasma glucose levels are usually between 3.7 and 4.6 mM [46].

Lactate is produced by anaerobic ATP production (glycolysis) when oxygen is unavailable for cells during aerobic metabolism. The peak of plasma lactate during stress events ranges from 6.4 to 13.3 mM (normal condition 3 to 5 mM). Cortisol in plasma stimulates anaerobic metabolism that results in lactic acid accumulation. In normal conditions, most lactate is produced by the erythrocytes; however, during intense physical activity, the muscle produces large amounts of lactate because of insufficient muscle oxygenation [47]. Salamanca et al. (2022) showed the reduction of lactate in *Sparus aurata* subjected to chronic stress and fed with a Phe-enriched diet [41]. Additionally, Herrera et al., 2020 showed the same effect of Trp-enriched diets in *Argyrosomus regius* [38] (Table 2). Finally, Longbaf Dezfouli et al. (2019) showed the reduction of plasmatic glucose, increased weight gain, final weight, and feed intake in *Lates calcarifer* subjected to freshwater as an abiotic stressor and including magnesium and selenium nanoparticles as alternative ingredients in the fish diet [48].

2.3. Oxidative Stress

Organisms subjected to stressors undergo physiological changes and activation of biochemical pathways mediated by the neuroendocrine system. Additionally, cellular stress includes oxidative stress caused by overproduced reactive oxygen species (ROS) in fish [49,50]. Oxygen is the primary biological acceptor of electrons with a vital role in cellular functions. However, it contributes to the undesirable formation of ROS, such as superoxide, hydrogen peroxide, and hydroxyl radical. ROS can be classified as free radicals or non-radicals [5]. Organisms have enzymatic and non-enzymatic antioxidant defenses to minimize the damaging effects of ROS. Enzymatic antioxidants include superoxide dismutase (SPD), catalase (CAT), glutathione peroxidase (GPx), glutathione reductase, and glutathione-S-transferases. In contrast, non-enzymatic antioxidants are vitamin E, vitamin C, β carotene, vitamin A, glutathione, flavonoids, thiols, coenzyme Q, and uric acid [5].

Cellular defense pathways in fish are triggered by oxidative stress due to acute and chronic unfavorable water conditions. The primary cellular markers during oxidative stress include the nuclear factor erythroid 2 (NRF2), HIF1α, heat shock protein 70 (HSP 70), tumor necrosis factor-alpha TNFα, and nuclear factor kappa-light-chain-enhancer of activated B cells [4]. Kulczykowska (2019) suggests that mucosal skin markers activate during oxidative stress in fish (e.g., melatonin (Mel) and its active metabolites) [51].

Research supports that dietary intervention with certain nutrients modulates chronic stress. Harsij et al. (2020) proved that diets enriched with selenium, vitamin C, and E for *Oncorhynchus mykiss* under ammonia exposure had positive effects on the antioxidant response as the levels of aspartate aminotransferase (AST), alanine transaminase (ALP), and alkaline phosphatase (ALT) were decreased [52]. Authors suggest that selenium protects fish against stress by forming seleno-proteins such as GPx and thioredoxin reductase. Additionally, juveniles of Wuchang bream (*Megalobrama amblycephala*) under ammonia stress fed with dietary selenium, yeast, and polyphenols showed decreased levels of cortisol, TNFα, IL-1β, GPx, and increased CAT [53].

Agavine, a bioactive compound from agave agro-industrial wastes, effectively functioned in juvenile Nile Tilapia (*Oreochromis niloticus*) stressed by high-density stress (63 kg/m^3) for 20 days. Stressed fish fed with agavine presented low plasmatic cortisol and high levels of SOD and CAT, promoting the antioxidant effect of fructants that produce fermented metabolites [54]. Cruz-Marín et al. (2023) used inulin, a nondigestive additive, in native fish tropical gar (*Atractosteus tropicus*) larvae. Fish fed with 2.5% inulin showed diminished intestinal activities of GPx, SOD, and CAT, which resulted in improved antioxidant status. Additionally, the larvae survival increased as the concentration of inulin increased [55]. In Nile Tilapia (*Oreochromis niloticus*), similar effects of dietary inulin were observed under 16 hypersaline-stressed conditions for eight weeks, similar to a chronic state as CAT and GAST diminished in the gut [56].

More examples of how dietary interventions with certain additives modulate chronic stress are shown in Table 2.

2.4. Immune Response

There is much information about the effects of additives on fish immune responses to prevent health care by dietary manipulation. This strategy is supported by substantial evidence that nutrition is an essential modulator of the fish immune system [57]. In this sense, in aquaculture practices, knowing those dietary additives that positively affect stressed fish at a chronic level is crucial and allows the possibility to cope with any threat. Acute stress stimulates the fish immune response by showing an activating phase that significantly enhances innate responses, such as recruiting subsets of immune cells and the presence of cytokines proinflammatory such as interleukin-8 (IL-8), C-C motif chemokine ligand 2 (CCL2), interleukin -2 (IL-2), interferon-gamma (IFN-γ), and TNF-α [6]. At the same time, cortisol in the blood is often observed under acute stress, and in conjunction with IL-1β and TNF-α is a critical mediator of acute inflammatory reactions to counteract external challenges [58]. After this process, anti-inflammatory cytokines down-regulate inflammation. However, chronic stress and its primary signaling mediator, cortisol, have traditionally been understood to have an overall suppressive effect on fish immunity, reducing antibody production, leukocyte mitosis, phagocytosis, and leukocyte trafficking. Thus, the continuous presence of cortisol promotes the expression of proinflammatory cytokines such as IL-1β, TNFα, and IL-6 [6,59–61].

Cytokines play an essential role in the innate adaptive and immune response. A series of sensitive molecular reactions regulate relative signal transduction from environmental stress to inflammatory response. SOCS (suppressor of cytokine signaling) family members function as feedback inhibitors of cytokine signaling by inhibiting various signal transduction pathways. SOCS1, the prototype of SOCS members, is deeply involved in negative regulations of cytokines through the Janus kinases (JAKs)—signal traducers and activators of the transcription (STATs) pathway. Elevated cortisol due to stress response can inhibit cytokine signaling by binding to glucocorticoid receptor (GR), suppressing the JAK/STAT3 pathway, enhancing the SOCS1 expression, and decreasing proinflammatory cytokine production [59]. Then, the stress response is mediated by the close interaction of hormones and cytokines [60]. Dai et al. (2023) noticed that 21 days of chronic stress in *Carassius gibel* promoted systemic inflammation, and fish showed a significant increase but low fold changes in IL-1β, IL-6, and IL-8 in the trunk kidney, spleen, intestine, and liver. The increased amplitude of all detected proinflammatory cytokines remained relatively low, suggesting sustained low-grade inflammation. Midgut samples manifested morphological changes, including damage in obliterated villi, decreased number of goblet cells, and loosened submucous layer. Microbiota also changed, reducing *Bifidobacterium*, *Lactobacillus*, *Faecalibacterium*, and *Flavobacterium* [62]. These results showed the detrimental effects of chronic stress in fish, the importance of microbiota, and that any damage in the gut might directly affect fish's immune status [63]. In the same way, low-grade inflammation differs from normal inflammation in that there are no typical signs of inflammation. However, it is also similar in involving the same inflammatory mediators and signaling pathways. Having unraveled the communication pathways between the brain and the gut, the gut microbiota has emerged as an essential component affecting all neuroimmune endocrine pathways. Microorganisms are vital in controlling stress and the gut–brain axis under stress conditions [64]. Increasing attention has been paid to the microbiota–gut–brain (MGB) axis, and its potential role in the immune system is a crucial communication pathway between the gut microbiota and the brain [63].

Nutritionists are working on immunomodulation to promote intestinal microbiota by enhancing protection through dietary supplementation using nondigestive ingredients. Prebiotics are non-digestible food ingredients that beneficially affect the host by selectively stimulating the growth and activity of health-promoting bacteria in the intestinal tract [65]. Fructooligosaccharides (FOS), mannooligosaccharides (MOS), and inulin are the most studied prebiotics in fish, while short-chain Fructooligosaccharides (scFOS), galactooligosaccharides (GOS) and xylooligosaccharides (XOS) have been much less studied [63,66]. Asencio-Alcudia et al. (2023) demonstrated the protective role of dietary α-Tocopherol, the most active form of vitamin E, in Longfin Yellowtail (*Seriola rivoliana*). Fish fed with 500 mg/kg α-Tocopherol show an enhanced activity of the anti-inflammatory process by over-expressing immune system genes IL-10 in the spleen due to the immunomodulation effect of additives [67].

Zhang et al. (2020) showed that administering FOS in chronically stressed mice diets can prevent intestinal barrier damage and neuroinflammation by expressing IL-10 anti-inflammatory cytokine, while the expression of TNF-α decreases in the gut [68]. The protective effect of FOS supplementation was demonstrated by reducing the pro-inflammatory response in goldfish (*Carassius auratus*) subjected to different LC50 concentrations of triphenyltin for short periods. Fish fed with FOS showed significant suppression levels of pro-inflammatory cytokines such as TNF-α, IL-6, and IL-1β [66]. *Megalobrama amblycephala* fed 0.4% FOS after six hours of 10 mg/L ammonia exposure presented an increment of nonspecific immunity through the lysozyme activity [69].

A mixture comprising live microorganisms and substrates selectively utilized by host microorganisms was an effective dietary symbiotic supplementation, as evaluated by Singh et al. (2019) [70]. *Bacillus circulans* PB7 (BCPB7) and FOS were proposed to improve the immune response of juveniles *Labeo rohita* under persistent stress to low pH for 60 days. Lysozyme and antioxidant activity were enhanced in fish fed with the synbiotic treatment compared to those organisms fed only with probiotic BCPB7 or FOS [71] (Table 2).

Table 2. Additives used in diets with positive effects on mitigating abiotic chronic stress on fish.

Nutrient	Supplementation	Specie	Chronic Stress Factor	Observation	Reference
Tryptophan	*Argyrosomus regius*	Crowding and netting stress for four months	0.25% Trp	No affects body weight, maintain glucose, lactate, and cortisol level ↓ Cortisol in mucus skin ↓ Plasma lactate	[38]
Tryptophan	*Oncorhynchus mykiss*	High density for 70 days	5 g Trp per kg diet	↑ Lysozyme ↑ Bactericidal activity ↓ CAT and MDA	[33]
Methionine, Lysine, Tryptophan, Threonine	*Oncorhynchus mykiss*	For six weeks, two times per week, handling stress 30 s of chasing, followed by capture in nets, removal from tanks, and 30 s of air exposure	0.17% DL-Methionine 0.81% L-Lysine sulfate 0.22% L-Tryptophan 0.20% L-Threonine		
Phenylalanine	*Sparus aurata*	Confinement and netting/chasing stress 5 min 3 times a day	5% Phe	↓ Plasma lactate	[41]
Magnesium and selenium nanoparticles	*Lates calcarifer*	Freshwater	4 mg NanoSe/kg diet 500 mg NanoMg/kg diet	↑ Final weight, weight gain, SGR, feed intake ↑ IgM and ACH50 ↑ ALT, AST ↓ Glucose	[48]
Nucleotides	Hybrid stripe bass *Morone chrysops* × *Morone saxatilis*	Salinity of 15 g/L for four weeks	0.5% de Adenosine 5′-monophosphate	↓ Blood glucose ↑ Weight gain ↑ Lysozyme activity, anti-protease activity	[72]
Vitamin E and NanoSe	*Oncorhynchus mykiss*	Dense stocking density (80 kg/m^3) 60 days	500 mg/kg Vit E and 1 mg/kg nano Se	Final Weight ↑ SGR ↑ FCR ↑ Cortisol ↓ Lactate ↓ ALT ↓ AST ↓ ALP ↓	[73]
NanoSe, Vitiamin C and E	*Oncorhynchus mykiss*	Sublethal concentration of ammonia exposure (0.024 mg/L)	0.2 mg/kg NanoSe, 200 mg/kg Vitamin C and 60 mg/kg Vitamin E	↑ final weight, ↓ FCR ↓ AST and ALP ↑ ALT, TAG, IMg Lysozyme	[52]
DHA/EPA	*Salmo sala*	Three weeks with an unpredictable chronic stress (UCS) protocol	25 g/kg EPA 14.2 g/kg total fatty acids	↑ weight gain ↓ mucosal fold height, enterocyte height, and vacuolization Gene expression related to environmental information processing	[74]

Table 2. *Cont.*

Nutrient	Supplementation	Specie	Chronic Stress Factor	Observation	Reference
n-6/n-3	*Salmo salar*	Hypoxia 3 times per week, during four weeks	Diet 6 46.7% LA 18:2n-6 0.1%ARA 20:4n-6 3.1%18:3n-3 2.0% EPA 20:5n-3 1.8% DHA 22:6n-3 3.8% EPA + DHA	Suppressed cortisol response. ↑ Level of eicosanoid PGD2 in liver ↑ level or leukotrienes LTB4 1 h after acute stress, LTB4 was the eicosanoid with the highest concentration before the acute stressor. ↓ IGF-1 was significantly lower	[75]
S. cerevisiae	Nile tilapia (*Oreochromis niloticus*)	Heat and hypoxia, Dynamic heat stress 0.01 °C per min up to 40 °C Static heat stress from 40 °C to 28 °C for 90 min Exposure to glyphosate and/or malathion Hypoxia stress all consume of oxygen (0 mg/L) for 24 h	50–70% *S. cerevisiae*	↑ survival	[76]
Selenium yeast supplementation	Nile tilapia (*Oreochromis niloticus*)	60 days against the harmful effects of glyphosate and/or malathion chronic toxicity	Selenium yeast supplementation 3.3 mg/kg diet (2.36 mg/kg selenomethionine and 0.94 mg organic selenium)	↑ survival ↑ growth ↑ SGR, ↓ FCR ↑ protection for free radicals	[77]
Spent oleaginous yeast	Juvenile red sea bream (*Pagrus major*)	Low salinity water (0.2%) The test was terminated when all the fish died.	2.5%, 25 g/kg supplement spent oleaginous yeast	↑ FW, SGR ↓ FCR ↑ SOD, GPx, IgM, and Lysozyme activity Maintain values of MDA	[78]
Dietary selenium yeast and tea-polyphenols	Juvenile Wuchang bream (*Megalobrama amblycephala*)	Ammonia stress 22.5 mg/L ammonia	Dietary selenium yeast and tea-polyphenols	↓ Cortisol, TNFα, IL-1β, ↓ GPx ↑ CAT	[53]
Bacillus circulans PB7 (BCPB7) and Fructoligosaccharide	Juveniles *Labeo rohita*	Low pH for 60 days		↑ WG, SGR ↓FCR ↑ Lysozyme activity ↓ Cortisol ↓ HSP70	[71]
B. coagulants	Common carp (*Cyprinus carpio*)	Long-term exposure to Cd^{2+} 30 and 60 days	2.0×10^8 CFU/g of *B. coagulants* in diet	Activation of Nrf gene family for resistance to oxidative stress and immune response	[79]
Clostridium autoethanogenum	Largemouth bass (*Micropterus salmoides*)		*Clostridium autoethanogenum*		[80]

3. Discussion

3.1. Feeding Intervention with Alternative Ingredients in Fish

Considering the issues that arise from the scarcity of suitable protein sources (e.g., fishmeal in animal farming) in the aquaculture industry, increasing prices of feeding, and environmental concerns, farmers are forced to prove alternative protein-rich ingredients to be replaced in conventional fish meal [81]. However, in the frontier of science in aquaculture, alternative ingredients are considered to replace fish meal as they confer fish health benefits during environmental changes that cause chronic stress. In this section, we discussed using alternative ingredients such as animal by-products, bacteria, fungi, and insects' meals in fish feeding under chronic stress and their benefits in fish growth performance, metabolic and cellular markers, and oxidative stress.

3.2. Animal By-Products Meal

Animal by-product meal (ABPM) consists of the discarded parts of farmed animals and represents a viable alternative to protein sources to replace FM in aquafeed [82]. ABPM is the group of ingredients that include meat and bone meal (MBM), bone meal, hydrolyzed feather meal, and poultry by-product meal [83,84]. Products from the non-ruminant and poultry processing industry suffer physicochemical transformation to obtain high aggregate value ingredients [83,84]. A disadvantage of ABPM as an FM protein replacement is reduced digestibility and feed efficiency at high dietary inclusion levels [85]. ABPM biotransformation with yeast or enzymes to obtain hydrolysates that increase digestibility and feed efficiency is under research [85]. ABPM is a good protein source to replace FM [86]. However, further studies are necessary to establish an adequate inclusion to avoid adverse effects on growth and improve fish condition under chronic stress abiotic. The work with positive results in fish related to stressful conditions and the partial substitution of FM with MBM occurs under acute stress. Zare et al. (2023) recorded netting results as a stressor and FM replacement with MBM and their interactions on growth, blood chemistry, immune responses, antioxidant system, and stress response in the ornamental fish *Astronotus ocellatus* [87].

3.3. Bacteria

The growing bacterial protein industry has gained much attention as an alternative protein source for animal diets, as no land to grow crops is used [88]. Bacterial protein meal (BPM) contains about 70% crude protein, 10% crude lipid (dry matter), a well-balanced amino acid profile, and vitamins similar to fish. BPM improves some aquatic animals' digestive capacity, meal taste, and immune function [80]. BPM that exert beneficial effects on fish health are *Lactobacillus*, *Shewanella*, and *Bacillus* [89].

In Salmo salar, BPM replaced up to 25% of the amino acids from FM, but fish growth performance and survival were negatively affected when reaching 50% of BPM. Tang et al., 2020 proved that dietary supplementation with a host-derived *Bacillus subtilis* strain in *Oreochromis niloticus* under hypersaline stress for eight weeks resulted in a *significantly* higher growth rate (4.8%/d) and weight gain (1400%). Regarding immune response, the brackish water and *B. subtilis* treatment presented the highest hemolytic complement (CH50) level and the lowest immunoglobulin M (IgM compared to the control. The presence of the complement system (CH50) is directly related to killing microorganisms. A complement system is a group of proteins that, when activated, lead to target cell lysis and facilitate pathogen microorganisms' phagocytosis through opsonization during immune response [90], and the absence of IgM could be a response to severe injury by hypersaline stress. On the contrary, fish control treatment without *B. subtilis* feeding and freshwater or brackish water showed low levels of CH50 and high levels of IgM. Then, *B. subtilis* can be implemented as a dietetic strategy to help fish cope with stressful environments [91]. Lactic acid bacteria (LAB) have been widely used in aquaculture as feed-supplements probiotics. *Lactobacillus plantarum* is a probiotic that enhances fish growth, health, antioxidants, and immunity [92].

3.4. Fungal (Yeast, Filamentous Fungi, and Mushroom)

Fungal biomass is gaining attention due to the emerging need for suitable and sustainable novel feed ingredients for aquaculture. Fungi comprise a large amount of morphological diversity, from unicellular yeasts to large fruiting bodies [93]. Yeast protein biomass is a potentially sustainable ingredient because it converts low-value non-food products from the agricultural industry into high-value feed with less dependence on arable land and water [94]. Yeast composition presents approximately 50% of protein content, trace minerals, and vitamins, including B-group [95]. In searching for strategies to introduce yeast in fish diets, yeast hydrolysate (YH) has high protein content and nutrients and is listed in the Chinese "feed material catalog" [96]. Yuan et al. (2017) obtained significant results in juveniles of *Cyprinus carpio* (var. Jian) fed with a yeast hydrolysate-enriched diet

(3% YH) by showing the fish's highest final weight (178.60 ± 8.42 g). Fish subjected to biotic stress with *Aeromonas hydrophila* displayed a modulated inflammatory signal, indicating that an adequate replacement level may help avoid the inflammatory response [97].

Species such as crustaceans have also shown the functionality of YH-enriched diets to manage abiotic stress. Environmental stress triggers the overproduction of ROS, producing cell damage. In crustaceans, antioxidant defense, especially the specialized antioxidant enzymes SOD and GPx, plays a crucial role in eliminating ROS [98]. In response to abiotic stress, (Chen et al., 2020) described positive results in *Litopenaeus vannamei* fed for eight weeks with 0.5% YH (*Rhodotorula mucilaginosa*) and 0.1% *Bacillus licheniformis* (BL). In this work, shrimps subjected to a high ammonia level (150 g/L NH_4CL into 100 L) and fed with a combination of 0.5 YH/0.1 BL and only 0.5% HY treatment (60%) showed the highest survival (80% and 60%, respectively). In addition, shrimps fed with a combination of YH and BL showed an enhanced antioxidant system, the levels of CAT, GPx, and SOD [99]. CAT, GPx, and SOD are the enzymes that reduce free radicals and prevent cellular damage. Jin et al. (2018) evaluated the effects of dietary YH and brewer yeast supplementations on growth performance and ammonia nitrogen stress resistance (24 h solution of 28 mg/L ammonia) of *Litopenaeus vannamei*. Shrimp fed with 1% yeast hydrolysate supplementation showed the highest final growth and the lower inflammatory-related gene expression of TNF-α and IL-1β in the intestine, and surprisingly, the levels of mTOR signal pathway genes were significantly up-regulated in those fed with 1% supplemented yeast hydrolysate [100].

3.4.1. Filamentous Fungus

Filamentous fungi generate multicellular colonies, or mycelia, through hyphae cells' extension and repeated branching [101]. When hyphae aggregate, more complex three-dimensional structures emerge. The most complex structures in the fungal kingdom are the multicellular sexual fruiting bodies, with distinct fungal tissues and multiple cell types [102]. Fruit bodies, including the familiar mushrooms of basidiomycetes, are multicellular reproductive organs. Mushroom formation involves the coordinated growth of millions of hyphae [102]. Filamentous fungi in the diet offer benefits to fish, including growth promotion, inhibition of pathogen colonization, improvement of nutrient digestion, and enhancement of reproduction. However, there are very few insights on chronic alleviation [103]. *Aspergillus* sp. V2 grown on vinasse can be used for fish feed. Proximate analysis revealed a total protein content of 34%, fat of 4.7%, ash of 15.8%, crude fiber of 4%, moisture of 0.3%, and carbohydrate of 45.2%, all within acceptable levels for commercial fish diets. Its composition is analogous to soybean meal, containing all the essential amino acids for fish feeding and permissible aflatoxin levels for feedstuff [104]. *Aspergillus niger* is a promising ingredient for formulating cost-effective shrimp feed. This fungus has been reported to produce numerous hydrolytic enzymes for degrading the limiting factors of plant proteins, such as antinutrients and fiber. *A. niger* enhances the quantity and quality of proteins and amino acids, especially methionine and lysine, helping to replace dietary fishmeal up to 80% in penaeid shrimps [105].

Dawood et al. (2020) used two levels of supplemented *Aspergillus oryzae* (ASP 1×10^6 CFU/g and ASP 2 1×10^8 CFU/g) in Nile tilapia (*Oreochromis niloticus*) under hypoxia, resulting in enhanced oxidative status, heat shock protein, and immune-related gene expression. The hypoxia stress was achieved by reducing the tank water volume until the symptoms in the dorsal fin were evident. Fish with both diets presented significantly higher levels of final body weight (61.40 ± 0.58 g with ASP and 63.25 ± 1 with ASP2) than fish control (56.77 ± 0.26). In terms of blood chemistry, antioxidant status, and immunological response after hypoxia, fish with supplemented diets showed lower glucose and cortisol concentrations (ASP 2 148 mg/dL and 0.65 mM/mL, respectively) in comparison with the fish control diet (160 mg/dL and 0.9 mM/mL). The enzymatic activity in fish fed with supplementation ASP 2 had higher levels of SOD (26 IU/L), GPX (25 IU/L), and lower MDA (14 nmol/mL), which are antioxidant defense network compounds, indicating reduced cell damage compared with fish control diet (SOD 21.5 IU/L, GPX 15.7 IU/L; MDA 17.5 nmol/mL) [106].

3.4.2. Mushroom Meal

Macroscopic filamentous fungi that form large fruiting bodies are known as mushrooms [107]. Edible mushrooms are the higher fungi in the *Ascomycetes* and *Basidiomycetes* classes [108]. Great attention was given to the polysaccharides produced by numerous fungi, especially mushrooms, because of their various biological properties, such as antioxidant, antimicrobial, and immunostimulant activities [109].

Mushrooms have been consumed as a healthy nutritional food and medicine for decades due to their significant antioxidant, anti-inflammatory, anti-tumor, and immune-stimulating properties [110]. By-products of king oyster mushroom *Pleurotus eryngii* (KOME) and *Lactobacillus plantarum* (LP) were incorporated into the white shrimp *Litopenaeus vannamei* diet. Shrimp fed with 5 g/kg KOME and 1×10^8 CFU/kg LP had significantly higher weight gain and total final weight than the control [110].

Pleurotus ostreatus, also known as the oyster mushroom, has anticancer, immunomodulatory, antiviral, antibiotic, anti-inflammatory, and cholesterol-lowering properties. A diet supplemented with this mushroom for *Oreochromis mossambicus* showed a positive effect on survival rate when fish were subjected to septicemia [111]. *Agaricus blazei* is an edible mushroom native to Brazil that has gained interest in various fields of medical research, as it has shown remarkable immunomodulating effects. The number of by-products and waste substrate from mushroom production has increased drastically as the annual global mushroom yield is estimated to be over 25 million tons. P.T. Lee, Wu, Tseng, Lu, and Lee described efforts to recycle and reuse waste mushroom compost, taking advantage of its immunomodulatory compounds, highlighting that waste mushroom substrate from *A. blazei* cultivation is a cost-effective feed additive for *Oreochromis niloticus* that protects fish from *S. agalactiae* infection [112].

A Taiwanese medicinal mushroom, *Antrodia cinnamomea* (AC), possesses triterpenoids, polysaccharides, superoxide dismutase (SOD), adenosine, succinic and maleic acid derivates, proteins, vitamins, nucleic acid, lectin, ergosterol, and lignin. The raw materials of this mushroom have bioactive compounds. Zebrafish were fed with an AC-supplemented diet for studying fish heat resistance (increased temperature from 28 °C to 39 °C in 100 min) and cold tolerance (decreased temperature from 28 °C to 11.5 °C in 100 min). Although this work describes acute stress conditions, the results for survival, growth performance, and immunology responses were improved. Fish fed with 10% AC presented almost 95% survival under heat and 80% under cold. After 30 days, the group of 10% AC/39 °C showed a significantly increased body weight gain (0.107 ± 0.018 g) compared to the 0.2% AC/39 °C group (0.086 ± 0.024 g). Fish fed with 10% AC presented the highest specific growth rate (1.084 ± 0.082%/day) compared to the 0.2% AC group (0.859 ± 0.075%/day) [113].

Enhancing the antioxidant defense status by supplementing with natural antioxidant-rich sources is the most convenient alternative to protect fish from stress. Disposing of undesirable mushroom parts, such as mushroom stalk waste (MSW), increases exponentially along with its production. As mushrooms are a rich source of polysaccharides and antioxidants, raw polysaccharides (RPs) from MSW were supplemented to fish feed to evaluate the protective effect on stress caused by pH fluctuation. The 5 and 10 g RP/kg feeds showed the highest survival for *Oreochromis niloticus*. Fish post-stress conditions in RP 5 g/kg feeds were improved, with enhanced SOD and CAT activity for pH 5.5, 8.5, and 10.5 during one week [114].

3.5. Insect

Insect meal has been suggested as a promising and natural alternative to FM and FO, as it contains a large number of amino acids, lipids, vitamins, minerals, and crude protein, where the latter represents between 50% and 82% of its dry mass, depending on the insect species and processing method [115]. Given these promising features, several types of insects, such as Silkworm (*Bombyx mori*), black soldier fly (*Hermetia illucens*), Housefly (*Musca domestica*), Yellow mealworm (*Tenebrio molitor*), Lesser mealworm (*Alphitobius diaperinus*), House cricket (*Acheta domesticus*), Banded cricket (*Gryllodes sigillatus*), and Jamaican field cricket (*Gryllus assimilis*) have been considered to elaborate meal [116]. Although the impact of insect meal on chronic stress has not been studied, some investigations have proven benefits in fish health. For example, an immunostimulant effect was described in rainbow trout *Oncorhynchus mykiss* with black solider fly and poultry by-product combination [117], in Pacific white shrimp (*Litopenaeus vannamei*) [118] and mandarin fish (*Siniperca scherzeri*) with Yellow mealworm (*Tenebrio molitor*) [119], in Atlantic salmon (*Salmo salar*) with black solider fly [120], and Nile tilapia (*Oreochromis niloticus*) with black soldier fly larvae [121]. Taufek et al. (2018) observed the effect of the cricket meal diet (*Gryllus bimaculatus*) on African catfish (*Clarias garienpinus*). After 40 days with experimental diets, fish were subjected to bacterial stress with *Aeromonas hydrophila* [122]. The results showed that fish fed with 40% of *G. bimaculatus* presented elevated levels of white blood cells ($19.2 \pm 0.13 \times 10^3$ mm^3) and lower with FM ($13.6 \pm 0.86 \times 10^3$ mm^3), higher lysozyme activity (22.2 ± 2.54 U/mL) compared to FM (8.4 ± 1.16 U/mL), and more elevated globulin in serum (59 ± 6.08 mg/dL) compared to FM (51.5 ± 5.49 mg/dL). Moreover, after 12 days of *A. hydrophila* infection, fish fed with cricket meal presented a 66.7% relative survival percentage, indicating a positive immunostimulant effect.

BSFL meals have been tested in different studies with different inclusion levels, ranging from 25% to 100%, meal quality, fish species, and diet formulations. Kortner et al. (2020) reviewed the effect of an insect meal-based diet (60% BSFL) on the gut health of Atlantic salmon in freshwater and seawater conditions. The results showed that fish from freshwater (pre-smoltification) presented gene expression related to immune tolerance (foxp3), stress response (hsp70), and detoxification activity (CYP1A1, MTA, SOD, AND CAT); additionally, no enterocyte steatosis in the proximal intestine was observed. Enterocyte steatosis is a metabolism disorder in enterocytes, which, in severe cases, may be accompanied by accumulations of lipidic materials inside the gut lumen [123]. The effect of BSFL meal was supplemented in a diet for juvenile rainbow trout (Oncorhynchus mykiss) reared at different stocking densities [124]. At 4.9 kg/m^3, fish fed with a diet of 50 g/kg BSFL meal and 10 g/kg BSFL oil showed an increase in final growth (27.12 ± 0.42 g) compared to control diet (without supplementation 26.07 ± 0.42 g). In terms of plasma antioxidant metabolites, immune and stress-related parameters of rainbow trout at the same stocking density showed the highest values of IGF-I (450 pg/mL), SOD (9 ng/mL), IL-1β (21 pg/mL), Lyz (30 U/mL), and lower values of HSP-70 (7.9 ng/mL) and Cortisol (30 ng/mL).

3.6. Non-Conventional Plant Meal

The most important groups of seed crops used for food include cereals, pseudocereals, pulses, oilseeds, and nuts, with a loss or waste through the food chain. Seed loss refers to materials that do not reach consumers. In contrast, seed wastes are all those parts of the seeds not intended for human consumption, including hull, husk, or shattered cotyledon, often produced in dehulling and milling processes. This seed coating contains bioactive compounds [125]. Corn husk is part of the corn stover, which has the highest concentration of phenolic compounds with ferulic and ρ-coumaric acids [126]. Galeana-López et al. (2021) used a 25 g corn husk meal/kg diet in Nile Tilapia (*Orechromis niloticus*) exposed to hypoxia at 1.5 mg/L of oxygen dissolved for a short period of 5 h, resulting in a significant increase in CAT activity in tilapias exposed to stress [127].

One problem with employing plant sources as alternative ingredients in aquafeeds is the anti-nutritional factors that impede the digestive process in fish. A solution uses biological strategies that degrade these compounds, such as solid-state fermentation (SSF), which involves the microbial fermentation of a substrate where subsequent by-products may provide residual bioactive components [114]. Amaral et al. (2023) evaluated the effect of SSF on plant feedstuff (soybean, rapeseed, sunflower, and rice bran) fermented with *Aspergillus niger* in stressed European Seabass (*Dicentrarchus labrax*). However, the results of fish-fed test diets subjected to different temperatures and salinity oscillation indicated that fermented products did not mitigate the impact of environmental stress [128].

4. Perspectives

Novel protein sources are functional and may exert a positive modulatory effect in fish subjected to chronic stress. Fish growth can be promoted by inhibiting cortisol hormone, potentializing antioxidant capacity, and reducing metabolic stress markers with adequate nutrition. More studies are needed to achieve the level of inclusion for each species.

Animal by-product meals are rendered products from farm animals that can be used as valuable feedstuffs for fish feed, displacing traditional fishmeal and promoting a circular economy system by reducing waste and reintroducing them in the economy as high-value products. These ingredients have shown positive effects in fish under stress conditions. Yeast hydrolysate is rich in peptides and is used as an additive because it stimulates the antioxidant system in fish and prepares the organism's defense to cope with infections. Including yeast hydrolysate in the diet might prevent outbreaks when fish are subject to stress. The fungal meal significantly affects dietary interventions and is a promising feed source that can be added to animal feed to replace familiar protein sources such as fishmeal. Valuable filamentous fungal biomass contains high protein concentrations composed of functional amino acids. Mushrooms and mycelia provide protein, lipids, fatty acids, vitamins, and antioxidants. However, research on using novel ingredients for dietary interventions has been performed in most acute rather than chronic stress situations.

Insects are one of the most promising alternative ingredients to supply a protein source. Their nutritional composition includes long-chain polyunsaturated fatty acids (VLPUFAs), Vitamin C, and Vitamin D, which promote immune stimulation, and fatty acids that allow fish maintenance besides environmental stress. Then, insects can be used because of their high amount of protein and their beneficial effects on fish health. Interestingly, amino acids (e.g., Trp and Phe), vitamins E and C, and some minerals (e.g., Se) are functional nutrients that alleviate abiotic chronic stress. More studies must be performed to evaluate novel meal ingredients to optimize the nutritional composition of fish feeding. On the other hand, investigations must be performed to establish the level of inclusion by species. Additionally, the combination and pretreatment of ingredients, such as hydrolysates, solid fermentations, and metabolites extraction, potentialize the effects of feeding for chronically stressed welfare and, not less importantly, the performance of aquaculture production systems. Therefore, more studies are needed to elucidate this novel generation of diets aiming to accomplish sustainable and resilient aquaculture.

5. Conclusions

Imminent intensive production, global warming, and the severity of infections directly affect fish integrity under farming conditions. Abiotic stress takes visibility due to negative impacts on the organism, resulting in diminished production. Novel ingredients for protein sources possess functional nutrients that promote a set of fish responses to cope with adverse environmental conditions. The frontier of science in fish nutrition is focused on developing a novel generation of diets to make aquaculture more resilient from molecular, individual, and shoal levels, with benefits to livestock units.

Author Contributions: Conceptualization, investigation, writing—original draft preparation, visualization J.S.-V. Conceptualization, investigation, resources, writing—original draft preparation, visualization, supervision, project administration, funding acquisition: H.A.-B. and G.A.P.-H. All authors have read and agreed to the published version of the manuscript.

Funding: This research was funded by CONAHCYT through the scholarship of J. Sánchez Velázquez during doctoral studies, grant number 632434, and by the fund for the development of knowledge FONDEC-UAQ 2019 (END202008).

Institutional Review Board Statement: Not applicable.

Informed Consent Statement: Not applicable.

Data Availability Statement: Not applicable.

Acknowledgments: The National Council of Science and Technology (CONAHCYT) partially funded this work through J. Sánchez-Velázquez's scholarship during doctoral studies (632434).

Conflicts of Interest: The authors declare no conflicts of interest.

References

1. Shoko, A.P.; Limbu, S.M.; Mrosso, H.D.J.; Mgaya, Y.D. A comparison of diurnal dynamics of water quality parameters in Nile tilapia (*Oreochromis niloticus*, Linnaeus, 1758) monoculture and polyculture with African sharp tooth catfish (*Clarias gariepinus*, Burchell, 1822) in earthen ponds. *Int. Aquat. Res.* **2014**, *6*, 56. [CrossRef]
2. Mateus, A.P.; Power, D.M.; Canário, A.V.M. *Stress and Disease in Fish*; Elsevier: Amsterdam, The Netherlands, 2017; ISBN 9780128045640.
3. Ahmed, I.; Reshi, Q.M.; Fazio, F. The influence of the endogenous and exogenous factors on hematological parameters in different fish species: A review. *Aquac. Int.* **2020**, *28*, 869–899. [CrossRef]
4. Cossins, A.R.; Crawford, D.L. Fish as models for environmental genomics. *Nat. Rev. Genet.* **2005**, *6*, 324–333. [CrossRef]
5. Chowdhury, S.; Saikia, S.K. Oxidative Stress in Fish: A Review. *J. Sci. Res.* **2020**, *12*, 145–160. [CrossRef]
6. Guo, H.; Dixon, B. Understanding acute stress-mediated immunity in teleost fish. *Fish Shellfish Immunol. Rep.* **2021**, *2*, 100010. [CrossRef] [PubMed]
7. Jerez, I. Bienestar Animal en el Cultivo de Dorada (Sparus aurata) Mecanismos de Atenuación del Estrés. Ph.D. Thesis, Universidad de Cádiz, Tesis para grado Doctor en Recursos Marinos (Acuicultura) de la Escuela Internacional de Doctorado en Estudios del Mar (EIDEMAR) de la Universidad de Cádiz, Cádiz, España, 2019. Volume 53.
8. Hoem, K.S.; Tveten, A. Current approaches in decoding the molecular mechanisms of long-term stress in adult farmed Atlantic salmon (*Salmo salar*). *Rev. Aquac.* **2019**, *12*, 1708–1720. [CrossRef]
9. Ugwemorubong, U.G.; Akinrotimi, O.A. *Management of Stress in Fish for Sustainable Aquaculture Development*; Researcher 2011; Marsland Press: New York, NY, USA, 2011; Volume 3, ISBN 1553-9865.
10. Boonstra, R. Reality as the leading cause of stress: Rethinking the impact of chronic stress in nature. *Funct. Ecol.* **2013**, *27*, 11–23. [CrossRef]
11. Abisha, R.; Krishnani, K.K.; Sukhdhane, K.; Verma, A.K.; Brahmane, M.; Chadha, N.K. Sustainable development of climate-resilient aquaculture and culture-based fisheries through adaptation of abiotic stresses: A review. *J. Water Clim. Chang.* **2022**, *13*, 2671–2689. [CrossRef]
12. Pawlak, P.; Burren, A.; Seitz, A.; Glauser, G.; Pietsch, C. Differential effects of acute eustress and distress on gene regulation patterns in the carp (*Cyprinus carpio* L.) brain. *Aquac. Res.* **2022**, *53*, 5075–5096. [CrossRef]
13. Ataallahi, M.; Nejad, J.G.; Park, K.-H. Selection of appropriate biomatrices for studies of chronic stress in animals: A review. *J. Anim. Sci. Technol.* **2022**, *64*, 621–639. [CrossRef]
14. Schreck, C.B.; Tort, L. The concept of stress in fish. In *Fish Physiology: Biology of Stress in Fish*; Elsvier: Amsterdam, The Netherlands, 2016; Volume 2016.
15. Balasch, J.C.; Tort, L. Netting the Stress Responses in Fish. *Front. Endocrinol.* **2019**, *10*, 62. [CrossRef]
16. Hanke, I.; Hassenrück, C.; Ampe, B.; Kunzmann, A.; Gärdes, A.; Aerts, J. Chronic stress under commercial aquaculture conditions: Scale cortisol to identify and quantify potential stressors in milkfish (*Chanos chanos*) mariculture. *Aquaculture* **2020**, *526*, 735352. [CrossRef]
17. Lee, C.S.; Lim, C.; Gatlin, D.M.; Webster, C.D. *Dietary Nutrients, Additives, and Fish Health*; Wiley-Blackwell: Chennai, India, 2015; ISBN 9781119005568.
18. Gupta, A.; Gupta, S.K.; Priyam, M.; Siddik, M.A.B.; Kumar, N.; Mishra, P.K.; Gupta, K.K.; Sarkar, B.; Sharma, T.R.; Pattanayak, A. Immunomodulation by dietary supplements: A preventive health strategy for sustainable aquaculture of tropical freshwater fish, *Labeo rohita* (Hamilton, 1822). *Rev. Aquac.* **2021**, *13*, 2364–2394. [CrossRef]
19. Yılmaz, S.; Ergün, S.; Yiğit, M.; Yılmaz, E. An Extensive Review on the Use of Feed Additives against Fish Diseases and Improvement of Health Status of Fish in Turkish Aquaculture Sector. *Aquac. Stud.* **2022**, *22*, AQUAST710. [CrossRef]
20. Mugwanya, M.; Dawood, M.A.O.; Kimera, F.; Sewilam, H. Anthropogenic temperature fluctuations and their effect on aquaculture: A comprehensive review. *Aquac. Fish.* **2022**, *7*, 223–243. [CrossRef]
21. Kumar, P.; Thirunavukkarasu, A.R.; Subburaj, R.; Thiagarajan, G. Concept of Stress and Its Mitigation in Aquaculture. In *Advances in Marine and Brackishwater Aquaculture*; Springer: New Delhi, India, 2015; pp. 95–100.
22. Ciji, A.; Akhtar, M.S. Stress management in aquaculture: A review of dietary interventions. *Rev. Aquac.* **2021**, *13*, 2190–2247. [CrossRef]
23. Eissa, N.; Wang, H. Transcriptional stress responses to environmental and husbandry stressors in aquaculture species. *Rev. Aquac.* **2014**, *8*, 61–88. [CrossRef]
24. Dawood, M.A.O.; Noreldin, A.E.; Sewilam, H. Blood biochemical variables, antioxidative status, and histological features of intestinal, gill, and liver tissues of African catfish (*Clarias gariepinus*) exposed to high salinity and high-temperature stress. *Environ. Sci. Pollut. Res.* **2022**, *29*, 56357–56369. [CrossRef]

25. Sadoul, B.; Alfonso, S.; Cousin, X.; Prunet, P.; Bégout, M.-L.; Leguen, I. Global assessment of the response to chronic stress in European sea bass. *Aquaculture* **2021**, *544*, 737072. [CrossRef]
26. Li, M.; Wang, X.; Qi, C.; Li, E.; Du, Z.; Qin, J.G.; Chen, L. Metabolic response of Nile tilapia (*Oreochromis niloticus*) to acute and chronic hypoxia stress. *Aquaculture* **2018**, *495*, 187–195. [CrossRef]
27. Abdel-Tawwab, M.; Monier, M.N.; Hoseinifar, S.H.; Faggio, C. Fish response to hypoxia stress: Growth, physiological, and immunological biomarkers. *Fish Physiol. Biochem.* **2019**, *45*, 997–1013. [CrossRef]
28. Islam, M.J.; Kunzmann, A.; Slater, M.J. Responses of aquaculture fish to climate change-induced extreme temperatures: A review. *J. World Aquac. Soc.* **2022**, *53*, 314–366. [CrossRef]
29. Lemos, L.S.; Angarica, L.M.; Hauser-Davis, R.A.; Quinete, N. Cortisol as a Stress Indicator in Fish: Sampling Methods, Analytical Techniques, and Organic Pollutant Exposure Assessments. *Int. J. Environ. Res. Public Health* **2023**, *20*, 6237. [CrossRef]
30. Sadoul, B.; Mathilakath, M. 5. Stress and Growth. In *Biology of Stress in Fish*; Elsevier Inc.: Amsterdam, The Netherlands, 2016; Volume 35, pp. 167–205. ISBN 9780128027288.
31. Johnston, I.A.; Bower, N.I.; Macqueen, D.J. Growth and the regulation of myotomal muscle mass in teleost fish. *J. Exp. Biol.* **2011**, *214*, 1617–1628. [CrossRef]
32. Fuentes, E.N.; Valdés, J.A.; Molina, A.; Björnsson, B.T. Regulation of skeletal muscle growth in fish by the growth hormone—Insulin-like growth factor system. *Gen. Comp. Endocrinol.* **2013**, *192*, 136–148. [CrossRef]
33. Hoseini, S.M.; Taheri Mirghaed, A.; Ghelichpour, M. Effects of dietary tryptophan levels and fish stocking density on immunological and antioxidant responses and bactericidal activity against *Aeromonas hydrophila* in rainbow trout (*Oncorhynchus mykiss*). *Aquac. Res.* **2020**, *51*, 1455–1463. [CrossRef]
34. Tejpal, C.S.; Pal, A.K.; Sahu, N.P.; Ashish Kumar, J.; Muthappa, N.A.; Vidya, S.; Rajan, M.G. Dietary supplementation of l-tryptophan mitigates crowding stress and augments the growth in *Cirrhinus mrigala* fingerlings. *Aquaculture* **2009**, *293*, 272–277. [CrossRef]
35. Yin, Y. Tryptophan metabolism in animals important roles in nutrition and health. *Front. Biosci.* **2011**, *S3*, 152. [CrossRef] [PubMed]
36. Lepage, O.; Tottmar, O.; Winberg, S. Elevated dietary intake of L-tryptophan counteracts the stress-induced elevation of plasma cortisol in rainbow trout (*Oncorhynchus mykiss*). *J. Exp. Biol.* **2002**, *205*, 3679–3687. [CrossRef]
37. Akhtar, M.S.; Pal, A.K.; Sahu, N.P.; Ciji, A.; Meena, D.K.; Das, P. Physiological responses of dietary tryptophan fed Labeo rohita to temperature and salinity stress. *J. Anim. Physiol. Anim. Nutr.* **2013**, *97*, 1075–1083. [CrossRef]
38. Herrera, M.; Fernández-alacid, L.; Sanahuja, I.; Ibarz, A.; Salamanca, N.; Morales, E. Physiological and metabolic effects of a tryptophan-enriched diet to face up chronic stress in meagre (*Argyrosomus regius*). *Aquaculture* **2020**, *522*, 735102. [CrossRef]
39. Salamanca, N.; Giráldez, I.; Morales, E.; de La Rosa, I.; Herrera, M. Phenylalanine and Tyrosine as Feed Additives for Reducing Stress and Enhancing Welfare in Gilthead Seabream and Meagre. *Animals* **2020**, *11*, 45. [CrossRef]
40. Herrera, M.; Herves, M.A.; Giráldez, I.; Skar, K.; Mogren, H.; Mortensen, A.; Puvanendran, V. Effects of amino acid supplementations on metabolic and physiological parameters in Atlantic cod (*Gadus morhua*) under stress. *Fish Physiol. Biochem.* **2017**, *43*, 591–602. [CrossRef]
41. Salamanca, N.; Moreno, O.; Giráldez, I.; Morales, E.; de la Rosa, I.; Herrera, M. Effects of Dietary Phenylalanine and Tyrosine Supplements on the Chronic Stress Response in the Seabream (*Sparus aurata*). *Front. Physiol.* **2022**, *12*, 775771. [CrossRef]
42. Petitjean, Q.; Jean, S.; Gandar, A.; Côte, J.; Laffaille, P.; Jacquin, L. Stress responses in fish: From molecular to evolutionary processes. *Sci. Total Environ.* **2019**, *684*, 371–380. [CrossRef]
43. Aedo, J.E.; Zuloaga, R.; Boltaña, S.; Molina, A.; Valdés, J.A. Membrane-initiated cortisol action modulates early pyruvate dehydrogenase kinase 2 (pdk2) expression in fish skeletal muscle. *Comp. Biochem. Physiol. Part A Mol. Integr. Physiol.* **2019**, *233*, 24–29. [CrossRef] [PubMed]
44. Wu, C.-Y.; Lee, T.-H.; Tseng, D.-Y. Glucocorticoid Receptor Mediates Cortisol Regulation of Glycogen Metabolism in Gills of the Euryhaline Tilapia (*Oreochromis mossambicus*). *Fishes* **2023**, *8*, 267. [CrossRef]
45. Nipu, N.; Antomagesh, F.; Faught, E.; Vijayan, M.M. Glucocorticoid receptor activation reduces food intake independent of hyperglycemia in zebrafish. *Sci. Rep.* **2022**, *12*, 15677. [CrossRef] [PubMed]
46. Svendsen, E.; Føre, M.; Økland, F.; Gräns, A.; Hedger, R.D.; Alfredsen, J.A.; Uglem, I.; Rosten, C.M.; Frank, K.; Erikson, U.; et al. Heart rate and swimming activity as stress indicators for Atlantic salmon (*Salmo salar*). *Aquaculture* **2021**, *531*, 735804. [CrossRef]
47. da Fonseca, L.A.; Fagundes, V.; Girardi, F.M.; Dornelas, L.R.S. Blood values of cortisol, glucose, and lactate in healthy green turtle (*Chelonia mydas*) and affected by fibropapillomatosis. *Comp. Clin. Path.* **2020**, *29*, 1099–1105. [CrossRef]
48. Longbaf Dezfouli, M.; Ghaedtaheri, A.; Keyvanshokooh, S.; Salati, A.P.; Mousavi, S.M.; Pasha-Zanoosi, H. Combined or individual effects of dietary magnesium and selenium nanoparticles on growth performance, immunity, blood biochemistry and antioxidant status of Asian seabass (*Lates calcarifer*) reared in freshwater. *Aquac. Nutr.* **2019**, *25*, 1422–1430. [CrossRef]
49. Nakano, T.; Kameda, M.; Shoji, Y.; Hayashi, S.; Yamaguchi, T.; Sato, M. Effect of severe environmental thermal stress on redox state in salmon. *Redox Biol.* **2014**, *2*, 772–776. [CrossRef]
50. Menon, S.V.; Kumar, A.; Middha, S.K.; Paital, B.; Mathur, S.; Johnson, R.; Kademan, A.; Usha, T.; Hemavathi, K.N.; Dayal, S.; et al. Water physicochemical factors and oxidative stress physiology in fish, a review. *Front. Environ. Sci.* **2023**, *11*, 1240813. [CrossRef]
51. Kulczykowska, E. Stress Response System in the Fish Skin—Welfare Measures Revisited. *Front. Physiol.* **2019**, *10*, 72. [CrossRef] [PubMed]

52. Harsij, M.; Gholipour Kanani, H.; Adineh, H. Effects of antioxidant supplementation (nano-selenium, vitamin C and E) on growth performance, blood biochemistry, immune status and body composition of rainbow trout (*Oncorhynchus mykiss*) under sub-lethal ammonia exposure. *Aquaculture* **2020**, *521*, 734942. [CrossRef]
53. Guo, H.; Lin, W.; Wang, L.; Zhang, D.; Wu, X.; Li, L.; Li, D.; Tang, R.; Yang, L.; Qiu, Y. The supplementation of dietary selenium yeast and green tea-derived polyphenols improves antioxidant capacity and immune response in juvenile Wuchang bream under ammonia stress. *Aquac. Res.* **2020**, *51*, 3790–3803. [CrossRef]
54. Flores-Méndez, L.C.; Lizárraga-Velázquez, C.E.; Sánchez-Gutiérrez, E.Y.; Arrizon, J.; Leyva-López, N.; Hernández, C. Study of the Effect of Dietary Agavin Supplementation in Blood Parameters and Antioxidant Enzymes of Juvenile Nile Tilapia (*Oreochromis niloticus*) under Stress Conditions. *Fishes* **2022**, *7*, 340. [CrossRef]
55. De La Cruz-Marín, E.; Martínez-García, R.; López-Hernández, J.F.; Méndez-Marín, O.; De la Rosa-García, S.C.; Peña-Marín, E.S.; Tovar-Ramírez, D.; Sepúlveda-Quiroz, C.A.; Pérez-Jiménez, G.M.; Jiménez-Martínez, L.D.; et al. Inulin Supplementation in Diets for Tropical Gar (*Atractosteus tropicus*) Larvae: Effects on Growth, Survival, and Digestive and Antioxidant Enzyme Activities. *Aquac. J.* **2023**, *3*, 43–55. [CrossRef]
56. Zhou, L.; Zhang, J.; Yan, M.; Tang, S.; Wang, X.; Qin, J.G.; Chen, L.; Li, E. Inulin alleviates hypersaline-stress induced oxidative stress and dysbiosis of gut microbiota in Nile tilapia (*Oreochromis niloticus*). *Aquaculture* **2020**, *529*, 735681. [CrossRef]
57. Caipang, C.M.A.; Lazado, C.C. Nutritional impacts on fish mucosa: Immunostimulants, pre- and probiotics. In *Mucosal Health in Aquaculture*; Elsevier: Amsterdam, The Netherlands, 2015; pp. 211–272.
58. Su, M.; Zhang, R.; Liu, N.; Zhang, J. Modulation of inflammatory response by cortisol in the kidney of spotted scat (*Scatophagus argus*) in vitro under different osmotic stresses. *Fish Shellfish Immunol.* **2020**, *104*, 46–54. [CrossRef]
59. Tort, L. Stress and immune modulation in fish. *Dev. Comp. Immunol.* **2011**, *35*, 1366–1375. [CrossRef] [PubMed]
60. Yada, T.; Tort, L. Stress and Disease Resistance: Immune System and Immunoendocrine Interactions. *Fish Physiol.* **2016**, *35*, 365–403. [CrossRef]
61. Tort Bardolet, L.; Balasch Alemany, J.C.; Mackenzie, S. Fish health challenge after stress: Indicators of immunocompetence. *Contrib. Sci.* **2004**, *2*, 443–454.
62. Dai, C.; Zheng, J.; Qi, L.; Deng, P.; Wu, M.; Li, L.; Yuan, J. Chronic stress boosts systemic inflammation and compromises antiviral innate immunity in Carassius gibel. *Front. Immunol.* **2023**, *14*, 1105156. [CrossRef] [PubMed]
63. Guerreiro, I.; Couto, A.; Machado, M.; Castro, C.; Pousão-Ferreira, P.; Oliva-Teles, A.; Enes, P. Prebiotics effect on immune and hepatic oxidative status and gut morphology of white sea bream (*Diplodus sargus*). *Fish Shellfish Immunol.* **2016**, *50*, 168–174. [CrossRef] [PubMed]
64. Youn, H.S.; Kim, J.-H.; Lee, J.S.; Yoon, Y.Y.; Choi, S.J.; Lee, J.Y.; Kim, W.; Hwang, K.W. Lactobacillus plantarum Reduces Low-Grade Inflammation and Glucose Levels in a Mouse Model of Chronic Stress and Diabetes. *Infect. Immun.* **2021**, *89*, e0061520. [CrossRef] [PubMed]
65. Hoseinifar, S.H.; Soleimani, N.; Ringø, E. Effects of dietary fructo-oligosaccharide supplementation on the growth performance, haemato-immunological parameters, gut microbiota and stress resistance of common carp (*Cyprinus carpio*) fry. *Br. J. Nutr.* **2014**, *112*, 1296–1302. [CrossRef]
66. Zhang, C.; Wang, J.; Qi, Q.; Yang, L.; Sun, P.; Yuan, X. Modulatory effect of fructooligosaccharide against triphenyltin-induced oxidative stress and immune suppression in goldfish (*Carassius auratus*). *Ecotoxicol. Environ. Saf.* **2021**, *212*, 111966. [CrossRef]
67. Asencio-Alcudia, G.G.; Sepúlveda-Quiroz, C.A.; Pérez-Urbiola, J.C.; Rodríguez-Jaramillo, M.d.C.; Teles, A.; Salas-Leiva, J.S.; Martínez-García, R.; Jiménez-Martínez, L.D.; Galaviz, M.; Tovar-Ramírez, D.; et al. Stress-Protective Role of Dietary α-Tocopherol Supplementation in Longfin Yellowtail (*Seriola rivoliana*) Juveniles. *Fishes* **2023**, *8*, 526. [CrossRef]
68. Zhang, Y.; Wang, P.; Xia, C.; Wu, Z.; Zhong, Z.; Xu, Y.; Zeng, Y.; Liu, H.; Liu, R.; Liao, M. Fructooligosaccharides supplementation mitigated chronic stress-induced intestinal barrier impairment and neuroinflammation in mice. *J. Funct. Foods* **2020**, *72*, 104060. [CrossRef]
69. Zhang, C.-N.; Li, X.-F.; Tian, H.-Y.; Zhang, D.-D.; Jiang, G.-Z.; Lu, K.-L.; Liu, G.-X.; Liu, W.-B. Effects of fructooligosaccharide on immune response, antioxidant capability and HSP70 and HSP90 expressions of blunt snout bream (*Megalobrama amblycephala*) under high ammonia stress. *Fish Physiol. Biochem.* **2015**, *41*, 203–217. [CrossRef] [PubMed]
70. Pawar, N.A.; Prakash, C.; Kohli, M.P.S.; Jamwal, A.; Dalvi, R.S.; Devi, B.N.; Singh, S.K.; Gupta, S.; Lende, S.R.; Sontakke, S.D.; et al. Fructooligosaccharide and Bacillus subtilis synbiotic combination promoted disease resistance, but not growth performance, is additive in fish. *Sci. Rep.* **2023**, *13*, 11345. [CrossRef]
71. Singh, S.K.; Tiwari, V.K.; Chadha, N.K.; Munilkumar, S.; Prakash, C.; Pawar, N.A. Effect of dietary synbiotic supplementation on growth, immune and physiological status of Labeo rohita juveniles exposed to low pH stress. *Fish Shellfish Immunol.* **2019**, *91*, 358–368. [CrossRef] [PubMed]
72. de Cruz, C.R.; Yamamoto, F.Y.; Castillo, S.; Gatlin, D.M. Establishing the optimal adenosine 5′-monophosphate level for hybrid striped bass *Morone chrysops* × *Morone saxatilis*: Effects on growth performance, nutrient digestibility, and immune modulation during acute and chronic stress. *Aquaculture* **2020**, *520*, 734668. [CrossRef]
73. Naderi, M.; Keyvanshokooh, S.; Salati, A.P.; Ghaedi, A. Effects of dietary vitamin E and selenium nanoparticles supplementation on acute stress responses in rainbow trout (*Oncorhynchus mykiss*) previously subjected to chronic stress. *Aquaculture* **2017**, *473*, 215–222. [CrossRef]

74. Løvmo, S.D.; Sundh, H.; Whatmore, P.; Nordvi, M.F.; Sigholt, T.; Madaro, A.; Bardal, T.; Olsen, R.E. Intestinal health in Atlantic salmon post-smolt (*Salmo salar*) when fed low- and high HUFA diets. *Aquaculture* **2022**, *557*, 738318. [CrossRef]
75. Hundal, B.K.; Liland, N.S.; Rosenlund, G.; Höglund, E.; Araujo, P.; Stubhaug, I.; Sissener, N.H. Increasing the dietary n-6/n-3 ratio alters the hepatic eicosanoid production after acute stress in Atlantic salmon (*Salmo salar*). *Aquaculture* **2021**, *534*, 736272. [CrossRef]
76. Abass, D.A.; Obirikorang, K.A.; Campion, B.B.; Edziyie, R.E.; Skov, P.V. Dietary supplementation of yeast (*Saccharomyces cerevisiae*) improves growth, stress tolerance, and disease resistance in juvenile Nile tilapia (*Oreochromis niloticus*). *Aquac. Int.* **2018**, *26*, 843–855. [CrossRef]
77. Hassan, M.A.; Hozien, S.T.; Abdel Wahab, M.M.; Hassan, A.M. Ameliorative effect of selenium yeast supplementation on the physio-pathological impacts ofchronic exposure to glyphosate and or malathion in *Oreochromis niloticus*. *BMC Vet. Res.* **2022**, *18*, 159. [CrossRef]
78. Shadrack, R.S.; Manabu, I.; Koshio, S.; Waqalevu, V. Physiological condition, digestive enzyme, blood haemato-biochemistry, antioxidant, immune and stress response of juvenile red sea bream (*Pagrus major*) fed diets containing spent oleaginous yeast. *Aquac. Rep.* **2021**, *21*, 100913. [CrossRef]
79. Jiang, Z.; Li, X.; Dong, C. Effect of Feed Supplementation with Bacillus coagulans on Nrf Gene Family Expression in Common Carp (*Cyprinus carpio*) under Long-Term Exposure to Cd^{2+}. *Fishes* **2022**, *7*, 48. [CrossRef]
80. Zhu, S.; Gao, W.; Wen, Z.; Chi, S.; Shi, Y.; Hu, W.; Tan, B. Partial substitution of fish meal by Clostridium autoethanogenum protein in the diets of juvenile largemouth bass (*Micropterus salmoides*). *Aquac. Rep.* **2022**, *22*, 100938. [CrossRef]
81. Karimi, S.; Ferreira, J.A.; Taherzadeh, M.J. Filamentous fungi as animal and fish feed ingredients. In *Current Developments in Biotechnology and Bioengineering*; Elsevier: Amsterdam, The Netherlands, 2023; pp. 399–433.
82. Lewis, M.J.; Francis, D.S.; Blyth, D.; Moyano, F.J.; Smullen, R.P.; Turchini, G.M.; Booth, M.A. A comparison of in-vivo and in-vitro methods for assessing the digestibility of poultry by-product meals using barramundi (lates calcarifer); impacts of cooking temperature and raw material freshness. *Aquaculture* **2019**, *498*, 187–200. [CrossRef]
83. Leiva, A.; Granados-Chinchilla, F.; Redondo-Solano, M.; Arrieta-González, M.; Pineda-Salazar, E.; Molina, A. Characterization of the animal by-product meal industry in Costa Rica: Manufacturing practices through the production chain and food safety. *Poult. Sci.* **2018**, *97*, 2159–2169. [CrossRef]
84. Rimoldi, S.; Terova, G.; Ascione, C.; Giannico, R.; Brambilla, F. Next generation sequencing for gut microbiome characterization in rainbow trout (*Oncorhynchus mykiss*) fed animal by-product meals as an alternative to fishmeal protein sources. *PLoS ONE* **2018**, *13*, e0193652. [CrossRef]
85. Dawson, M.R.; Alam, M.S.; Watanabe, W.O.; Carroll, P.M.; Seaton, P.J. Evaluation of Poultry By-Product Meal as an Alternative to Fish Meal in the Diet of Juvenile Black Sea Bass Reared in a Recirculating Aquaculture System. *N. Am. J. Aquac.* **2018**, *80*, 74–87. [CrossRef]
86. Gaylord, T.G.; Sealey, W.M.; Barrows, F.T.; Myrick, C.A.; Fornshell, G. Evaluation of ingredient combinations from differing origins (fishmeal, terrestrial animal and plants) and two different formulated nutrient targets on rainbow trout growth and production efficiency. *Aquac. Nutr.* **2017**, *23*, 1319–1328. [CrossRef]
87. Zare, M.; Kazempour, M.; Hosseini, H.; Hosseini Choupani, S.M.; Akhavan, S.R.; Rombenso, A.; Esmaeili, N. Fish Meal Replacement and Early Mild Stress Improve Stress Responsiveness and Survival of Fish after Acute Stress. *Animals* **2023**, *13*, 1314. [CrossRef]
88. Sharif, M.; Zafar, M.H.; Aqib, A.I.; Saeed, M.; Farag, M.R.; Alagawany, M. Single cell protein: Sources, mechanism of production, nutritional value and its uses in aquaculture nutrition. *Aquaculture* **2021**, *531*, 735885. [CrossRef]
89. Islam, F.; Salam, M.A.; Rahman, M.A.; Paul, S.I.; Das, T.R.; Rahman, M.M.; Shaha, D.C.; Gupta, D.R.; Alam, M.S.; Islam, T. Plant endophytic yeasts Pichia fermentans and Meyerozyma caribbica improve growth, biochemical composition, haematological parameters and morphology of internal organs of premature *Barbonymus gonionotus*. *Aquac. Rep.* **2021**, *19*, 100575. [CrossRef]
90. Costabile, M. Measuring the 50% Haemolytic Complement (CH50) Activity of Serum. *J. Vis. Exp.* **2010**, *29*, 1923. [CrossRef]
91. Tang, S.; Liu, S.; Zhang, J.; Zhou, L.; Wang, X.; Zhao, Q.; Weng, W.; Qin, J.G.; Chen, L.; Li, E. Relief of hypersaline stress in Nile tilapia *Oreochromis niloticus* by dietary supplementation of a host-derived Bacillus subtilis strain. *Aquaculture* **2020**, *528*, 735542. [CrossRef]
92. Abou-El-Atta, M.E.; Abdel-Tawwab, M.; Abdel-Razek, N.; Abdelhakim, T.M.N. Effects of dietary probiotic *Lactobacillus plantarum* and whey protein concentrate on the productive parameters, immunity response and susceptibility of Nile tilapia, *Oreochromis niloticus* (L.), to Aeromonas sobria infection. *Aquac. Nutr.* **2019**, *25*, 1367–1377. [CrossRef]
93. Taylor, J.W.; Ellison, C.E. Mushrooms: Morphological complexity in the fungi. *Proc. Natl. Acad. Sci. USA* **2010**, *107*, 11655–11656. [CrossRef] [PubMed]
94. Agboola, J.O.; Øverland, M.; Skrede, A.; Hansen, J.Ø. Yeast as major protein-rich ingredient in aquafeeds: A review of the implications for aquaculture production. *Rev. Aquac.* **2021**, *13*, 949–970. [CrossRef]
95. Jach, M.E.; Serefko, A.; Ziaja, M.; Kieliszek, M. Yeast Protein as an Easily Accessible Food Source. *Metabolites* **2022**, *12*, 63. [CrossRef] [PubMed]
96. Yang, X.; Chi, S.; Tan, B.; Nie, Q.; Hu, J.; Dong, X.; Yang, Q.; Liu, H.; Zhang, S. Yeast hydrolysate helping the complex plant proteins to improve the growth performance and feed utilization of Litopenaeus vannamei. *Aquac. Rep.* **2020**, *17*, 100375. [CrossRef]

97. Yuan, X.-Y.; Liu, W.-B.; Liang, C.; Sun, C.-X.; Xue, Y.-F.; Wan, Z.-D.; Jiang, G.-Z. Effects of partial replacement of fish meal by yeast hydrolysate on complement system and stress resistance in juvenile Jian carp (*Cyprinus carpio* var. Jian). *Fish Shellfish Immunol.* **2017**, *67*, 312–321. [CrossRef]
98. Wang, X.; Shen, Z.; Wang, C.; Li, E.; Qin, J.G.; Chen, L. Dietary supplementation of selenium yeast enhances the antioxidant capacity and immune response of juvenile Eriocheir Sinensis under nitrite stress. *Fish Shellfish Immunol.* **2019**, *87*, 22–31. [CrossRef]
99. Chen, M.; Chen, X.-Q.; Tian, L.-X.; Liu, Y.-J.; Niu, J. Enhanced intestinal health, immune responses and ammonia resistance in Pacific white shrimp (*Litopenaeus vannamei*) fed dietary hydrolyzed yeast (*Rhodotorula mucilaginosa*) and *Bacillus licheniformis*. *Aquac. Rep.* **2020**, *17*, 100385. [CrossRef]
100. Jin, M.; Xiong, J.; Zhou, Q.-C.; Yuan, Y.; Wang, X.-X.; Sun, P. Dietary yeast hydrolysate and brewer's yeast supplementation could enhance growth performance, innate immunity capacity and ammonia nitrogen stress resistance ability of Pacific white shrimp (*Litopenaeus vannamei*). *Fish Shellfish Immunol.* **2018**, *82*, 121–129. [CrossRef]
101. Money, N.P. Fungal Cell Biology and Development. In *The Fungi*; Elsevier: Amsterdam, The Netherlands, 2016; pp. 37–66.
102. Kües, U.; Khonsuntia, W.; Subba, S. Complex fungi. *Fungal Biol. Rev.* **2018**, *32*, 205–218. [CrossRef]
103. Poirier, M.; Hugot, C.; Spatz, M.; Da Costa, G.; Lapiere, A.; Michaudel, C.; Danne, C.; Martin, V.; Langella, P.; Michel, M.-L.; et al. Effects of Five Filamentous Fungi Used in Food Processes on In Vitro and In Vivo Gut Inflammation. *J. Fungi* **2022**, *8*, 893. [CrossRef]
104. Rulli, M.M.; Villegas, L.B.; Barcia, C.S.; Colin, V.L. Bioconversion of sugarcane vinasse into fungal biomass protein and its potential use in fish farming. *J. Environ. Chem. Eng.* **2021**, *9*, 106136. [CrossRef]
105. Jannathulla, R.; Dayal, J.S. Beneficial effects, challenges and opportunities of the filamentous fungus, *Aspergillus niger* with special reference to the shrimp feed industry—A review. *Rev. Aquac.* **2023**, *15*, 1311–1334. [CrossRef]
106. Dawood, M.A.O.; Eweedah, N.M.; Moustafa, E.M.; Farahat, E.M. Probiotic effects of *Aspergillus oryzae* on the oxidative status, heat shock protein, and immune related gene expression of Nile tilapia (*Oreochromis niloticus*) under hypoxia challenge. *Aquaculture* **2020**, *520*, 734669. [CrossRef]
107. Strong, P.J.; Self, R.; Allikian, K.; Szewczyk, E.; Speight, R.; O'Hara, I.; Harrison, M.D. Filamentous fungi for future functional food and feed. *Curr. Opin. Biotechnol.* **2022**, *76*, 102729. [CrossRef] [PubMed]
108. Amara, A.A.; El-Baky, N.A. Fungi as a Source of Edible Proteins and Animal Feed. *J. Fungi* **2023**, *9*, 73. [CrossRef] [PubMed]
109. Mohan, K.; Ravichandran, S.; Muralisankar, T.; Uthayakumar, V.; Chandirasekar, R.; Seedevi, P.; Rajan, D.K. Potential uses of fungal polysaccharides as immunostimulants in fish and shrimp aquaculture: A review. *Aquaculture* **2019**, *500*, 250–263. [CrossRef]
110. Prabawati, E.; Hu, S.-Y.; Chiu, S.-T.; Balantyne, R.; Risjani, Y.; Liu, C.-H. A synbiotic containing prebiotic prepared from a by-product of king oyster mushroom, *Pleurotus eryngii* and probiotic, *Lactobacillus plantarum* incorporated in diet to improve the growth performance and health status of white shrimp, Litopenaeus vannamei. *Fish Shellfish Immunol.* **2022**, *120*, 155–165. [CrossRef]
111. Shyamala, V.; UmaMaheswar, N. *Immunostimulation of Dietary Prebiotic Mushroom and Disease Resistant of Oreochromis mossambicus against Septicemia*; India Journal of Natural Sciences 2021; TamilNadu Scientific Research Organisation: Pudukkottai, Tamil Nadu, India, 2021; Volume 12.
112. Lee, P.-T.; Wu, Y.-S.; Tseng, C.-C.; Lu, J.-Y.; Lee, M.-C. Dietary *Agaricus blazei* Spent Substrate Improves Disease Resistance of Nile Tilapia (*Oreochromis niloticus*) against Streptococcus agalactiae In Vivo. *J. Mar. Sci. Eng.* **2022**, *10*, 100. [CrossRef]
113. Chang, C.-C.; Lu, Y.-C.; Wang, C.-C.; Ko, T.-L.; Chen, J.-R.; Wang, W.; Chen, Y.-L.; Wang, Y.-W.; Chang, T.-H.; Hsu, H.-F.; et al. Antrodia cinnamomea Extraction Waste Supplementation Promotes Thermal Stress Tolerance and Tissue Regeneration Ability of Zebrafish. *Molecules* **2020**, *25*, 4213. [CrossRef] [PubMed]
114. Ahmed, M.; Abdullah, N.; Shuib, A.S.; Abdul Razak, S. Influence of raw polysaccharide extract from mushroom stalk waste on growth and pH perturbation induced-stress in Nile tilapia, *Oreochromis niloticus*. *Aquaculture* **2017**, *468*, 60–70. [CrossRef]
115. Rema, P.; Saravanan, S.; Armenjon, B.; Motte, C.; Dias, J. Graded Incorporation of Defatted Yellow Mealworm (*Tenebrio molitor*) in Rainbow Trout (*Oncorhynchus mykiss*) Diet Improves Growth Performance and Nutrient Retention. *Animals* **2019**, *9*, 187. [CrossRef] [PubMed]
116. Alfiko, Y.; Xie, D.; Astuti, R.T.; Wong, J.; Wang, L. Insects as a feed ingredient for fish culture: Status and trends. *Aquac. Fish.* **2022**, *7*, 166–178. [CrossRef]
117. Cardinaletti, G.; Di Marco, P.; Daniso, E.; Messina, M.; Donadelli, V.; Finoia, M.G.; Petochi, T.; Fava, F.; Faccenda, F.; Contò, M.; et al. Growth and Welfare of Rainbow Trout (*Oncorhynchus mykiss*) in Response to Graded Levels of Insect and Poultry By-Product Meals in Fishmeal-Free Diets. *Animals* **2022**, *12*, 1698. [CrossRef] [PubMed]
118. Motte, C.; Rios, A.; Lefebvre, T.; Do, H.; Henry, M.; Jintasataporn, O. Replacing Fish Meal with Defatted Insect Meal (Yellow Mealworm *Tenebrio molitor*) Improves the Growth and Immunity of Pacific White Shrimp (*Litopenaeus vannamei*). *Animals* **2019**, *9*, 258. [CrossRef]
119. Sankian, Z.; Khosravi, S.; Kim, Y.-O.; Lee, S.-M. Effects of dietary inclusion of yellow mealworm (*Tenebrio molitor*) meal on growth performance, feed utilization, body composition, plasma biochemical indices, selected immune parameters and antioxidant enzyme activities of mandarin fish (*Siniperca scherze*). *Aquaculture* **2018**, *496*, 79–87. [CrossRef]
120. Belghit, I.; Waagbø, R.; Lock, E.-J.; Liland, N.S. Insect-based diets high in lauric acid reduce liver lipids in freshwater Atlantic salmon. *Aquac. Nutr.* **2019**, *25*, 343–357. [CrossRef]

121. Tippayadara, N.; Dawood, M.A.O.; Krutmuang, P.; Hoseinifar, S.H.; Doan, H.V.; Paolucci, M. Replacement of Fish Meal by Black Soldier Fly (*Hermetia illucens*) Larvae Meal: Effects on Growth, Haematology, and Skin Mucus Immunity of Nile Tilapia, *Oreochromis niloticus*. *Animals* **2021**, *11*, 193. [CrossRef]
122. Taufek, N.M.; Simarani, K.; Muin, H.; Aspani, F.; Raji, A.A.; Alias, Z.; Razak, S.A. Inclusion of cricket (*Gryllus bimaculatus*) meal in African catfish (*Clarias gariepinus*) feed influences disease resistance. *J. Fish.* **2018**, *6*, 623–631. [CrossRef]
123. Li, Y.; Kortner, T.M.; Chikwati, E.M.; Belghit, I.; Lock, E.-J.; Krogdahl, Å. Total replacement of fish meal with black soldier fly (*Hermetia illucens*) larvae meal does not compromise the gut health of Atlantic salmon (*Salmo salar*). *Aquaculture* **2020**, *520*, 734967. [CrossRef]
124. Cho, J.-H.; Bae, J.; Hwang, I.J. Effects of black soldier fly (*Hermetia illucens*) pre-pupae meal on the growth, stress, and immune responses of juvenile rainbow trout (*Oncorhynchus mykiss*) reared at different stocking densities. *Aquac. Rep.* **2022**, *25*, 101202. [CrossRef]
125. Martinez-Soberanes, E.E.; Mustafa, R.; Reaney, M.J.T.; Zhang, W.J. Seed Hull Utilization. In *Food Wastes and By-Products*; Wiley: Hoboken, NJ, USA, 2020; pp. 291–326.
126. Galeana-López, J.A.; Hernández, C.; Leyva-López, N.; Lizárraga-Velázquez, C.; Sánchez-Gutiérrez, E.; Basilio Heredia, J. Corn husk extracts as an antioxidant additive in diets for Nile tilapia (*Oreochormis niloticus*) fingerlings: Effect on growth performance, feed intake and toxicity. *Biotecnia* **2020**, *22*, 147–154. [CrossRef]
127. Galeana-López, J.A.; Lizárraga-Velázquez, C.E.; Hernández, C.; Leyva-López, N.; Heredia, J.B. Corn Husk Phenolics Modulate Hepatic Antioxidant Response in Nile Tilapia (*Oreochromis niloticus*) Exposed to Hypoxia. *Molecules* **2021**, *26*, 6161. [CrossRef]
128. Amaral, D.; Filipe, D.M.; Cavalheri, T.F.; Vieira, L.; Magalhães, R.P.; Belo, I.; Peres, H.; Ozório, R.O.d.A. Solid-State Fermentation of Plant Feedstuff Mixture Affected the Physiological Responses of European Seabass (*Dicentrarchus labrax*) Reared at Different Temperatures and Subjected to Salinity Oscillation. *Animals* **2023**, *13*, 393. [CrossRef]

animals

MDPI

Review

The Role of Insects in Sustainable Animal Feed Production for Environmentally Friendly Agriculture: A Review

Csaba Hancz [1], Sadia Sultana [1], Zoltán Nagy [2] and Janka Biró [2,*]

[1] Kaposvár Campus, Hungarian University of Agriculture and Life Sciences, Guba S. 40., 7400 Kaposvár, Hungary; sadiazoology@nstu.edu.bd (S.S.)
[2] Research Center for Fisheries and Aquaculture, Institute of Aquaculture and Environmental Safety, Hungarian University of Agriculture and Life Sciences, Anna-liget 35., 5540 Szarvas, Hungary
* Correspondence: nagyne.biro.janka@uni-mate.hu

Simple Summary: This paper provides an overview of the key issues concerning the role of edible insects in sustainable feed production and environmentally friendly agriculture. The use of insect-derived feeds for animal production is presented in detail and discussed for the major terrestrial livestock and aquaculture groups.

Abstract: The growing demand for animal protein, the efficient use of land and water, and the limitations of non-renewable energy sources highlight the global importance of edible insects. This paper provides an overview of the key issues regarding the role of edible insects in sustainable feed production and environmentally friendly agriculture. The indispensable ecological services provided by insects are discussed, as well as the farming, products, and nutritional value of edible insects. A representative selection of the literature reviewing major insect species' chemical compositions and nutritional value is also presented. The use of insect-derived feeds for animal production is presented in detail and discussed for the major terrestrial livestock and aquaculture groups.

Keywords: insects; sustainability; ecosystem services; insect products

1. Introduction

Sustainability, circularity, and environmental friendliness are widely acknowledged to be essential components of contemporary human endeavors. Simultaneously, food production for a rapidly expanding global population remains a pressing concern in the face of scarce opportunities for expanding arable land. To ensure high levels of sustainable agricultural production, it is crucial to incorporate all available knowledge. Thus, the significance of insects must be re-evaluated in this context. Dunkel and Payne [1] offer a comprehensive overview of the global significance of edible insects in their introduction to "Insects as Sustainable Food Ingredients," highlighting the growing demand for animal-based protein, the efficient use of land and water, and the limitations of non-renewable energy sources. Highlighting the increasing concern for sustainability, Guiné et al. [2] present important findings on the efficiency of insects in comparison to other farm animals. They note that the environmental impact of insect production encompasses feed conversion, land use, and water consumption. Insects require the least amount of feed, land, and water, followed by chickens, pigs, and cows. In contrast, cows require over five times as much feed, land, and water as insects. Calculations for protein efficiency, expressed in terms of protein concentration, reveal that beef, pork, and poultry have protein values of 190 g/kg, 50 g/kg, and 200 g/kg, respectively, while crickets have values of 154 g/kg and 205 g/kg for nymphs and adults, respectively [2].

Insects represent the largest and most diverse group of animals, with over a million described species, accounting for more than half of all known living organisms, i.e., over 90% of the animal life forms on Earth. The roles of insects are fundamental and diverse.

They play a crucial role in providing ecosystem services, including pollination, biological control, the provision of food, and organic matter recycling. Insects represent a dominant component of biodiversity across most terrestrial ecosystems and are critical to nutrient cycling and overall ecosystem functioning [3,4]. Although herbivorous insects cause damage to about 18% of global agricultural production, fewer than 0.5% of all known insect species are considered pests [5]. Aquatic insects are also highly diverse, with over 100,000 species inhabiting freshwater ecosystems, and they play crucial ecological roles, such as decomposing organic matter and transferring energy between trophic levels. These hexapods serve as links in the nutrient cycle, and their biological interactions have substantial impacts on the community structure within the freshwater ecosystem [6].

Insects possess the potential to tackle contemporary global challenges and advance the Sustainable Development Goals (SDGs) owing to their significant influence on various ecosystem services [7]. Ecosystem services are frequently undervalued; hence, a lucid illustration of this, in the context of insects, would be valuable (see Figure 1). However, Schowalter et al. [8] introduced the concept of ecosystem disservices, which serves as an important counterpoint to insect ecosystem services and warrants due consideration. Insect larvae can provide atypical ecosystem services, such as plastic degradation and biofuel production, which have the potential to aid in mitigating the effects of climate change, as suggested by Morimoto and Harris [9], and could become a priority in the future.

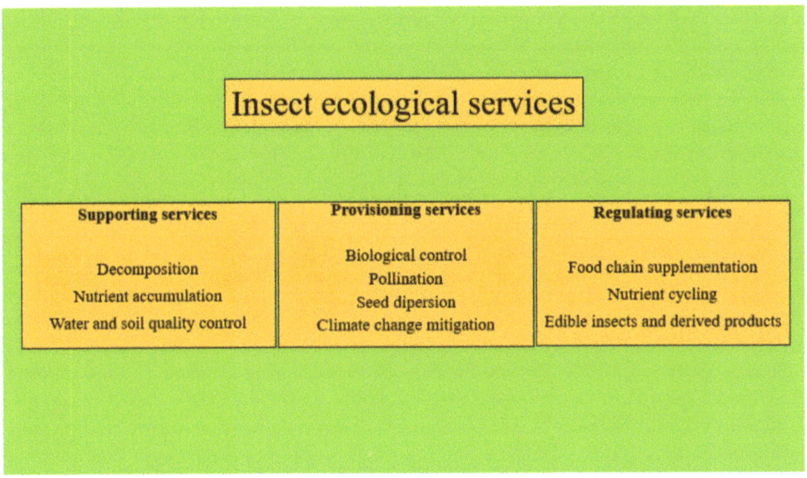

Figure 1. Ecological service categories for insects. Sources: [4,10].

In addition to ecosystem services, another fundamental benefit of insects is their mass production based on agricultural and other organic wastes. Insects efficiently bioconvert organic waste into new products, mirroring the principle of a circular economy, providing benefits such as reduced waste management costs and resource use in protein and fat production through insect-derived feed [11].

Insects have served as food for humans since the earliest times, and their consumption (entomophagy) is included in traditions in many areas of the world. It is impossible to find reliable data on the amount of direct human consumption, but currently, according to Govorushko [12], 2 billion people in 113 countries across Africa, Asia, Central and South America, and Australia consume insects, although Western attitudes towards this type of food are certainly ambiguous. Liceaga et al. [13] provide a nicely illustrated review of the current situation regarding the potential use of insects and insect products for human consumption.

However, in their stimulating and thorough review, Akhtar and Isman [14] emphasize that in situations where there is an inadequate supply of conventional animal protein sources such as beef, pork, and chicken, alternative sources such as insects need to be found. Edible insects have great potential as an environmentally friendly choice for future food systems, primarily because of their high nutritional content. In addition, insects have greater food conversion efficiency and produce lower levels of greenhouse gas emissions (GHGEs) while requiring less water and land than vertebrates, a serious consideration for present and future generations.

A decade ago, the FAO called for increased use of forage insects in feed production to conserve the environment and resources and ensure food and feed security in the face of a growing world population [15]. An overview of recent developments in the use of insects as food and feed was provided by van Huis [16], who reviewed the literature on edible insects in recent years and noted that attention shifted from insect harvesting to insect farming about a decade ago. In addition, most production is for pet food but will soon turn to aquafeed. (As discussed below, other major animal production sectors are also increasingly using insect meal in feeds.) In addition to the obvious economic and environmental considerations, the insect food and feed sector is developing rapidly thanks to an increasingly supportive regulatory framework. This latter aspect has recently been highlighted by Sogari et al. [17], who, among other things, summarized the pertinent EU regulatory framework, according to which insects for human consumption fall under the category of so-called novel foods (NFs) (as of 2018). Chia et al. [18] discuss another important aspect of insect farming: it has the potential to promote inclusive business opportunities for smallholder farmers within the agribusiness value chain while also contributing to solving socio-economic and environmental problems in developing countries in ways that align with the United Nations' Sustainable Development Goals.

Finally, this section provides an overview of worldwide market statistics and expectations related to edible insects. According to market research, the Edible Insects Market is projected to reach USD 9.60 billion by 2030, with a Compound Annual Growth Rate (CAGR) of 28.3% during the forecast period of 2022–2030. In terms of volume, the market is expected to reach 3,139,035.10 tonnes by 2030, with a CAGR of 31.1% during the forecast period 2022–2030. The edible insect market is growing due to increased greenhouse gas emissions from the livestock and poultry industries, the high nutritional value insects offer, the environmental advantages of eating them, the rising demand for insect-derived protein in animal feed, and the low risk of acquiring zoonotic diseases when consuming them. Based on products, the whole segment is projected to make up the greatest share of the edible insect market in 2022. This is primarily due to the readily available and lower cost of whole insects compared to processed insects. Additionally, the lack of insect-processing facilities in certain regions of the world and the increasing demand for insects in the animal feed industry contribute to the growth of the whole insect segment. Regarding the types of insect, black soldier fly production is expected to experience significant growth during the 2020–2030 forecast period. This growth is being driven by the expanding aquaculture industry, rising demand for alternative proteins in the animal feed sector due to increased soy meal prices, government backing of insect meal use in livestock feed, and increased investment by major players in the black soldier fly industry. Based on its end use, the segment of human consumption is anticipated to rapidly grow during the 2020–2030 forecast period. This growth is attributed to several factors, including the increasing demand for insect-based foods to feed a growing global population, the high nutritional value insects can provide to humans, and the rising demand for environmentally friendly protein alternatives. Based on geography, the Asia–Pacific region is expected to account for the biggest portion of the edible insect market in 2022 [19].

2. The Role of Insects in Agriculture

Intensified agricultural production systems use huge external inputs to increase yields, but they also destabilize ecosystems [20]. Biodiversity is the key to sustainable agricultural production and, more broadly, a livable planet. Understanding the role of insects in ecosystems will help in achieving this goal. As briefly mentioned above, insects also play an important role in agroecology by acting as drivers of ecosystem functions. Ecologically sustainable management of agricultural systems includes the use of beneficial organisms [21]. According to Jankielsohn [5], the main functions of insects in the ecological processes of terrestrial ecosystems are nutrient cycling, seed dispersal, bioturbation [22,23], pollination [24,25], and biological pest control [26,27]. Cock et al. [28] group important invertebrates in agricultural ecosystems into soil invertebrates, biological control agents (BCAs), and pollinators. Follett et al. [29] provide a new perspective on BCAs, including with respect to environmentally friendly methods such as intercropping and insect habitat modification. In a review of mass-produced insects for biological control, most of which are used for agricultural purposes, Francuski and Beukeboom [30] list 33 species. Also worth mentioning is their subcategory of sterile insect technology, which includes nine species.

As the mass rearing of insects for food and feed is undergoing full industrial development as an effective alternative to animal protein production, its by-product, insect excreta (frass), must also be considered for use as an organic fertilizer to replace the use of artificial ones, contributing to sustainable agriculture and a circular economy [31]. Insect frass and exuviae, which are mostly composed of chitin, exert beneficial effects in a variety of ways, such as on plant growth, resistance, and reproduction, or by stimulating the soil microbial community [32]. For these reasons, they could certainly be a valuable alternative to conventional fertilizers and pesticides [17].

3. Edible Insects, Farming, Products, and Nutritional Value

Francuski and Beukeboom [30] provided an excellent review of the literature on insect mass production. They list a total of 62 species in the following categories: biological control (33), food (6), feed (5), pollination (4), industrial production (3), medicine and cosmetics (4), waste management (2), and research (5). According to Nolan et al. [33], about 2300 edible insects have been reported, belonging to a large number of different phyla and species and having diverse sets of morphological characteristics, showing great variation in composition and nutritional aspects. Factors affecting the digestibility and bioavailability of the digested proteins were studied, and the effect of the digested edible insect protein on the gut microbiota was summarized in their work; additionally, a critical evaluation of the positive and negative aspects of edible insect protein consumption was given.

Cortes Ortiz et al. [34] provided an overview of all aspects of insect mass production, covering its history, which began with pest control objectives in the 1970s; the species used and their respective applications; production techniques; and environmental control. Like many others, they emphasize that developments in insect mass production will be directed toward food and feed production. The benefits of insect production and its current status in Europe are well illustrated by Thrastardottir et al. [35], whose review focuses on the most popular insects farmed in Europe, the yellow mealworm, *Tenebrio molitor*, and the black soldier fly (BSF), *Hermetia illucens*, together with the main obstacles and risks. According to Hong et al. [36], *T. molitor* larvae are being mass-produced as feed for pets, zoo animals, and production animals. These larvae are a highly sustainable protein source, containing high-quality protein present in high quantities, and can replace soybean meal or fishmeal. Supplementation with *T. molitor* larvae improved the growth performance of broiler chickens without having negative effects on traits [37]. In studies involving swine, supplementation of *T. molitor* larvae improved performance and protein utilization for growing pigs [38].

The main species involved in mass production worldwide are listed in Table 1.

For the time being, progress has reached a level where the effect of strain on the overall performance of yellow mealworms (*T. molitor*) has been studied. Rumbos et al. [39] found that an Italian strain had the best growth performance in terms of survival, larval biomass production, development time, and feed utilization. The results of their study highlight the need to select strains that can increase the overall productivity of insect farming systems.

Table 1. Major insect species farmed for food and/or feed.

Scientific Name	Common Name	Developmental Stage	Food	Feed
Acheta domesticus	House cricket	Adult	x	x
Tenebrio molitor	Mealworm	Larvae	x	x
Gryllus bimaculatus	Mediterranean field cricket	Adult		x
Bombyx mori	Silkworm	Larvae, pupae	x	x
Galleria mellonella	Wax worm	Larvae	x	
Apis mellifera	European honeybee	Adult	x	
Musca domestica	Common housefly	Larvae		x
Lucilia sericata	Common green bottle fly	Larvae (maggot)		x
Rhynchophorus ferrugineus	Red palm weevil	Larvae, pupae	x	
Rhynchophorus phoenicis	Palm weevil	Larvae	x	
Pachnoda marginata	Sun beetle	Larvae	x	x
Hermetia illucens	Black soldier fly (BSF)	Larvae	x	x

Sources: [34,40].

The metabolic flexibility and efficiency of insects are certainly the most important aspects of mass production for food and feed. Ramos-Elorduy [41] emphasizes that this is mainly based on the well-known fact that insects are poikilothermic. According to her data, the energy content of edible insects varies between species and regions, but in general, Coleoptera and Lepidoptera species provide more energy. The energy values for livestock are 165–705 kcal/100 g, and those for vegetables 308–352 kcal/100 g, while edible insects provide 217–777 kcal/100 g, and insects raised on organic waste provide 288–575 kcal/100 g. Waldbauer [42] provided perhaps the first comprehensive review of insect food consumption and utilization, analyzing differences among species, diets, and environmental factors. According to Maino and Kearney [43], insects are characterized by their small size, large numbers, impressive reproductive output, and rapid growth that follows a qualitatively different trajectory than that of many other animals. Their model of insect growth predicts that energy reserves per biomass increase with age, implying higher production efficiency and biomass energy density in later instars. This model also suggests that insects achieve greater production efficiency and higher growth rates by increasing specific assimilation and increasing energy reserves per biomass, which are less costly to maintain than structural biomass. A long path of research and development led to the study of Halloran et al. [44], who were already able to analyze the environmental impacts of different insect production systems based on life cycle inventories. In addition, the need for some genetic work to increase the productivity of insect farming systems has also emerged to take advantage of the given metabolic characteristics [39].

There is an abundance of research discussing the chemical composition and nutritional value of insects. Weru et al. [45], in their review on edible insects as a suitable source for human nutrition, screened hundreds of articles. Their data showed a large variation in nutritional value, even within species, according to sex, geographical origin, and growth stage. It has to be mentioned that the quality of much of the published data was poor because of the small number of samples analyzed.

Table 2 contains mean data on crude protein and crude fat content, constituting the main characteristics of the nutritional value of some edible insects. Some references are given where the standard deviation shows rather high values, demonstrating that the chemical composition of insects is highly variable due to several factors, such as species, diet, sex, developmental stage, and processing method.

Table 2. The crude protein (CP) and crude fat (CF) content of some important edible insect species (dry weight, %).

Scientific Name	Common Name	CP	CF	References
Orthoptera	Crickets, locusts, grasshoppers (adult)	61.3	13.4	[46]
		66.1	21.9	[47]
	house crickets	63.3 ± 5.7	17.3 ± 6.3	[48]
	grasshoppers	50.5	15.3	[49]
Tenebrio molitor	Mealworm larvae defatted meal	47.1–49.4	35.2–38.1	[46]
		58.4	41.9	[47]
		52.8 ± 4.2	36.1 ± 4.1	[48]
		55.8	25.2	[49]
		56.5	6.2	[50]
Zophobas morio	Superworm	46.8	42.0	[47]
Bombyx mori	Silkworm spent pupae larvae pupae meal	58.0	35.0	[46] [48]
		53.8–69.8	8.1–9.5	
		60.7 ± 7.0	25.7 ± 9.0	
Galleria mellonella	Wax worm	34.0	60.0	[47]
Musca domestica	Common housefly	64.0	24.3	[46]
	larvae	63.1	15.5	[48]
	pupae	50.4	18.9	[48]
Hermenia illucens	Black soldier fly (BSF) defatted meal	45.1	36.1	[47]
		56.1	25.2	[49]
		52.5	9.3	[50]

The nutritional value of insect meal cannot be assessed without its amino acid profile and digestibility, constituting exactly what can be found in the Makkar et al.'s [48] paper, which also examined the latter according to consumer species. According to Sánchez-Muros et al. [51], who provided a long list of the chemical compositions of species, insects can have an adequate amino acid profile, depending on the insect species. The most common limiting amino acids are histidine, lysine, and tryptophan, which can be incorporated into the diet.

The other component of the insect body to consider is chitin. Chitin has both beneficial functions (e.g., prebiotic effects) and putative negative effects on digestibility [47]. However, chitin is a relatively small component of most insects; e.g., Sándor et al. [50] found its proportions to be 9.6 and 5.5% in dry matter in BSF and mealworm meal, respectively. On the other hand, some animal species have chitinases in their gastrointestinal tracts. Where this is not the case, the inclusion of insect meal in feed must certainly be limited [52].

In addition to nutritional value, insects also contain bioactive compounds that serve as health stimulators in livestock animals. Animal researchers regularly focus on three categories of bioactive compounds present in insects: antimicrobial peptides (AMPs), fatty acids (especially lauric acid), and polysaccharides such as chitin and chitosan. Each of them demonstrates antimicrobial activity through various mechanisms, such as forming or destroying membrane pores, impeding intracellular processes, or supporting the host's immune system. Additionally, the antioxidant capabilities of insect proteins may protect against oxidative tissue damage. The combination of AMPs, which destroy the bacterial cell envelope with fatty acids, and chitin or chitosan may provide a solution to antimicrobial resistance [53].

4. Use of Insect-Derived Feeds in Animal Production

Insect meal is considered a viable solution to mitigating the limitations imposed by natural resource depletion, climate change, and the consumption rivalry between food, feed, and fuel. Due to their increasing demand and resulting cost escalation, primary protein sources for animal nutrition such as soybeans, peas, and fish meal are no longer sustainable for long-term use. Insects like black soldier fly larvae, crickets, and mealworms can effectively supplement feed sources by offering valuable energy, protein, and fat for an animal's diet [54].

The costs of conventional feed resources such as soybean meal and fishmeal are very high, and their availability will be limited in the future. Insect farming could therefore be part of the solution. The nutritional quality of black soldier fly larvae, house fly maggots, mealworms, locusts/grasshoppers/crickets, and silkworm meal and their use as replacements for soybean meal and fishmeal in the diets of poultry, pigs, fish species, and ruminants were discussed by Makkar et al. [48]. Their main findings are as follows: the protein content of these alternative resources is as high as 42 to 63%, and so is their lipid content, which can be extracted and used for various applications, even for biodiesel production. The concentration of unsaturated fatty acids is also high but varies between species. According to the studies discussed, the palatability of these alternative diets is good for the animals, and these resources can replace 25–100% of soybean meal or fishmeal, depending on the species. Digestibility analyses of dog food containing BSF larvae meal as the sole source of protein (36.5% inclusion) showed values that off meat-based diet, indicating its suitability as a sustainable protein source for pet food [55]. However, most types of insect meal are deficient in calcium, so its supplementation is necessary, especially for growing animals and laying hens. Gasco et al. [56], besides reviewing the factors affecting the decision-making process regarding using insect-based products, also provide a comprehensive summary of the field, ranging from the species involved to processing methods and questions of safety, cost, availability, and the consistency of supply and legislation. Important aspects of product quality such as digestibility, palatability, and even pelletability are also discussed.

Insect-derived feeds are part of many animal feeding programs because they provide a sustainable way of recycling waste into nutrient-rich ingredients. Secondly, since many animals naturally consume insects, their inclusion in feeding programs can actually improve animal welfare. In addition, the nutrient composition and availability of insect-derived ingredients are generally very high relative to requirements and formulation needs, and components such as chitin, proteins, and fatty acids with antimicrobial activity provide additional benefits, as Koutsos [57] summarized. Last but not least, serious economic considerations will also appear soon with regard to using them to replace other animal protein sources, especially fishmeal. However, due to low produced quantities and high production costs, the price of insect-based proteins is still high and not competitive when compared to that of fishmeal or soybean meal, although continued efforts to upscale production will increase product availability and quality/consistency and reduce costs [58]. However, the data provided by Pinotti et al. [59] deserve attention. They compared the prices of different insect meals with those of soybean and fish meal and found that the defatted meal of BSF and crickets were 3 and 7–12 times more expensive, but if the comparison was made on 100 g of protein base, these values were 2 and 6–9. Shah et al. [49] made a similar comparison based on a literature review that included housefly maggots, mealworms, and BSF. According to their evaluation, housefly maggots and mealworms showed very favorable values both concerning product price and that calculated for CP, while the values of BSF were much higher. However, it should be noted that these data for BSF came from only two review articles, so their accuracy may be questionable.

The main and most important insect-derived products are meal and oils, which are used as feed ingredients in increasing quantities and proportions for a growing number of species [60]. Although insects are a rich source of fat that could be very well used in animal feeds, facilitating even pelleting, this practice is not widespread due to this product's short shelf life. Other beneficial effects of insect oil, namely, that of BSF larvae, were revealed by Prachumchai and Cherdthong [61], who found, in their study using in vitro gas production, that adding 4% black soldier fly larva oils (BSFOs) at different roughage (R)-to-concentrate (C) ratios increased propionate levels, decreased methane emission, and preserved dry matter (DM) degradability. In addition, insect oils have other applications, such as use in green cosmetics or biodiesel production [17].

5. The Use of Insect Meal in the Feeding of Terrestrial Farm Animals

5.1. Ruminants

While live insects and their products have been shown to be suitable sources of protein and fat that can be used in the diets of farmed monogastric animals, very little information is available on the effects of insect inclusion in ruminant diets on feed digestibility and performance. An important reason for this is that the use of insects in ruminant diets is currently banned in many countries worldwide due to the risk of BSE, although there is no evidence to date that insects carry and transmit prions [62]. Nevertheless, some in vitro studies have highlighted the relatively low digestibility of insects for ruminants, unless developing processing technologies can improve this aspect. Furthermore, in vivo feeding studies are needed to evaluate the impact of live insect products and by-products on the quantitative and qualitative aspects of ruminant production and health [63]. The main conclusion for the future, according to Renna et al. [64], is that a major research effort is needed in the coming years to achieve a thoughtful nutritional evaluation of insect-derived products (i.e., whole and defatted meals, oils, and other ingredients). There is also a need to establish specific recommended inclusion levels for ruminants, taking into account the wide variation in insect composition due to species and production systems. It would be advisable to evaluate the use of insect products in commercial ruminant diets, taking into account their economic and environmental impacts on different farming systems.

An example of the much-needed research in this area was published by Toral et al. [65], who compared the in vitro ruminal protein degradation of four insect species using three methods. They found that all the analyzed techniques appeared to provide similar rankings, with good correlations between methods, particularly between regression and in situ results. Irrespective of the methods, nitrogen from the four insects did not show high ruminal degradation (41–76%). However, among the species studied, *T. molitor* showed the lowest and highest values of ruminal N degradation and intestinal digestibility, respectively, so it may be the best option for replacing dietary soybean meal in ruminant diets.

5.2. Pigs

Although they primarily focus on the Australian context and BSF, as a protein and fat source, DiGiacomo and Leury's [62] literature review provides valuable insights into the use of insect meal in pig feeds. Their review indicates that while fat content affects palatability and digestibility, BSF larvae constitute a suitable ingredient for pig diets, as demonstrated by previous studies [66]. Since the inclusion level was 33% in this early experiment with 5-week-old barrows, the observed decrease in dry matter digestibility can be understood, as well as the increase in feed intake. In a subsequent study [67], where early-weaned pigs were fed with 0%, 50%, and 100% replacement rates of dried plasma with black soldier fly larvae, it was found that 50% replacement increased while 100% decreased pigs' performance. Replacing the soybean meal in growing pigs' diets with 50, 75, and 100% partially defatted BSF meal (61% CP and 14% lipid) had no adverse effect on pork quality or sensory parameters. Furthermore, the inclusion of BSF larvae improved juiciness in the supplemented groups. Pigs fed with BSF larvae produced back fat with a higher polyunsaturated fatty acid content, likely due to the larvae's high-fat content [68]. Biasato et al. [69] conducted an experiment where soybean meal was replaced with BSF larvae at rates of 0%, 30%, and 50% in the diets of weaned pigs. This study revealed a linear increase in daily feed intake; however, no effects on growth were observed. Jin et al. [38] replaced soybean meal with dried mealworms at lower rates, namely, 1.5%, 3%, 4.5%, and 6%, which caused a linear increase in growth and digestibility parameters in weaned female piglets.

Among edible insects, according to Hong et al. [70], the black soldier fly, yellow mealworm, and common housefly are viable protein source alternatives for pigs. Both can replace fish meal in diets for weaned pigs without having any adverse effects on growth performance or nutrient digestibility. Moreover, insect products exhibit higher standardized ileal digestibility values of amino acids compared to conventional animal proteins in growing pigs [70].

T. molitor larvae have been industrially produced as feed for pets, zoo animals, and even production animals. They represent a protein-rich alternative to soybean meal or fishmeal, boasting a high quality and quantity of protein and an optimal amino acid profile. Therefore, they are considered an environmentally sustainable protein source. In swine studies, supplementing mealworm larvae enhanced growth performance and protein utilization for weaning pigs. Additionally, growing pigs exhibited greater amino acid digestibility when fed 10% *T. molitor* larvae than when they were fed conventional animal proteins [70]. However, Hong et al. [70] suggest that further studies are required on the optimal inclusion level of insect products in every phase of pig diets, from weaned pigs to sows. This is because previous studies have predominantly been conducted on weaned pigs and using a limited number of insect products.

5.3. Poultry

De Marco et al. [37] discovered how many of the nutrients in two types of insect larval meals (*T. molitor* and *H. illucens*) were digested by broiler chickens. They also figured out how much energy the chickens acquired from the meals. Finally, they measured the amounts of amino acids that were digested in the gut. The study tested three diets: a basic diet and two diets made by replacing 250 g/kg (*w/w*) of the basic diet with either *T. molitor* meal or *H. illucens* meal. The coefficients of total tract apparent digestibility (CTTAD) of the nutrients for both insect larval meals did not differ significantly, though differences were found for the CTTAD of ether extract for apparent metabolizable energy (AME) or nitrogen-corrected apparent metabolizable energy (AMEn) (AME = 16.86 and 17.38 MJ/kg DM, respectively; AMEn = 16.02 and 16.60 MJ/kg DM, respectively). This study concluded that both *T. molitor* and *H. illucens* meal can be viable sources of AME and easily digestible amino acids (AAs).

Schiavone et al. [71] found that using fat derived from black soldier fly larvae (BSFL) as a replacement for soybean oil in broiler chickens' diets did not affect growth rates, feed preferences, blood characteristics, carcass features, or meat quality. Regardless of the inclusion of BSFL, different pieces of broiler chicken breast meat had similar levels of crude protein and fat content and showed comparable thawing loss. In addition, pH, color values, and drip loss were not affected by dietary treatments during refrigerated storage on day 0 and day 9. The fatty acid composition of broiler chicken breasts was significantly influenced by the amount of BSFL included in the diet. As the inclusion rate of BSFL increased, the proportion of saturated fatty acids increased, while the polyunsaturated fatty acid fraction decreased. On the other hand, the monounsaturated fatty acid fraction remained unchanged. The conclusion was reached that the inclusion of BSFL ensured satisfactory productivity, carcass characteristics, and overall meat quality. Therefore, BSFL has the potential to be a promising new chicken feed ingredient.

Józefiak et al. [72] assessed the impact of full-fat insect meal on performance and microbiota composition in the gastrointestinal tracts of broiler chickens. The trial incorporated *Gryllodes sigillatus*, *Shelfordella lateralis*, *Gryllus assimilis*, *T. molitor*, and *H. illucens* in variable proportions, ranging from 0.05% to 0.2%. This study found that the inclusion of insect meal in the broilers' diets had no effect on their growth performance, but an increase in feed intake was observed. The inclusion of insect feeds lowered the pH levels of the digesta in both the crop and caeca. The introduction of *H. illucens* as a supplement had the greatest impact on the microbiota populations in the crop, ileum, and caeca. However, as the amount of *S. lateralis* added to the broiler diets increased, the count of selected microbiota in the crop and ileum also increased. These results show that using insect full-fat meals in small quantities can alter the microbiota composition in the digestive tract of broiler chickens.

Ducks serve as a viable and unique option for livestock. Kovitvadhi et al. [73] conducted a study to identify insect species that could serve as potential protein sources in the duck diet via in vitro digestibility tests. Yellow mealworm larvae, giant mealworm larvae (*Zophobas morio*), lesser wax moth larvae (*Achroia grisella*), house fly larvae, mulberry silkworm pupae (*Bombyx mori*), and American cockroach nymphs (*Periplaneta americana*) demonstrated high digestibility and were thus deemed excellent alternative protein sources for ducks. However, further research is necessary to confirm the findings of this study on incorporating insects into the diets of ducks at varying proportions and to bolster sustainable duck farming.

Elahi et al. [74] provided an informative summary of insect meal as a potential feed ingredient for poultry, citing 133 references that cover the literature on various insect species such as BSF, mealworms, houseflies, silkworms, termites (*Macrotermes subhyalinus, Macrotermes bellicosus, Glyptotermes montanus*), the cricket/grasshopper/locust group, and bees, as well as earthworms, which appears to be an outlier in this context. They concluded that insect meal (BSFM, mealworm meal, and housefly meal) presents the most promising prospects for industrialization in broiler and laying hen diets. However, earthworms, silkworms, and locusts can also be effectively utilized in poultry feed. Given the centuries-long use of insects in medicine, it is reasonable to consider replacing antibiotics used in poultry diets with insects due to their antimicrobial properties. Insect meal can be utilized in low-protein diets for amino acid adjustment due to insects' high essential amino acid content.

Chickens naturally consume insects, which is why including insect products in their industrial feeds seems like a logical choice. However, it is important to educate consumers on the use of insects as a feed source to address fears and misconceptions that may arise. This objective was achieved by studies like Khalifah et al.'s [75], which explores the composition and nutritional benefits of insects, as well as their contribution to maintaining poultry farming and minimizing ecological hazards. Their analysis also assesses the pros and cons of insects and the potential of utilizing them as a primary nutrient source in upcoming years. Sajid et al.'s [76] review has similar content but that is obtained from different sources. They reference Raju et al.'s [77] paper, which summarizes two experiments on broiler chickens. The chickens were fed different levels (0, 0.25, 5, and 7.5%) of black soldier fly larva meal (BSFLM) from three sources, which had varying amounts of CP and CF. They discovered that bodyweight gain and food consumption increased consistently in the groups that were given BSFLM at up to 5.0% during the 0–3 weeks of age and remained similar throughout the rest of the period. However, the feed conversion rate (FCR) increased consistently over 3 to 5 weeks. When higher BSFLM levels ($\geq$7.5% in the diet) were used in the second experiment, the body weight gain and food intake decreased significantly, and the FCR increased. Furthermore, dressing yield, breast weight, and abdominal fat content increased linearly with the level of BSFLM in the diet. It was found that the inclusion of BSFLM in broiler chicken feed at up to 5% can have positive impacts on growth during the early stages.

6. Use of Insect Meal in Feeding of Aquatic Animals

In his book chapter, Riddick [78] clearly summarized the then-contemporary situation of using insects as protein sources in aquaculture. He reviewed the research on four key species—the BSF, common house fly, silkworm moth, and yellow mealworm—that served as model insects to highlight progress. The protein and fat content differences between these species are important considerations as they vary between species and developmental stages within a species. The main conclusions of his research were that insects in the form of meal or pellets could provide sufficient protein to partially replace standard fish meal in the diets of fish that are omnivorous, such as catfish and carp, rather than carnivorous (trout and salmon). There is a need to develop cost-effective, large-scale farming practices to meet the increasing demand for farmed fish.

According to Freccia et al. [79], aquafeeds are primarily composed of cereals, oilseeds, and marine-origin components. The demand for feedstuff from the terrestrial animal industries poses a challenge to the profitability of aquafeeds, necessitating the identification of complementary ingredients. Many studies have focused on alternative protein sources, but research on plant proteins, microorganism-based proteins, and various animal by-products is ongoing to address constraints such as antinutritional factors and unbalanced nutrient profiles. The use of insects as a potential alternative in the nutrition of aquatic animals is still being investigated. Researchers are examining the advantages of utilizing insects as animal feed, the nutritional characteristics of various insect-based diets, advancements in aquatic feed technology, and considerations regarding the obstacles and future prospects of insect usage.

Insects have started to serve a significant function in aquaculture as substitute sources of protein. Insects are abundant in most freshwater environments, whereas in seawater, only three genera exist, with one occurring in the open sea and two in coral reef and tide pool marine environments. This group constitutes a natural part of the diets of both carnivorous and omnivorous fish [80] as well as other farm animals. Specifically, during the larval and fingerling stages of fish rearing, various insect species are crucial components of fish diets. Insects provide a valuable source of protein for fish in their natural habitat due to their high protein content ranging from 9.3% to 76% [48,51] and fat content ranging from 7.9% to 40% [81,82]. These variations lead to differences in the content of fatty acids and amino acids that need to be taken into account [83]. It should also be considered that insects are not only meal replacements but also prebiotics due to the presence of chitin and AMPs, so the inclusion of insect meal in fish diets, even at relatively low levels, could improve the immune system of fish and enhance their performance, as previously shown in other livestock species. However, it is also important to remember that there are more than 200 species of farmed fish, and their dietary requirements are not well understood. In addition, the process of the elaboration of insect meal before its use in aquatic animal feed should be considered [83].

7. Conclusions

In today's world, with the increasing emphasis on sustainability and environmental awareness and with food production constituting a critical concern for a growing global population, a reassessment of the importance of insects is needed. Their ecological services are essential for sustainable agriculture, and edible insects offer a promising source of animal protein because they use land and water efficiently while recycling waste into nutrient-rich ingredients.

Insects are now being incorporated into feeding programs for land and aquatic animals due to their highly nutritious composition and availability. In addition, insect-derived components such as chitin, proteins, and fatty acids (with antimicrobial properties) offer other benefits. In the near future, there may be compelling economic reasons to replace other animal protein sources, particularly fishmeal, with insect-based feeds.

Low production volumes and high production costs render insect protein uncompetitive in terms of price compared to fishmeal or soybean meal. Nonetheless, this sector is witnessing notable growth concerning environmental and sustainability apprehensions, backed by comprehensive worldwide research efforts. Market statistics and forecasts reveal that this sector is expected to achieve a compound annual growth rate (CAGR) of 28.3% between 2022 and 2030.

In summary, the reviewed literature indicates that expanding insect production and utilization for food and feed offers the prospect of significantly bolstering agricultural sustainability, provided that ongoing extensive research is conducted to bridge knowledge gaps within this emerging field.

Author Contributions: Cs.H., conceptualization, writing—original draft preparation; S.S. and Z.N., writing—original draft preparation; and J.B., writing—review and editing. All authors have read and agreed to the published version of the manuscript.

Funding: This research received no external funding.

Institutional Review Board Statement: Not applicable.

Informed Consent Statement: Not applicable.

Data Availability Statement: No new data were created or analyzed in this study. Data sharing is not applicable to this article.

Conflicts of Interest: The authors declare no conflicts of interest.

References

1. Dunkel, F.V.; Payne, C. Chapter 1—Introduction to Edible Insects. In *Insects as Sustainable Food Ingredients*; Dossey, A.T., Morales-Ramos, J.A., Guadalupe Rojas, M., Eds.; Academic Press: London, UK, 2016; pp. 1–27. [CrossRef]
2. Guiné, R.P.F.; Correia, P.; Coelho, C.; Costa, C. The role of edible insects to mitigate challenges for sustainability. *Open Agric.* **2021**, *6*, 34–36. [CrossRef]
3. Weisser, W.W.; Siemann, E. *Insects and Ecosystem Function. Ecological Studies*; Springer-Verlag: Berlin/Heidelberg, Germany, 2004; Volume 173. [CrossRef]
4. Noriega, J.A.; Hortal, J.; Azcárate, F.M.; Berg, M.P.; Bonada, N.; Briones, M.J.I.; del Toro, I.; Goulson, D.; Ibanez, S.; Landis, D.A.; et al. Research trends in ecosystem services provided by insects. *Basic Appl. Ecol.* **2018**, *26*, 8–23. [CrossRef]
5. Jankielsohn, A. The Importance of Insects in Agricultural Ecosystems. *Adv. Entomol.* **2018**, *6*, 62–73. [CrossRef]
6. Baskar, K.; Gawade, S. Aquatic insects and their importance in assessing ecosystem health. *MOJ Eco. Environ. Sci.* **2021**, *6*, 136–137. [CrossRef]
7. Dangles, O.; Casas, J. Ecosystem services provided by insects for achieving sustainable development goals. *Ecosyst. Serv.* **2019**, *35*, 109–115. [CrossRef]
8. Schowalter, T.D.; Noriega, J.A.; Tscharntke, T. Insect effects on ecosystem services—Introduction. *Basic Appl. Ecol.* **2018**, *26*, 1–7. [CrossRef]
9. Morimoto, J. Addressing global challenges with unconventional insect ecosystem services: Why should humanity care about insect larvae? *People Nat.* **2020**, *2*, 582–595. [CrossRef]
10. Millenium Ecosystem Assessment. *Ecosystems and Human Well-Being: Wetlands and Water. Synthesis*; World Resources Institute: Washington, DC, USA, 2005.
11. Gasco, L.; Biancarosa, I.; Liland, N.S. From waste to feed: A review of recent knowledge on insects as producers of protein and fat for animal feeds. *Curr. Opin. Green Sustain. Chem.* **2020**, *23*, 67–79. [CrossRef]
12. Govorushko, S. Global status of insects as food and feed source: A review. *Trends Food Sci. Technol.* **2019**, *91*, 436–445. [CrossRef]
13. Liceaga, A.M.; Aguilar-Toalá, J.A.; Vallejo-Cordoba, B.; González-Córdova, A.F.; Hernández-Mendoza, A. Insects as an Alternative Protein Source. *Annu. Rev. Food Sci. Technol.* **2022**, *13*, 19–34. [CrossRef] [PubMed]
14. Akhtar, Y.; Isman, M.B. Insects as an Alternative Protein Source. In *Proteins in Food Processing*, 2nd ed.; Yada, R.Y., Ed.; Woodhead Publishing: Duxford, UK, 2018; pp. 263–288. [CrossRef]
15. FAO. The Contribution of Insects to Food Security, Livelihoods and the Environment. 2013. Available online: https://www.fao.org/documents/card/en/c/c499eb3c-58dc-52dc-ac64-5f2ff88dce8f (accessed on 1 October 2023).
16. Van Huis, A. Edible insects: Challenges and prospects. *Entomol. Res.* **2022**, *52*, 161–177. [CrossRef]
17. Sogari, G.; Amato, M.; Palmieri, R.; Hadj Saadoun, J.; Formici, G.; Verneau, F.; Mancini, S. The future is crawling: Evaluating the potential of insects for food and feed security. *Curr. Res. Food Sci.* **2023**, *6*, 100504. [CrossRef] [PubMed]
18. Chia, S.; Tanga, C.M.; van Loon, J.J.A.; Dicke, M. Insects for sustainable animal feed: Inclusive business models involving smallholder farmers. *Curr. Opin. Environ. Sustain.* **2019**, *41*, 23–30. [CrossRef]
19. Research and Markets, Global Edible Insects Market Report 2022–2030. 2022. Available online: https://www.globenewswire.com/en/news-release/2022/06/15/2462794/28124/en/Global-Edible-Insects-Market-Report-2022-2030-Environmental-Benefits-of-Edible-Insects-Consumption-Rising-Demand-for-Insect-Protein-in-the-Animal-Feed-Industry.html (accessed on 17 August 2023).
20. Robertson, G.P.; Swinton, S.M. Reconciling Agricultural Productivity and Environmental Integrity: A Grand Challenge for Agriculture. *Front. Ecol. Environ.* **2005**, *3*, 38–46. [CrossRef]
21. Pywell, R.F.; Heard, M.S.; Woodcock, B.A.; Hinsley, S.; Ridding, L.; Nowakowski, M.; Bullock, J.M. Wildlife Friendly Farming Increases Crop Yield: Evidence for Ecological Intensification. *Proc. R. Soc. B Biol. Sci.* **2015**, *282*, 20151740. [CrossRef] [PubMed]
22. Slade, E.M.; Riutta, T.; Roslin, T.; Tuomisto, H.L. The Role of Dung Beetles in Reducing Greenhouse Gas Emissions from Cattle Farming. *Sci. Rep.* **2016**, *6*, 18140. [CrossRef] [PubMed]
23. Nichols, E.; Spector, S.; Louzada, J.; Larsen, T.; Amezquita, S.; Favila, M.E. Ecological Functions and Ecosystem Services Provided by Scarabaeinae Dung Beetles. *Biol. Conserv.* **2008**, *141*, 1461–1474. [CrossRef]
24. Lonsdorf, E.; Kremen, C.; Ricketts, T.; Winfree, R.; Williams, N.; Greenleaf, S. Modelling Pollination Services across Agricultural Landscapes. *Ann. Bot.* **2009**, *103*, 1589–1600. [CrossRef] [PubMed]

25. Bartomeus, I.; Potts, S.G.; Steffan-Dewenter, I.; Vaissière, B.E.; Woyciechowski, M.; Krewenka, K.M.; Tscheulin, T.; Roberts, S.P.M.; Szentgyörgyi, H.; Westphal, C.; et al. Contribution of insect pollinators to crop yield and quality varies with agricultural intensification. *PeerJ* **2014**, *2*, e328. [CrossRef] [PubMed]
26. Landis, D.A.; Wratten, S.D.; Gurr, G.M. Habitat Management to Conserve Natural Enemies of Arthropod Pests in Agriculture. *Annu. Rev. Entomol.* **2000**, *45*, 175–201. [CrossRef] [PubMed]
27. Bell, J.R.; Traugott, M.; Sunderland, K.D.; Skirvin, D.J.; Mead, A.; Kravar-Garde, L.; Reynolds, K.; Fenion, J.S.; Symondson, W.O.C. Beneficial Links for the Control of Aphids: The Effects of Compost Applications on Predators and Prey. *J. Appl. Ecol.* **2008**, *45*, 1266–1273. [CrossRef]
28. Cock, M.; Biesmeijer, J.; Cannon, R.J.C.; Gerard, P.J.; Gillespie, D.; Jiménez, J.J.; Lavelle, P.M.; Raina, S.K. The positive contribution of invertebrates to sustainable agriculture and food security. *Biol. Cab Rev. Perspect. Agric. Vet. Sci. Nutr. Nat. Resour.* **2012**, *7*, 1–27. [CrossRef]
29. Follett, P.A.; Bruin, J.; Desneux, N. Insects in agroecosystems—An introduction. *Entomol. Exp. Appl.* **2020**, *168*, 3–6. [CrossRef]
30. Francuski, L.; Beukeboom, L.W. Insects in production: An introduction. *Entomol. Exp. Appl.* **2020**, *168*, 422–431. [CrossRef]
31. Poveda, J. Insect frass in the development of sustainable agriculture. A review. *Agron. Sustain. Dev.* **2021**, *41*, 5. [CrossRef]
32. Barragán-Fonseca, K.Y.; Nurfikari, A.; van de Zande, E.M.; Wantulla, M.; van Loon, J.J.A.; de Boer, W.; Dicke, M. Insect frass and exuviae to promote plant growth and health. *Trends Plant Sci.* **2022**, *27*, 646–654. [CrossRef] [PubMed]
33. Nolan, P.; Mahmoud, A.E.; Kavle, R.R.; Carne, A.; Bekhit, A.E.A.; Agyei, D. Edible insects: Protein composition, digestibility, and biofunctionalities. In *Processing Technologies and Food Protein Digestion*; Bhat, Z.F., Morton, J.D., Bekhit, A.E.A., Suleira, H.A.R., Eds.; Academic Press: London, UK, 2023; pp. 429–494. [CrossRef]
34. Cortes Ortiz, J.A.; Ruiz, A.T.; Morales-Ramos, J.A.; Thomas, M.; Rojas, M.G.; Tomberlin, J.K.; Yi, L.; Han, R.; Giroud, L.; Jullien, R.L. Insect Mass Production Technologies. In *Insects as Sustainable Food Ingredients*; Dossey, A.T., Morales-Ramos, J.A., Rojas, M.G., Eds.; Academic Press: London, UK, 2016; pp. 153–201. [CrossRef]
35. Thrastardottir, R.; Olafsdottir, H.T.; Thorarinsdottir, R.I. Yellow Mealworm and Black Soldier Fly Larvae for Feed and Food Production in Europe, with Emphasis on Iceland. *Foods* **2021**, *10*, 2744. [CrossRef] [PubMed]
36. Hong, J.; Han, T.; Kim, Y.Y. Mealworm (*Tenebrio molitor* Larvae) as an Alternative Protein Source for Monogastric Animal: A Review. *Animals* **2020**, *10*, 2068. [CrossRef] [PubMed]
37. De Marco, M.; Martínez, S.; Hernandez, F.; Madrid, J.; Gai, F.; Rotolo, L.; Belforti, M.; Bergero, D.; Katz, H.; Dabbou, S.; et al. Nutritional value of two insect larval meals (*Tenebrio molitor* and *Hermetia illucens*) for broiler chickens: Apparent nutrient digestibility, apparent ileal amino acid digestibility and apparent metabolisable energy. *Anim. Feed. Sci. Technol.* **2015**, *209*, 211–218. [CrossRef]
38. Jin, X.H.; Heo, P.S.; Hong, J.S.; Kim, N.J.; Kim, Y.Y. Supplementation of dried mealworm (*Tenebrio molitor* larva) on growth performance, nutrient digestibility and blood profiles in weaning pigs. *Asian–Australasian J. Anim. Sci.* **2016**, *29*, 979–986. [CrossRef] [PubMed]
39. Rumbos, C.; Adamaki-Sotiraki, C.; Gourgouta, M.; Karapanagiotidis, I.; Asimaki, A.; Mente, E.; Athanassiou, C. Strain matters: Strain effect on the larval growth and performance of the yellow mealworm, *Tenebrio molitor* L. *J. Insects Food Feed.* **2021**, *7*, 1195–1205. [CrossRef]
40. Varelas, V. Food Wastes as a Potential New Source for Edible Insect Mass Production for Food and Feed: A review. *Fermentation* **2019**, *5*, 81. [CrossRef]
41. Ramos-Elorduy, J. Energy supplied by edible insects from Mexico and their nutritional and ecological importance. *Ecol. Food Nutr.* **2008**, *47*, 280–297. [CrossRef]
42. Waldbauer, G.P. The Consumption and Utilization of Food by Insects. *Adv. Insect Physiol.* **1968**, *5*, 229–288. [CrossRef]
43. Maino, J.; Kearney, M. Testing mechanistic models of growth in insects. *Proc. R. Soc. B Biol. Sci.* **2015**, *282*, 20151973. [CrossRef]
44. Halloran, A.; Roos, N.; Eilenberg, J.; Cerutti, A.; Bruun, S. Life cycle assessment of edible insects for food protein: A review. *Agron. Sustain. Dev.* **2016**, *36*, 57. [CrossRef] [PubMed]
45. Weru, J.; Chege, P.; Kinyuru, J. Nutritional potential of edible insects: A systematic review of published data. *Int. J. Trop. Insect Sci.* **2021**, *41*, 2015–2037. [CrossRef]
46. Rumpold, B.A.; Schlüter, O.K. Nutritional composition and safety aspects of edible insects. *Mol. Nutr. Food Res.* **2013**, *57*, 802–823. [CrossRef] [PubMed]
47. Koutsos, L.; McComb, A.; Finke, M. Insect Composition and Uses in Animal Feeding Applications: A Brief Review. *Ann. Entomol. Soc. Am.* **2019**, *112*, 544–551. [CrossRef]
48. Makkar, H.P.S.; Tran, G.; Huezé, V.; Ankers, P. State-of-the-art on use of insects as animal feed. *Anim. Feed Sci. Tech.* **2014**, *197*, 1–33. [CrossRef]
49. Shah, A.A.; Totakul, P.; Maharach, M.; Cherdthong, A.; Harnboonsong, Y.; Wanapat, M. Nutritional composition of various insects and potential uses as alternative protein sources in animal diets. *Anim. Biosci.* **2022**, *35*, 317–331. [CrossRef] [PubMed]
50. Sándor, Z.J.; Banjac, V.; Vidosavljević, S.; Káldy, J.; Egessa, R.; Lengyel-Kónya, É.; Tömösközi-Farkas, R.; Zalán, Z.; Adányi, N.; Libisch, B.; et al. Apparent Digestibility Coefficients of Black Soldier Fly (*Hermetia illucens*), Yellow Mealworm (*Tenebrio molitor*), and Blue Bottle Fly (*Calliphora vicina*) Insects for Juvenile African Catfish Hybrids (*Clarias gariepinus* × *Heterobranchus longifilis*). *Aquac. Nutr.* **2022**, *2022*, 4717014. [CrossRef] [PubMed]
51. Sánchez-Muros, M.-J.; Barroso, F.G.; Manzano-Agugliaro, F. Insect meal as renewable source of food for animal feeding: A review. *J. Clean. Prod.* **2014**, *65*, 16–27. [CrossRef]

52. Kroeckel, S.; Harjes, A.G.E.; Roth, I.; Katz, H.; Wuertz, S.; Susenbeth, A.; Schulz, C. When a turbot catches a fly: Evaluation of a pre-pupae meal of the Black Soldier Fly (*Hermetia illucens*) as fish meal substitute—Growth performance and chitin degradation in juvenile turbot (*Psetta maxima*). *Aquaculture* **2012**, *364–365*, 345–352. [CrossRef]
53. Veldkamp, T.; Dong, L.; Paul, A.; Govers, C. Bioactive properties of insect products for monogastric animals—A review. *J. Insects Food Feed.* **2022**, *8*, 1027–1040. [CrossRef]
54. De Souza-Vilela, J.; Andrew, N.R.; Ruhnke, I. Insect protein in animal nutrition. *Anim. Prod. Sci.* **2019**, *59*, 2029–2036. [CrossRef]
55. Penazzi, L.; Schiavone, A.; Russo, N.; Nery, J.; Valle, E.; Madrid, J.; Martínez, S.; Hernández, F.; Pagani, E.; Ala, U.; et al. In vivo and in vitro Digestibility of an Extruded Complete Dog Food Containing Black Soldier Fly (*Hermetia illucens*) Larvae Meal as Protein Source. *Front. Vet. Sci.* **2021**, *8*, 653411. [CrossRef] [PubMed]
56. Gasco, L.; Bellezza Oddon, S.; Vandenberg, G.W.; Veldkamp, T.; Biasato, I. Factors affecting the decision-making process of using insect-based products in animal feed formulations. *J. Insects Food Feed.* **2023**, 1–12. [CrossRef]
57. Koutsos, L. Applications of Insect-derived Ingredients in Animal Diets. *J. Anim. Sci.* **2021**, *99*, 119. [CrossRef]
58. Pippinato, L.; Gasco, L.; Di Vita, G.; Mancuso, T. Current scenario in the European edible-insect industry: A preliminary study. *J. Insects Food Feed.* **2020**, *6*, 371–381. [CrossRef]
59. Pinotti, L.; Giromini, C.; Ottoboni, M.; Tretola, M.; Marchis, D. Review: Insects and former foodstuffs for upgrading food waste biomasses/streams to feed ingredients for farm animals. *Animal* **2019**, *13*, 1365–1375. [CrossRef] [PubMed]
60. Van Huis, A.; Gasco, L. Insects as feed for livestock production. *Science* **2023**, *379*, 138–139. [CrossRef] [PubMed]
61. Prachumchai, R.; Cherdthong, A. Black Soldier Fly Larva Oil in Diets with Roughage to Concentrate Ratios on Fermentation Characteristics, Degradability, and Methane Generation. *Animals* **2023**, *13*, 2416. [CrossRef]
62. DiGiacomo, K.; Leury, B.J. Review: Insect meal: A future source of protein feed for pigs? *Animal* **2019**, *13*, 3022–3030. [CrossRef] [PubMed]
63. Renna, M.; Rastello, L.; Gasco, L. Can insects be used in the nutrition of ruminants? *J. Insects Food Feed.* **2022**, *8*, 1041–1045. [CrossRef]
64. Renna, M.; Rastello, L.; Veldkamp, T.; Toral, P.G.; Gonzalez-Ronquillo, M.; Robles-Jimenez, L.E.; Gasco, L. Are insects a solution for feeding ruminants? Legislation, scientific evidence, and future challenges. *Anim. Front.* **2023**, *13*, 102–111. [CrossRef] [PubMed]
65. Toral, P.G.; Hervás, G.; González-Rosales, M.G.; Mendoza, A.G.; Robles-Jiménez, L.E.; Frutos, P. Insects as alternative feed for ruminants: Comparison of protein evaluation methods. *J. Animal Sci. Biotechnol.* **2022**, *13*, 21. [CrossRef] [PubMed]
66. Newton, G.L.; Booram, C.V.; Barker, R.W.; Hale, O.M. Dried *Hermetia illucens* larvae meal as a supplement for swine. *J. Anim. Sci.* **1977**, *44*, 395–400. [CrossRef]
67. Newton, L.; Sheppard, C.; Watson, D.W.; Burtle, G.; Dove, R. *Using the Black Soldier Fly, Hermetia Illucens, as a Value-Added Tool for the Management of Swine Manure*; Animal and Poultry Waste Management Centre, North Carolina State University: Raleigh, NC, USA, 2005.
68. Altmann, B.A.; Neumann, C.; Rothstein, S.; Liebert, F.; Mörlein, D. Do dietary soy alternatives lead to pork quality improvements or drawbacks? A look into micro-alga and insect protein in swine diets. *Meat Sci.* **2019**, *153*, 26–34. [CrossRef] [PubMed]
69. Biasato, I.; Renna, M.; Gai, F.; Dabbou, S.; Meneguz, M.; Perona, G.; Martinez, S.; Lajusticia, A.C.B.; Bergagna, S.; Sardi, L.; et al. Partially defatted black soldier fly larva meal inclusion in piglet diets: Effects on the growth performance, nutrient digestibility, blood profile, gut morphology and histological features. *J. Anim. Sci. Biotechnol.* **2019**, *10*, 1–11. [CrossRef] [PubMed]
70. Hong, J.; Han, T.; Kim, Y.Y. Insect as feed ingredients for pigs. *Anim. Biosci.* **2022**, *35*, 347–355. [CrossRef]
71. Schiavone, A.; Cullere, M.; De Marco, M.; Meneguz, M.; Biasato, I.; Bergagna, S.; Dezzutto, D.; Gai, F.; Dabbou, S.; Gasco, L.; et al. Partial or total replacement of soybean oil by black soldier fly larvae (*Hermetia illucens* L.) fat in broiler diets: Effect on growth performances, feed-choice, blood traits, carcass characteristics and meat quality. *Ital. J. Anim. Sci.* **2016**, *16*, 93–100. [CrossRef]
72. Józefiak, A.; Kierończyk, B.; Rawski, M.; Mazurkiewicz, J.; Benzertiha, A.; Gobbi, P.; Nogales-Merida, S.; Świątkiewicz, S.; Józefiak, D. Full-fat insect meals as feed additive–the effect on broiler chicken growth performance and gastrointestinal tract microbiota. *J. Anim. Feed. Sci.* **2018**, *27*, 131–139. [CrossRef]
73. Kovitvadhi, A.; Chundang, P.; Thongprajukaew, K.; Tirawattanawanich, C.; Srikachar, S.; Chotimanothum, B. Potential of Insect Meals as Protein Sources for Meat-Type Ducks Based on In Vitro Digestibility. *Animals* **2019**, *9*, 155. [CrossRef] [PubMed]
74. Elahi, U.; Xu, C.; Wang, J.; Lin, J.; Wu, S.; Zhang, H.; Qi, G. Insect meal as a feed ingredient for poultry. *Anim. Biosci.* **2022**, *35*, 332–346. [CrossRef] [PubMed]
75. Khalifah, A.; Abdalla, S.; Rageb, M.; Maruccio, L.; Ciani, F.; El-Sabrout, K. Could Insect Products Provide a Safe and Sustainable Feed Alternative for the Poultry Industry? A Comprehensive Review. *Animals* **2023**, *13*, 1534. [CrossRef] [PubMed]
76. Sajid, Q.U.A.; Asghar, M.U.; Tariq, H.; Wilk, M.; Płatek, A. Insect Meal as an Alternative to Protein Concentrates in Poultry Nutrition with Future Perspectives (An Updated Review). *Agriculture* **2023**, *13*, 1239. [CrossRef]
77. Raju, M.; Rao, S.R.; Paul, S.S.; Prakash, B.; Reddy, M.R.; Kannan, A.; Shanmugam, M.; Santosh Phani Kumar, P. Source Variation in Nutrient Profile and the Effects of Black Soldier Fly (*Hermetia illucens*) Larva Meal at Graded Levels in Diet on Performance, Carcass Traits, Serum Biochemical Profile and Cellular Immune Response in Broiler Chicken. *Int. J. Trop. Insect Sci.* **2023**, *preprint*. [CrossRef]

78. Riddick, E.W. Insect Protein as a Partial Replacement for Fishmeal in the Diets of Juvenile Fish and Crustaceans. In *Mass Production of Beneficial Organisms*; Morales-Ramos, J.A., Guadalupe Rojas, M., Shapiro-Ilan, D.I., Eds.; Academic Press: London, UK, 2014; pp. 565–582. [CrossRef]
79. Freccia, A.; Bee Tubin, J.S.; Nishioka, A.R.; Coelho, M.G.E. Insects in Aquaculture Nutrition: An Emerging Eco-Friendly Approach or Commercial Reality. In *Emerging Technologies, Environment and Research for Sustainable Aquaculture*; Lu, Q., Serajuddin, M., Eds.; IntechOpen: London, UK, 2020; pp. 119–132. [CrossRef]
80. Henry, M.; Gasco, L.; Piccolo, G.; Fountoulaki, E. Review on the use of insects in the diet of farmed fish: Past and future. *Anim. Feed. Sci. Technol.* **2015**, *203*, 1–22. [CrossRef]
81. Finke, M.D. Complete nutrient content of four species of commercially available feeder insects fed enhanced diets during growth. *Zoo Biol.* **2015**, *34*, 554–564. [CrossRef] [PubMed]
82. Meneguz, M.; Schiavone, A.; Gai, F.; Dama, A.; Lussiana, C.; Renna, M.; Gasco, L. Effect of rearing substrate on growth performance, waste reduction efficiency and chemical composition of black soldier fly (*Hermetia illucens*) larvae. *J. Sci. Food Agric.* **2018**, *98*, 5776–5784. [CrossRef] [PubMed]
83. Nogales-Mérida, S.; Gobbi, P.; Józefiak, D.; Mazurkiewicz, J.; Dudek, K.; Rawski, M.; Kierończyk, B.; Józefiak, A. Insect meals in fish nutrition. *Rev. Aquac.* **2018**, *11*, 1080–1103. [CrossRef]

animals MDPI

Review

Factors Affecting Yeast Digestibility and Immunostimulation in Aquatic Animals

Sadia Sultana [1], Janka Biró [2,*], Balázs Kucska [1] and Csaba Hancz [1]

1 Kaposvár Campus, Hungarian University of Agriculture and Life Sciences, Guba S. 40., 7400 Kaposvár, Hungary; sadiazoology@nstu.edu.bd (S.S.)
2 Research Center for Fisheries and Aquaculture, Hungarian University of Agriculture and Life Sciences, Anna-liget u. 35, 5540 Szarvas, Hungary
* Correspondence: nagyne.biro.janka@uni-mate.hu

Simple Summary: This paper provides an overview of the key issues concerning the use of yeast products as an alternative protein source and feed additives to develop a more sustainable and environmentally friendly aquaculture. The use of yeast is presented in detail, with a particular focus on its use as a primary protein source and as a supplement (such as probiotics and immune stimulants). Additionally, the digestibility of yeasts and their use in live aquafeed culture are discussed in detail.

Abstract: The aquafeed industry increasingly relies on using sustainable and appropriate protein sources to ensure the long-term sustainability and financial viability of intensive aquaculture. Yeast has emerged as a viable substitute protein source in the aquaculture sector due to its potential as a nutritional supplement. A substantial body of evidence exists to suggest that yeast has the potential to act as an effective immune-stimulating agent for a range of aquaculture fish species. Furthermore, the incorporation of yeast supplements and feed additives has the potential to bolster disease prevention, development, and production within the aquaculture sector. Except for methionine, lysine, arginine, and phenylalanine, which are typically the limiting essential amino acids in various fish species, the various yeast species exhibit amino acid profiles that are advantageous when compared to fishmeal. The present review considers the potential nutritional suitability of several yeast species for fish, with particular attention to the various applications of yeast in aquaculture nutrition. The findings of this study indicate that the inclusion of yeast in the diet resulted in the most favorable outcomes, with improvements observed in the overall health, growth performance, and nutritional condition of the fish. Digestibility, a key factor in sustainable feed development, is discussed in special detail. Additionally, this review addresses the utilization of yeast as an immunostimulating agent for fish and its digestion in fish. Furthermore, the research emphasizes the necessity of large-scale production of yeast as a substitute for fishmeal in aquaculture.

Keywords: yeast; fish; growth performance; aquafeeds; digestibility; immunostimulation

1. Introduction

Aquaculture is an essential component of the global animal production system to meet the long-term protein needs of the world's population [1]. In recent years, the global aquaculture industry has seen an increase in the importance of using cutting-edge nutritional techniques to improve fish health and disease resistance [2]. Aquaculture must continue to expand sustainably to meet the future demand for animal protein and support the continued growth of the global population [3]. The sector has encountered obstacles to its rapid growth, including diseases and restricted access to high-quality protein sources for fish feed. In aquatic species, protein represents the primary dry matter component of growth [4]. Aquafeeds are formulated to include all the essential nutrients that fish require to maintain optimal health. Fishmeal (FM) and fish oil (FO), which are derived primarily from wild-caught small pelagics like anchovies, sardines, and menhaden, have historically been a significant component of aquaculture diets. According to IFFO, approximately 69% of FM and 75% of FO are utilized in aquaculture production [5]. In particular, the proportions of fishmeal and fish oil have significantly decreased and have been substituted with components derived from plants. Despite changes in feed composition to reduce marine ingredient dependency, the sustainable development of aquaculture faces threats from stagnant fishmeal and fish oil supply [6]. Furthermore, fishmeal and plant-based meals are currently utilized as primary sources of protein. Consequently, sustainability concerns have necessitated the identification of novel sustainable components for aquafeeds [7]. Compared to animal-sourced feedstuffs, plant-sourced feedstuffs frequently exhibit reduced dry matter contents of amino acids (AAs) and crude protein (CP) [4,8]. The reduction in fish stocks and the elevated cost of fishmeal have created significant challenges for the aquaculture industry in meeting its consumption requirements. It is therefore a global priority to identify alternative protein sources that can be used to replace fishmeal in aquafeeds [9]. Over the next ten years, it is predicted that fisheries' trimmings and byproducts will provide increasing fishmeal and oil production [10]. However, they will likely not be sufficient to meet anticipated demand by 2050 [11], which may also have an impact on the nutritional value of farmed fish. In response to these growing concerns, large-scale research has recently been conducted to identify acceptable fishmeal substitutes that can be included in the diets of a variety of aquatic animal species. To address these challenges, novel aquafeed materials have been developed, including macroalgae, insects, genetically modified crops, and single-cell proteins derived from microalgae, bacteria, or yeasts [12]. Furthermore, single-cell protein sources, particularly yeast protein products, may serve as a supplementary solution to offset the adverse effects of plant proteins [13].

Recent studies have demonstrated the potential of yeast as a viable alternative protein source in aquaculture, offering a promising solution to replace fishmeal [7,14]. Research findings have shown that yeast can be a sustainable and high-quality protein source in fish diets, with favorable amino acid composition and health benefits [15]. The latest advances in yeast research have focused on its potential as a functional ingredient and nutritional supplement that can enhance fish immune systems and gut health [3]. Recent advances in the study of yeast have focused on its potential as a functional component and dietary supplement that strengthens the immune systems and gut health of fish. In aquaculture, *Saccharomyces cerevisiae* and its byproducts are the most often utilized functional feed additives [16,17]. Yeast supplements have improved immunity and enhanced aquaculture water quality, leading to better productivity and disease protection [18]. Dietary yeast extracts have also been found to promote the growth of beneficial bacteria and inhibit some pathogenic species in fish [19]. For instance, supplementing the diet of largemouth bass (*Micropterus salmoides*) with brewer's yeast hydrolysate inhibited bacterial members of the genus *Mycoplasma* and significantly increased *Cetobacterium* in their intestines [19]. Furthermore, yeast-derived cell wall fractions have been shown to have immunological and health benefits for fish [15]. Several scientific reviews have explained the health benefits of these cell wall components in different species over the years, but little is known about the function of yeast as a macro-ingredient in fish feeds [20]. However, supplementing with *S. cerevisiae* as a probiotic has improved the growth, immunity, and disease resistance of various fish species [21]. Probiotic fermentation can help remove toxicity from dietary ingredients used as alternative protein sources to replace fishmeal [22]. In addition, yeasts have been used to develop artemia and rotifers or to ferment feedstock following artificial or natural colonization in the gastrointestinal tracts of the hosts [18,23]. Other findings underscore the value and sustainability of yeast products as a nutritional source in aquaculture. Their probiotic effects also enhance water quality [24].

This review aims to present an overview of current knowledge regarding the use of yeasts as a primary source of protein, supplements, and probiotics in aquaculture systems.

In particular, it will emphasize the effects of these organisms on the growth performance, digestibility, physiology, and immune system of aquatic species.

2. Yeast Cell Wall Components and Their Nutritional Properties

Yeast, a single-celled eukaryote with membranous organelles such as mitochondria, nucleus, and endomembrane system, is a protein source in aquafeeds for shrimp and marine fish. Many yeast species are frequently utilized, such as Saccharomyces cerevisiae, *Kluyveromyces*, *Torulaspora*, *Saccharomyces*, and *Torulopsis* [14]. The yeast cell wall provides structural support and is made up of 10–15% proteins and 85–90% polysaccharides [25].

The mucosal immune responses of rainbow trout were evaluated when supplementing an experimental formulated feed with multi-strain yeast fraction product of *Saccharomyces cerevisiae* and *Cyberlindnera jardinii* and found that the extracts enhanced and strengthened the ability of innate immune cells of mucosal-associated lymphoid tissues of fish [26].

Another study aimed to characterize the biochemical and molecular properties of four proprietary yeast cell wall extracts from *S. cerevisiae* about their potential effect on the intestinal immune responses when given orally to zebrafish (*Danio rerio*), concluding that these characteristics give a chance to correctly evaluate the immune potential of the tested products. Additionally, this study offered novel perspectives on the development of specific fractions derived from *S. cerevisiae* for use in precision animal feeds [27].

The estimated polysaccharide content of the S. cerevisiae cell wall is 30–60% b-1, 3-glucan and 1, 6-glucan, 25–30% mannan, and 5–10% chitin [25]. Table 1 shows that yeast *S. cerevisiae* contains approximately 32–62% protein [28]. On average, 93–97% of dry yeast is composed of dry matter, which includes 40–60% crude nitrogen and poultry byproducts protein, 5–9% lipids, and 35–45% carbohydrates [28,29].

Table 1. Chemical composition of *Saccharomyces cerevisiae* as yeast meal and fishmeal.

Composition	*Saccharomyces cerevisiae* from Beer Fermentation	Menhaden Fishmeal
Dry matter%	93	91.2–92
ME (kcal/kg)	1990	3370
Crude protein%	44.4	59–68.5
Crude fat%	1	9.1–10.4
Crude fiber%	2.7	0.9
Ca%	0.12	4.87–5.34
P%	1.4	2.93–3.05

Source: [28–30].

Additionally, it is also stated that yeast species such as *S. cerevisiae* have unsaturated fatty acids, and linoleic and alpha-linolenic acids, which can also be formed by fatty acid desaturase [28]. Low-value and non-food cellulosic materials are fermented to produce dry yeast products, via the pre-treatment of biomass by the extraction of lignin; hydrolysis to release cellulose and hemicellulose; enzymatic hydrolysis to release hemicellulose and cellulose into C5 and C6 sugars; fermentation of sugars, ammonia, phosphorus, and other nutrients; and downstream processing into dry yeast products for use as a source of protein in fish feed [15,18].

It can be stated in general that the characteristics of yeast products make them ideal feed ingredients, serving as valuable protein sources. Additionally, their role as pre- and provitamins is of significant importance.

3. Use of Yeast in Fish Feeding

As mentioned above, fishmeal is a highly digestible feed ingredient for animals and a great source of high-quality protein, fatty acids, and readily available minerals and vitamins [31]. Finding local raw resources is necessary since the high cost of imported raw materials drives up the price of locally produced feed [32]. Fishmeal inclusion levels in aquafeed must also be lowered to save costs and save the environment [12]. Current research on fish nutrition and aquafeed production primarily focuses on assessing the potential of non-conventional protein sources. These sources include plant-derived proteins, processed animal byproducts, and single-cell proteins [33,34].

As a high-nutritional eukaryotic organism, yeast is an appropriate candidate for use as animal feed [35]. One of the most significant qualities of yeast as a potential feed ingredient is its capacity to transform industrial biomass, forestry waste, and less valuable, inedible food byproducts into high-quality food or feed ingredients with little to no dependence on arable land and water. Furthermore, the use of yeast as a feed ingredient results in a net reduction in climate change [36,37].

Furthermore, the utilization of yeast as a fish feed represents a potentially cost-effective and environmentally sustainable approach within the domain of aquaculture biotechnology, which could diminish reliance on fishmeal [38]. Fishmeal or plant proteins can be substituted with the affordable yeast *S. cerevisiae*, which has been shown to have no detrimental effects on the growth and nutritional performance of rainbow trout [39], Nile tilapia [40], and Arctic charr [41]. A study on alternate sources of protein also reveals that *S. cerevisiae* is an effective natural resource of protein in the replacement of fishmeal in tilapia feed [42]. Similarly, another study found that the 15% inclusion of yeast had increased the growth performance without reducing the quality of the end product [43]. The productivity, immunity, and overall health of Nile tilapia (*Oreochromis niloticus*) have all been improved by the use of fermented yeast products [17]. Recent advances in yeast research are more focused on their potential as nutritional supplements and functional properties with beneficial effects on the immune responses and gut health in fish [3].

The use of yeast as an aquatic feed ingredient is undoubtedly aligned with the developmental objectives of sustainability, resulting in both economic and environmental benefits. The extensive literature overview encompasses a vast array of products, doses applied, species fed on, and effects evaluated. To facilitate a summary, a division was applied as seen below. However, it must be acknowledged that this division may appear somewhat forced at times, given that multiple effects were measured and evaluated in the majority of the studies discussed here.

4. Use of Yeast as the Primary Protein Source in Aquafeed

Yeast cells have high protein content (approximately 40–55%) and other bioactive compounds that are essential for fish development and growth performance [7,39]. A multitude of research investigations have demonstrated the beneficial impacts of incorporating whole yeast and/or its byproducts into the diet on Nile tilapia growth performance, hematological indices, antioxidant defense systems, immunological response, and disease resistance [17,44]. Research suggests that a diet enriched with *Saccharomyces cerevisiae* (SC) can improve the growth performance, intestinal morphology, redox homeostasis, and immune response of *S. aurata*. The most effective concentration was found to be 4 g/kg [45]. Candida species are commonly used in rainbow trout diets, and up to 40% of fishmeal can be effectively substituted without any reduction in production or efficiency [14]. In a study, it was revealed that by increasing the feed conversion ratio and body nitrogen retention, a 15% protein-rich yeast fraction inclusion could enhance the growth performance of trout fed on plant-based diets [13].

Brewer's yeast, mainly the *Saccharomyces cerevisiae* strain, has been used as nitrogen-rich ingredients in aquaculture feeds from the beginning of the 1990s [46]. A study was conducted to assess the suitability of brewer's spent yeast (BSY) in aquafeed for *Labeo rohita* growth, hemato-biochemical parameters, and enzyme activities, and fishmeal was gradually replaced at an increased level [47]. The study found that replacing 30% of fishmeal with brewer's wasted yeast considerably improves growth and immunity in *L. rohita*. In addition, another study revealed that replacing 45% of fishmeal with brewer's yeast can improve the growth performance and immune response of the Thai Panga (*Pangasianodon hypophthalmus* × *Pangasius bocourti*) [48]. In addition, it was suggested that dried yeast could replace up to 30 and 50% of the protein provided by FM/SBM (50/50%), respectively, without affecting the performance of red drum (*Sciaenops ocellatus*) [49]. In the research, it was concluded that the growth, immunity, and disease resistance of channel catfish were enhanced by replacing fish feed with yeast culture, and the intestinal microbiota of the fish, which was induced by the yeast culture, was a key factor in these outcomes [50]. Furthermore, dietary fermented yeast could be used as a strategic option to maintain tilapia production [17]. Moreover, yeast can convert low-value biomass leftovers from agriculture and forests into high-value feed components, making it a sustainable and environmentally friendly element [37]. Consequently, it follows that fishmeal can be substituted with single-cell yeast protein at certain levels to reduce the aquafeed cost and preserve the sustainable output of aquaculture for the mentioned aquatic species. The effects of yeast products on different fish species are summarized in Table 2.

Table 2. Yeast species used in aquaculture as a primary protein source to enhance growth performance, disease resistance, and immune system of different fish species.

Product Name	Fish Species	Inclusion Levels	Effects	Source
Brewer's yeast	Largemouth bass (*Micropterus salmoides*)	20, 40, 60 g/kg	The supplementation of *S. cerevisiae* at 10 g kg^{-1} did not significantly reduce fishmeal in the diet.	[50]
Dietary yeast polysaccharides	Channel catfish (*Ictalurus punctatus*)	20 g/kg	The study found that fish fed experimental diets for 12 weeks significantly improved growth performance compared to the control.	[50]
Yeast, *Saccharomyces cerevisiae* (SC)	Gilthead sea bream (*Sparus aurata*)	1, 2, and 4 g/kg	The study showed that at a dose of 4 g/kg, a meal high in SC dramatically improved growth performance, intestinal morphology, redox homeostasis, and immunological response.	[45]
Protein-rich yeast fraction	Rainbow trout (*Oncorhynchus mykiss*)	0, 5, 10, or 15%	The use of yeasts in 60% of fishmeal protein can cause haemolytic anemia in rainbow trout, potentially restricting their inclusion in farmed fish diets.	[13]
Dietary yeast hydrolysate	Nile tilapia (*Oreochromis niloticus*)	1 and 3%	1% yeast hydrolysate supplementation enhances growth performance, feed utilization, and antioxidant status.	[51]
Baker's yeast	Freshwater catfish gulsa tengra (*Mystus cavasius*)	0.5, 1, and 1.5 g/kg	Yeast supplement is a potentially effective growth promoter that could replace antibiotics in the treatment of *M. cavasius* infections.	[42]
Dietary hydrolyzed yeast	Pacific white shrimp (*Litopenaeus vannamei*)	1%	The study found no significant effect on growth performance or body composition, but it did affect intestinal health, immunological responses, and ammonia resistance.	[52]
Autolyzed dried yeast	Gilthead sea bream (*Sparus aurata*)	0 and 5%	Following a 92-day feeding trial, there were no discernible variations between the dietary groups in terms of fish growth rate, mortality, or feed efficiency.	[53]
Dietary yeast hydrolysate	Largemouth bass (*Micropterus salmoides*)	0, 1.5, 3.0, 4.5%	Dietary yeast hydrolysate has been found to enhance the antioxidant capacity and immune response of largemouth bass without any adverse growth effects.	[54]
Yeast (*S. cerevisiae*)	Sea cucumber, (*Apostichopus japonicus*)	5%	The study aimed to enhance the growth, digestive enzyme activity, nutritional value, and immune system of sea cucumber juveniles.	[55]

Table 2. *Cont.*

Product Name	Fish Species	Inclusion Levels	Effects	Source
Yeast, *Saccharomyces cerevisiae*, extract	Juvenile pikeperch (*Sander lucioperca L.*)	2, 4, and 6%	The study found that pikeperch growth is stimulated by the lowest analyzed dose of yeast, which is 2% yeast extract.	[56]
Yeast culture (YC)	Gibel carp (*Carassius auratus gibelio CAS III*)	0, 20, 40, and 60%	The study suggests that yeast culture could be a suitable fishmeal alternative for Gibel carp diets, with a dietary inclusion of 4 g of yeast culture per 100 g diet.	[57]
Yeast hydrolysate (YH)	Juvenile Jian carp (*Cyprinus carpio var.* Jian)	1, 3, 5, and 7%	The study found that 3% YH was the most effective substitution for fishmeal (FM) in enhancing innate immunity and growth performance.	[58]
Dried yeast	Red drum (*Sciaenops ocellatus*)	20, 30, 40, 50%	Without impairing the red drum's functionality, the material may replace between 30 and 50% of the protein supplied by fishmeal.	[49]
Brewer's yeast (*Saccharomyces cerevisiae*)	Thai Panga (*Pangasianodon hypophthalmus* × *Pangasius bocourti*)	30, 45, 60, 75%	By substituting brewer's yeast for 45% of the fishmeal, one can improve the immune response and growth performance of the Thai Panga.	[48]
Brewer's yeast	Goldfish (*Carassius auratus*)	0, 15, 25, 35, and 45%	Fish fed a diet with yeast replacing 35% of their meal showed better weight gain, SGR, FCR, and protein efficiency ratio compared to other diets.	[59]
Dietary yeast cell wall (YCW)	Japanese seabass (*Lateolabrax japonicus*)	0, 250, 500, 1000, 2000, and 20,000 mg/kg	The study found that the optimal dose of YCW is 2000 mg/kg, with a 10 × safety margin.	[60]
Yeast (*Saccharomyces cerevisiae*)	Juvenile channel catfish (*Ictalurus punctatus*)	25, 50, 75, 100, and 125 g/kg	The study found that adding up to 100 g kg^{-1} of dried yeast without affecting growth performance is possible.	[61]
Baker yeast, *Saccharomyces cerevisiae* (SC)	Nile tilapia (*Oreochromis niloticus*) fingerlings	1 and 2 g/100 g	Nile tilapia fingerlings fed a yeast mixture for 119 days showed improved growth performance, feed efficiency, and hematological indices.	[62]
Ethanol yeast (EY)	Sunshine bass (female white bass *Morone chrysops* × male striped bass *M. saxatilis*)	0, 7.5, 15%	Reductions in carcass protein and increase in carcass lipid was the results of replacing fishmeal entirely with EY. It also caused changes in whole-body composition.	[63]
Brewer's yeast	Sea bass (*Dicentrachus labrax*)	0, 10, 20, 30, or 50%	Brewer's yeast can replace 50% of fishmeal protein without negative effects on fish performance and improve feed efficiency by up to 30% in the diet.	[64]
Brewer's yeast (*Saccharomyces cerevisiae*)	Beluga sturgeon (*Huso huso*) juveniles	1 and 2%	Brewer's yeast can enhance growth performance and modify intestinal microbiota in beluga sturgeon without negatively impacting basic hematological parameters.	[65]
Yeast-based, certified organic protein source	Cobia (*Rachycentron canadum*)	25, 50, 75, and 100%	The data indicates that at least 25% of dietary protein can be provided by yeast-based protein in diets for cobia, despite negative effects on production characteristics.	[66]
Brewer's yeast (*Saccharomyces cerevisiae*)	Hybrid striped bass (*Morone chrysops* × *M. saxatilis*)	1, 2, and 4%	The growth performance, feed efficiency, and infection resistance of hybrid striped bass were found to be greatly improved by brewer's yeast.	[67]

The results of the experiments demonstrate that the majority of animals exhibited good growth performance when provided with yeast products as a primary protein source. In instances where this trait did not outperform the control, other beneficial effects were observed. However, given the significant diversity in cultured species and age groups, as well as the range of products utilized, a continuous optimization of feed formulations is essential.

5. Use of Yeast as Supplement in Aquafeed

Yeast supplements and feed additives improve fish health and production, leading to increased growth in the aquaculture industry [18]. Yeast-based diets are high in protein, lipids, attractants, and other nutrients. *Candida* sp., *Kluveromyces* sp., and *Phaffia* sp. are three yeast species commonly utilized as fishmeal alternatives. Several studies have studied the effects of partial and complete substitution of fishmeal with brewer's yeast on the growth, body composition, feed consumption, and digestibility of juvenile tilapia [23]. Another study revealed that yeast supplement is a promising growth promoter and might be an alternative method to antibiotics for disease prevention of *Mystus cavasius*) [23]. Selenium is a crucial dietary mineral for immune system function, cell growth regulation, and stress response (Se). It has been observed to improve growth performance, feed utilization, and immunocompetence in a variety of fish species, including Wuchang bream (*Megalobrama amblycephala*) [68], hybrid striped bass (*Morone chrysops* × *Morone saxatilis*) [69], and rainbow trout (*Oncorhynchus mykiss*) [70]. Some studies reported that the dietary fishmeal inclusion level could be reduced to 14% with SPC (Soy Protein Concentrate) [71] or a blend of soybean meal and poultry byproduct meal [72] serving as the fishmeal substitute. In a recent study, it was proven that the dietary fishmeal inclusion level for golden pompano could be reduced to 240 $g.kg^{-1}$ with soy protein as a fishmeal alternative and 1 $g\ kg^{-1}$ of Se-yeast supplementation. The loss in growth in golden pompano fed a high soy protein-based diet could be due to selenium (Se) deficiency rather than low feed intake and utilization [73]. However, there hasn't been enough research conducted on how dietary yeast affects the replacement of meals.

A study revealed that the inclusion of yeast in fermented poultry byproduct meal (pbm) in the diet of common carp (*Cyprinus carpio*) at 15–20% level increased digestive enzyme activities, immune function, and the growth of the fish [74]. In contrast, it has been shown that common carps can utilize a high percentage of their dietary protein requirement from the yeasts *Candida tropicalis*, *Candida utilis*, and *Candida lipolytica* with better results than those obtained with soybean or meat and bone meal. Furthermore, it was revealed that various levels of dried yeast in fish feed results in significant impact on body weight of fish as indicated by 20% greater growth by the dietary yeast inclusion in the replacement of fishmeal up to 40% [43]. Likewise, many yeast species have a high impact on the growth performance, feed efficiency, and body composition of different aquaculture species, as shown in Table 3. Therefore, to lower the cost of aquafeed for the sustainable production of aquaculture for these species, it can be deduced that fishmeal should subsequently be substituted at these levels with single-cell yeast protein [75].

The findings of the literature review indicate that the evaluated yeast products effectively functioned as prebiotics and probiotics, and the animals fed them exhibited favorable growth, feed conversion, and survival rates. However, due to the ongoing advancement in methodology for measuring and evaluating the impact of immunostimulation and microbiome diversity enhancement, direct comparisons of the results are challenging.

Table 3. Yeast products used in aquaculture as supplementary protein sources to enhance growth performance, disease resistance, and the immune system of several species.

Product Name	Fish Species	Inclusion Levels	Effects/Outcome	Source
Torula Yeast (*C. utilis*)	Rainbow trout (*Oncorhynchus mykiss*)	10 and 20%	The findings of this study suggested that torula yeast (*C. utilis*) may promote more robust rainbow trout growth when given a diet free of fishmeal.	[6]
Selenium yeast	Golden pompano (*Trachinotus ovatus*)	1 g/kg	The study suggests that golden pompano can be reduced in dietary fishmeal to 1 g/kg Selenium-yeast.	[73]
Saccharomyces cerevisiae-fermented poultry byproduct meal (pbm)	Common carp (*Cyprinus carpio*)	0, 5, 10, 15, 20%	The inclusion of 15–20% yeast and fermented poultry byproduct meal in the common carp's diet has been proven to enhance digestive enzyme activity, immunological function, and growth.	[74]
Dietary inclusion of fermented poultry byproduct meal (FPBM)	Nile tilapia (*Oreochromis niloticus*)	10, 20, 30, and 40%	To improve tilapia health and growth, 11.17–25.14% of FPBM can be added to their diets in an efficient manner.	[22]
Yeast cell wall	Juvenile Persian sturgeon (*Acipenser persicus*)	0.5 and 1%	The growth metrics and feeding performance are not significantly affected by the administration of 0.5 and 1% Immunowall.	[76]
Selenium yeast (Se)	Meagre (*Argyrosomus regius*) fingerlings	0.77, 1.51, 2.97, and 3.98 mg/kg	To improve the development performance, feed utilization, liver and kidney histology, and economic benefits of meagre juveniles, a meal intake of 3.98 mg Se kg−1 (3 mg Se-yeast) is advised.	[77]
Yeast-fermented sunflower meal (YFSFM)	Nile tilapia (*Oreochromis niloticus*)	0, 25, 50, and 75%	The varying levels of YFSFM had no significant effect on the fish's dry matter, lipid, crude protein, or ash content.	[78]
Selenium yeast	Siberian Sturgeon *Acipenser baerii*	5, 10, 15 g/kg	A food supplement of Sel-Plex at levels ≤15 g kg−1 increased the growth performance of Siberian sturgeons even though the hematological indicators improved.	[79]
Substituting soybean meal with yeast (*Sacharomyces cerevisae*) meal	African catfish (*Clarias gariepinus*)	0, 10, 20, 30, 40, 50, and 100	The diet with a 50% yeast inclusion was deemed optimal due to its enhanced nutritional status, improved blood parameters, and enhanced fish health.	[80]
Grain distillers dried yeast (GDDY)	Juvenile rainbow trout (*Oncorhynchus mykiss*)	0, 25, 37.5, 50, 62.5, 75, 87.5, and 100%	Rainbow trout development and feed conversion were highly affected by high GDDY incorporation rates, but feed intake was not affected.	[81]
Yeast-fermented canola meal	Nile tilapia (*Oreochromis niloticus*)	0, 25, 50, 75, and 100%	The study found that replacing fish with 75 and 100% levels significantly reduced their protein efficiency ratio and nutrient digestibility compared to lower levels.	[82]
Yeast, Metschnikowia sp. in combination with Rhodotorula	Juvenile of sea cucumber (*Apostichopus japonicus*)	104, 105, and 106 CFU g−1	Boost the immune system, boost growth, increase the synthesis of digestive enzymes, and provide nourishment.	[83]
Yeast-fermented canola meal	Asian sea bass (*Lates calcarifer*)	25, 50, 75, 100%	The study revealed that yeast-fermented canola meal can replace 50% of fishmeal in the Asian sea bass diet without affecting growth.	[84]

Table 3. *Cont.*

Product Name	Fish Species	Inclusion Levels	Effects/Outcome	Source
Fry Fed Organic Diets Containing Yeast Extract (YE)	Nile tilapia *(Oreochromis niloticus)*	0, 15, 30, and 45%	Fry fed a control diet with 20% FM and a diet with 45% YE/36%SBM with amino acid supplementation showed no significant differences in final weight, weight gain, and specific growth rate.	[85]
Ethanol yeast	Sunshine bass *(Morone chrysops × M. saxatilis)*	0, 7.5, 15, and 22%	According to the study, sunshine bass should be fed ethanol yeast-based diets with an FM level of between 7.5% and 15%.	[63]
Dietary supplementation with brewer's yeast	Red drum *(Sciaenops ocellatus)*	10 g/kg	The study suggests that a ten-dose dose of several prebiotics is sufficient to enhance the feed efficiency and disease resistance of red drums.	[86]
Yeast protein *Saccharomyces cerevisiae* supplemented with biogenic L-carintine	Nile tilapia *(Oreochromis niloticus) fingerlings*	25, 50, 75, and 100%	Tilapia fed a diet containing 7.14 and 10.71% yeast, supplemented with 100 mg/100 g diet, showed optimal growth performance, feed and protein utilization.	[87]
Yeast extract powder (YEP)	Rohu *(Labeo rohita)* fingerlings	0.1, 0.2, 0.3, 0.4, and 0.5%	Up to a 0.2% margin, the fish fed diets enriched with YEP grew more rapidly than the control group.	[88]
Brewer's grains with yeast (BGY)	Australian red claw crayfish *(Cherax quadricarinatus)*	10, 20, and 30%	The study suggests that soybean meal and BGY can be completely replaced with fish and shrimp meal in the diets of juvenile red claw crayfish.	[89]

5.1. Use of Yeast as Probiotics in Aquafeed

Aquaculture is expanding worldwide, yet adverse circumstances of fish culture can cause disease and financial loss. Probiotics are being promoted as a safe and sustainable alternative to traditional aquaculture [90]. In addition, probiotics are considered as live or dead microorganisms, or microalgae or yeast that give health advantages when supplied in sufficient doses [91]. Additionally, probiotics can improve fish health by enhancing feed conversion and digestion, reducing stress sensitivity, and raising overall health [92]. Additionally, probiotics stimulate the secretion of intestinal mucus, promote the growth of microvilli, and create barriers against invading pathogens [21]. The scientific community is searching for alternatives to lessen the misuse of antibiotics because they impair fish immunity, harm the aquatic ecosystem, and are present in fish tissue, which can be harmful to consumers' health [93]. Additives with prebiotic properties from yeast byproducts are used in animal diets to promote growth, improve digestion, enhance microorganism activity, and enhance immune system functionality [94]. Yeasts are resistant to antibacterial medications, making them useful as probiotic supplements during antibiotic treatment. *Saccharomyces cerevisiae* and *S. cerevisiae var. boulardii* (or *S. boulardii*) are widely explored as probiotic yeasts from various sources [95–97]. The addition of supplements containing yeast probiotics to fish feed has been shown by several researchers to enhance growth rate [18,40,42], feed digestion [18,23,39], stress tolerance [40], the immune system [18,23], and disease control [40,42,98]. Studies have shown the individual and combinational effects of probiotics (*B. subtilis, B. licheniformis and S. cerevisiae*) on the growth, body composition, digestive enzymes, and intestinal morphology of *A. persicus* fingerling and also suggested that the bacterial probiotics might be more effective than the yeast [99]. Probiotics are an environmentally benign way to boost fish health and growth in tilapia farming, as they can replace antibiotics. The study suggests that the use of probiotics, prebiotics, and synbiotics in tilapia aquaculture can significantly enhance growth and immunity, requiring an optimal dosage and administration time [100]. In addition, probiotics and prebiotics have emerged as ecologically benign alternatives to antibiotic treatments in shrimp farming, as a result of the increased interest in investigating alternative approaches [101]. In addition, research into whiteleg shrimp *Penaeus vannamei* has demonstrated that the use of mixed probiotic cultures improves viability, feed conversion, and the final yield of farmed shrimp [102].

In a study, the effect of probiotics was analyzed on survival, welfare, growth indices, and blood composition in Siberian sturgeon (*Acipenser baerii*) raised in a recirculating system. The results revealed that adding yeasts as probiotics to the meals of *Acipenser baerii* fed in a circulating system increased the fish's growth, survival, and well-being [92]. In addition, it was suggested that probiotic-rich diets can enhance the immunological response and growth of Siberian sturgeons [103]. It has been demonstrated that probiotics positively impact the growth, survival, and health of aquatic animals. Probiotics have been included in aquaculture methods to address the requirement for greater disease resistance, enhanced aquatic organism growth, and feed efficiency [104,105]. The individual and combined effects of probiotics in Persian sturgeon (*Acipenser persicus*) were assessed by measuring growth performance, proximate body composition, digestive enzymes, and intestinal morphology. The findings contributed to a growing body of literature about the individual and combined effects of probiotics with *S. cerevisiae* on growth, body composition, digestive enzymes, and intestinal architecture of *Acipenser persicus* fingerling [99]. Furthermore, commercial probiotic products are currently being made from several species of yeast, such as *S. cerevisiae*, as indicated in Table 2. It can be concluded that the use of yeasts as probiotics is expanding, and it is advantageous to consider these yeasts as a potential substitute for improving development, digestibility, survival rate, feed efficiency, and capacity to fortify the immune systems of fish larvae and juveniles. A summary of the effects of yeast products as probiotics can be found in Table 4.

Table 4. Yeast products used in aquaculture as probiotics to enhance growth performance, disease resistance, and immune system of different species.

Fish Species	Inclusion Levels	Effects	Source
Pacific white shrimp (*Litopenaeus vannamei*)	10 g/kg	According to the study, adding yeast byproduct additions to sugar byproducts may enhance development, immunity, histological changes, and resistance to the AHPND-causing *V. parahaemolyticus*.	[94]
Siberian Sturgeon (*Acipenser baerii*)	50% Lactic acid + 50% *Saccharomyces boulardii*	Growth indices, survival, and well-being of *Acipenser baerii* fish were dramatically improved when probiotics, such as lactic acid bacteria and yeasts, were introduced to their meals.	[92]
Persian sturgeon (*Acipenser persicus*)	5 g	This study contributes to the growing literature on the effects of probiotics on the growth, body composition, digestive enzymes, and intestinal morphology of *A. persicus* fingerlings.	[99]
Gilthead sea bream (*Sparus aurata*)	0.55 or 1.1%	The study found that yeast, when consumed at a 0.55 or 1% basal diet, exhibited immunostimulant activity in gilthead seabream.	[106]
Rainbow trout, (*Oncorhynchus mykiss fry*)	1, 5, and 10%	Adding yeast to rainbow trout fry diets during early life stages is suitable, with a 5% concentration likely enhancing growth performance and feed efficiency ratio.	[107]

5.2. Use of Yeast as Immune-Stimulant for Fish

Yeast meal used in aquaculture has been proven to enhance growth, modulate gut microbiota, and boost fish immune systems. Yeast has emerged as a sustainable innovative element in aquafeed due to its potential benefits for immunostimulation and nutrition of several species in aquaculture [18]. To varied degrees of success, numerous studies have been conducted to determine the effects of these immunostimulants on fish development, intestinal morphology, and overall health. In addition to its probiotic effects, *Saccharomyces cerevisiae* can be utilized as an immunostimulant for fish raised in captivity that can benefit from *Saccharomyces cerevisiae's* ability to promote growth and disease resistance [108].

Research on the potential of yeast as a possible fish feed ingredient is scanty, with a lot of focus on *S. cerevisiae* [3]. Few studies assess the health benefits of using yeast as a source of protein, even though positive health effects have been reported when using entire *S. cerevisiae* yeast or yeast derivatives. Low levels of *S. cerevisiae* yeast have shown to enhance growth performance, immune responses, and/or protection against bacterial infection and increase disease resistance in several fish species, such as salmonids [109], common carp [74], Japanese seabass (*Lateolabrax japonicas*) [60], and Nile tilapia [110]. Yeasts such as *C. utilis* have also been shown to modulate immune responses in rainbow trout, [111] and *K. marxi-anus* has shown probiotic properties in an in vitro colonic model system [112]. Many researchers are interested in yeast cell wall derivatives because of their immunostimulatory qualities and, in part, since the use of chemotherapy in animal feed is prohibited. The majority of recent developments in the study of yeast cell walls, however, have focused more on polysaccharides devoid of the lipid moiety because it has also been noted that these molecules interact with immune system cells and chemicals related to humoral immunity [113]. Additionally, it has been reported that immunostimulatory polysaccharides promote the differentiation, proliferation, and improved function of macrophages by generating reactive oxygen species (ROS) [113].

Yeasts are a good source of β-glucan which is well known for its role in immunostimulation in fishes and other vertebrates [114]. β-glucan from yeast is crucial for fish development as it boosts the immune system and protects against infections in various ways. In addition, β-glucans facilitate several biological processes such as wound healing [115], stress tolerance [116], and antioxidant [117] and antibacterial processes [118]. The most commonly utilized β-glucans for studying mechanisms of action are those of commercial origin, with recommended doses from manufacturers. Nonetheless, their immunological activity is ineffective in other fish species, and increasing the dose may have negative consequences, including immune suppression [119]. β-glucans are considered as one kind of pathogen-associated molecular pattern (PAMP) [120]. They therefore cause fish to develop a signaling route, and the signaling pathway has been explained and exemplified in Figure 1 using the most recent fish β-glucan research. In addition, they are applied to farm animals to increase feed conversion ratios (FCR), replace antibiotics in the guts of fish, eradicate opportunistic bacteria, enhance intestinal morphology and function, and enhance both innate and acquired immune functions [121], a concept originated from the discovery that certain sugars, like mannose, could be used as inhibitors of pathogen adhesion to intestinal cells [20].

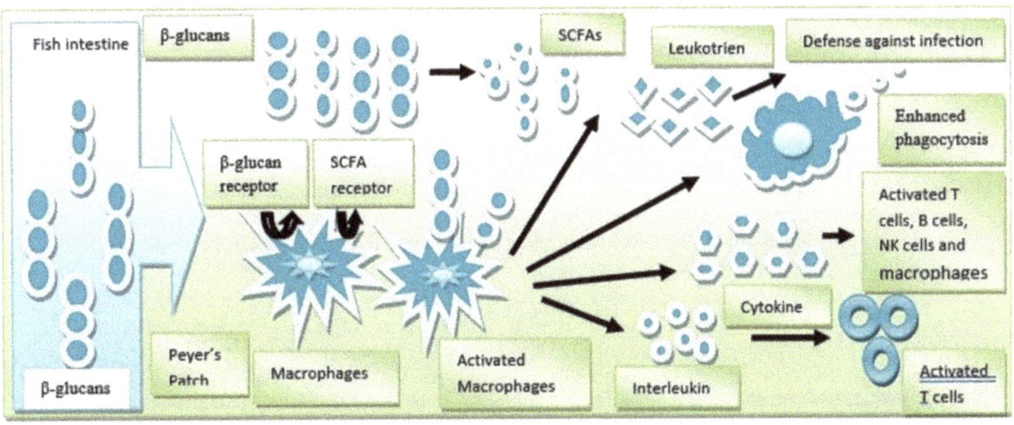

Figure 1. Signaling pathway for yeast β-glucans in the direction of fish immunization. Figure based on Mahdy et al. Source: [18].

Several studies have assessed the effects of probiotics on the immune status and gut microbiota of European sea bass, (*Dicentrarchus labrax* L., 1758). The investigated aspects range from antioxidant and digestive enzyme activities, growth performances, gut histology and immune-histochemistry, and gut microbiota to disease resistance and stress tolerance, the immunological and hematological response, body composition, malformations, and the survival rate [122–126]. However, there is a gap in the literature on the use of *S. boulardii* and its effects on the intestinal immune system of fish. These findings indicate that yeast products are thought to be more effective as immunostimulants than immunizations and antibiotics to reduce fish mortality.

The results of the literature review demonstrate that the evaluated yeast products effectively functioned as prebiotics and probiotics. The separation of the two previous chapters was solely for the purpose of providing an overview of the discussed questions. However, in experimental practice, these are inseparable, given the methodology that is typically applied

6. Digestibility of Yeast in Fish and Crustaceans

Digestibility is an important characteristic of yeast that makes it an economically competitive protein source for aquaculture, comparable to conventional sources in various fish species' diets [15,18]. Yeast cell walls represent 26–32% of the cell dry weight and contain varying proportions of mannan-oligosaccharides, β-glucan, chitin, and nucleic acids, depending on the species. In addition, the cell wall fraction contains bioactive substances like as β-glucan and mannan oligosaccharides, which promote protein solubility and digestibility during in vitro digestion [127]. Increasing replacement of fishmeal with *S. cerevisiae* in the latter study also resulted in slightly decreasing apparent digestibility of dry matter. Baker's yeast is a type of probiotic that can boost immunity, supply an enzyme for food digestion, and improve feed intake [108]. In a study, the effect of adding 10% or 20% torula yeasts to a fishmeal-free diet on the digestibility and growth of juvenile rainbow trout was evaluated. The results of this study showed that when fed a diet devoid of fishmeal, torula yeast (*C. utilis*) may help rainbow trout grow more robustly [6]. Brewer's yeast significantly decreased digestibility coefficients of dry matter, protein, and energy in sea bass juveniles when fishmeal was partially replaced with yeast [15]. Arctic char had lower energy and amino acid digestibility than extracted *S. cerevisiae*, but there was no significant difference in Eurasian perch (*Perca fluviatilis*) [41]. This demonstrates variances in digestive efficiency among species, probably related to varying gastrointestinal enzyme activity [128]. In studies with Atlantic salmon, the replacement of 40% of low temperature (LT) fishmeal with *C. utilis* or *K. marxianus* yeast had no significant effect on the digestibility of crude protein and energy, whereas the substitution with *S. cerevisiae* significantly reduced protein and energy digestibility [36]. Protein digestibility in Atlantic salmon was further enhanced by the improved solubility of proteins and β-glucans resulting from distinct downstream processing techniques of the yeast *S. cerevisiae* [127]. The digestibilities of individual amino acids were also unaffected by the dietary inclusion of *C. utilis* and *K. marxianus*, except for a reduced digestibility of methionine and cysteine in the *K. marxianus* diet. Conversely, the diet containing *S. cerevisiae* revealed significantly lower amino acid digestibility than the fishmeal control except for glycine and cysteine [36]. The latter study [15] indicated considerable differences in digestibility among different whole-cell yeast products. The enzymatic hydrolysis of cell wall rupture can improve the digestibility of yeast nutrients [109]. Moreover, by removing the cell wall material, which is a representation of the water-soluble cell contents, yeast can be extracted from lignocellulosic biomass and used as aquaculture feed. Since cell walls are removed and there is a greater amount of water-soluble low molecular weight proteins in the diet, adding more yeast extract in place of fishmeal in shrimp (*Litopenaeus vannamei*) diets increased the apparent digestibility of protein [129]. The extrusion of feed may cause partial disruption of yeast cell walls and potentially increase protein and amino acid digestibility [130]. Enhanced growth performance and health in fish fed on yeast might also be attributed to the improved digestive enzyme activity [111] and by supplying digestive enzymes that aid in the digestion of complex carbohydrates, [131] or by enhancing digestive capacity by improving gastrointestinal morphology and thereby increasing the absorptive surface [36,109,111].

7. Use of Yeast in Live Aquafeed Culture

The significant advancements in technology for larval rearing, along with the global proliferation of hatcheries, necessitate the use of appropriate rotifer culture techniques as a vital source of larval nourishment. Rotifers (*Brachionus* sp.) are supplied as a living capsule, providing the necessary nutrients for the proper growth and production of cultivated fish larvae [132]. In addition, rotifers are the most important live-feeding species utilized in the rearing of fish larvae because of their size, nutritional content, and behavior [18,133]. Furthermore, it has been proved by many researchers that a medium including yeast, starch, and albumen is suitable for the rotifer culture. In addition, by using yeasts in aquaculture, copepods, artemia, and rotifers can be produced along with live aquafeed [18]. Moreover, *Artemia nauplii* is now one of the most dependable live feeds for fish and is frequently utilized to nourish young aquatic species raised in aquaculture [134]. However, the amount of yeast used in rotifer culture varies greatly due to insufficient information on feeding rate, frequency, and composition range [135]. Through the use of a somewhat diet-dependent procedure with variable yeast amounts, the ideal feeding rate and frequency for *Brachionus plicatilis* was found [136]. The results showed that the feeding rate of the rotifers is sufficient at an average of 0.3 g of barker yeast per million rotifers. Additionally, increasing the feeding frequency can lead to a higher population growth rate and egg-bearing ratio, with an increase from twice to three times [136]. According to a study, yeast works better for Rotifer *Brachionus calyciflorus* cultures when combined with fresh *Chlorella* compared to powdered *Chlorella*. It is advised that better raising conditions will enable good rotifer development when the food contains both fresh *Chlorella* and yeast [133]. Furthermore, research has demonstrated that yeast increases the density of copepods in culture, suggesting that it could one day be used as a supplement to feed for additional copepod species [137]. According to other studies, the maximum relative population density of copepods has been reached when compared to marine microalgae when copepods are fed commercial *S. cerevisiae* baker's yeast with a soybean component [138]. Therefore, in the cultivation of live food, yeast is recommended as a supplement or protein alternative.

The application of live feed is a standard part of the production of some fish species, which is important from both a commercial and a nature conservation standpoint. The results of the experiments outlined above clearly demonstrate that yeast products are an excellent source of nutrition for rotifers and microcrustaceans used as live food.

8. Conclusions and Aspects of Future Research

In conclusion, the use of yeast as a primary protein-rich feed additive, supplement, and probiotic has the potential to provide the necessary solutions to meet current and future market demands and address issues associated with the production of several valuable species for the future of the aquaculture sector. Furthermore, the current investigation has demonstrated that the addition of yeast to the diets of several fish species can enhance their growth, immunity, and efficiency in utilizing nutrients. Our review of the literature indicates that the addition of yeast to fish feed, either directly or as a supplement, can facilitate growth, improve digestive efficiency, and bolster immune systems. Furthermore, the use of yeast as a probiotic can enhance immune systems and improve water quality, thereby increasing yield. However, to maximize growth, survival, and health outcomes, it is recommended to use probiotics in exploitable species. Furthermore, yeast is regarded as a viable option for live aquafeed and aquaculture. It is not recommended to substitute all other protein sources in aquafeed with yeast proteins, as this could have adverse effects on certain fish species and pose potential health risks to consumers.

However, the efficacy of yeast as a primary protein source in fish diets is still under investigation, and further research is necessary. Yeast products have been shown to have a protein digestibility that is similar to that of conventional protein sources when incorporated into the diets of different fish species. However, there is a lack of research comparing the digestibility of different yeasts in the same experiment. To compete with fishmeal in the aquaculture sector and become a reasonably priced alternative for fish farmers, large-scale investment is required for yeast production. This would also help to reduce their dependency on fishmeal. While yeast can be used as a protein source in aquafeed, there is currently limited knowledge regarding its safety. To prioritize the development of potential ways that address fish and environmental safety and human health, more scientific research on the use of yeast in the aquaculture industry is required. Future research should focus on understanding the specific nutritional needs of cultured aquatic species at different stages and developing affordable dry feeds that the industry can use.

Author Contributions: C.H., conceptualization, writing—original draft preparation; S.S. and B.K., writing—original draft preparation; J.B., writing—review and editing. All authors have read and agreed to the published version of the manuscript.

Funding: This research received no external funding.

Institutional Review Board Statement: Not applicable.

Informed Consent Statement: Not applicable.

Data Availability Statement: No new data were created or analyzed in this study. Data sharing is not applicable to this article.

Conflicts of Interest: The authors declare no conflicts of interest.

References

1. FAO. *The State of World Fisheries and Aquaculture (SOFIA)*; FAO: Rome, Italy, 2022; p. 266.
2. Reverter, M.; Tapissier-Bontemps, N.; Sarter, S.; Sasal, P.; Caruso, D. Moving towards more sustainable aquaculture practices: A meta-analysis on the potential of plant-enriched diets to improve fish growth, immunity and disease resistance. *Rev. Aquac.* **2021**, *13*, 537–555. [CrossRef]
3. Agboola, J.; Schiavone, M.; Øverland, M.; Morales-Lange, B.; Lagos, L.; Arntzen, M.Ø.; Lapeña, D.; Eijsink, V.G.H.; Horn, S.J.; Mydland, L.T.; et al. Impact of Down-Stream Processing on Functional Properties of Yeasts in Diets of Atlantic Salmon (*Salmo salar*): Implications for Gut Health. *Sci. Rep.* **2021**, *11*, 4496.
4. Jia, S.; Li, X.; He, W.; Wu, G. Protein-Sourced Feedstuffs for Aquatic Animals in Nutrition Research and Aquaculture. In *Advances in Experimental Medicine and Biology*; Dong, H., Radeke, H.H., Rezaei, N., Steinlein, O., Xiao, J., Rosenhouse-Dantsker, A., Gerlai, R., Eds.; Springer: Berlin/Heidelberg, Germany, 2021; Volume 1354. [CrossRef]
5. IFFO. Marine Ingredients Organization, Statistical Yearbook 2019. Marine Ingredients Organization. 2019. Available online: https://www.iffo.com/feeding-growing-population (accessed on 6 August 2024).
6. Ekmay, R.D.; Plagnes-juan, E.; Aguirre, P.; Surget, A.; Terrier, F.; Frohn, L.; Skiba-Cassy, S. Partially replacing plant protein sources with torula yeast in rainbow trout (*Oncorhynchus mykiss*) feed increases growth and factors related to immune status. *J. World Aquac. Soc.* **2024**, *55*, 169–186. [CrossRef]
7. Agboola, J.O.; Øverland, M.; Skrede, A.; Hansen, J.Ø. Yeast as major protein-rich ingredient in aquafeeds: A review of the implications for aquaculture production. *Rev. Aquac.* **2021**, *13*, 949–970. [CrossRef]
8. Hou, Y.; He, W.; Hu, S.; Wu, G. Composition of polyamines and amino acids in plant-source foods for human consumption. *Amino Acids* **2019**, *51*, 1153–1165. [CrossRef]
9. Dossou, S.; Koshio, S.; Ishikawa, M.; Yokoyama, S.; Dawood, M.A.O.; Basuini, M.F.E.; Olivier, A.; Zaineldin, A.I. Growth performance, blood health, antioxidant status and immune response in red sea bream (*Pagrus major*) fed *Aspergillus oryzae* fermented rapeseed meal (RM-Koji). *Fish Shellfish Immunol.* **2018**, *75*, 253–262. [CrossRef]
10. FAO. *The State of World Fisheries and Aquaculture 2018—Meeting the Sustainable Development Goals*; FAO Fisheries: Rome, Italy, 2018.
11. Froehlich, H.E.; Jacobsen, N.S.; Essington, T.E.; Clavelle, T.; Halpern, B.S. Avoiding the ecological limits of forage fish for fed aquaculture. *Nat. Sustain.* **2018**, *1*, 298–303. [CrossRef]
12. Hua, K.; Cobcroft, J.M.; Cole, A.; Condon, K.; Jerry, D.R.; Mangott, A.; Praeger, C.; Vucko, M.J.; Zheng, C.; Zenger, K.; et al. The future of aquatic protein: Implications for protein sources in aquaculture diets. *One Earth* **2019**, *1*, 316–329. [CrossRef]
13. Richard, N.; Costas, B.; Machado, M.; Fernández-Boo, S.; Girons, A.; Dias, J.; Corraze, G.; Terrier, F.; Marchand, Y.; Skiba-Cassy, S. Inclusion of a protein-rich yeast fraction in rainbow trout plant-based diet: Consequences on growth performances, flesh fatty acid profile and health-related parameters. *Aquaculture* **2021**, *544*, 737132. [CrossRef]

14. Glencross, B.D.; Huyben, D.; Schrama, J.W. The application of single-cell ingredients in aquaculture feeds—a review. *Fishes* **2020**, *5*, 22. [CrossRef]
15. Øverland, M.; Skrede, A. Yeast derived from lignocellulosic biomass as a sustainable feed resource for use in aquaculture. *J. Sci. Food Agric.* **2017**, *97*, 733–742. [CrossRef] [PubMed]
16. Dawood, M.A.O.; Koshio, S.; Ishikawa, M.; Yokoyama, S.; Basuini, E.; Hossain, M.S.; Nhu, T.H.; Moss, A.S.; Dossou, S.; Wei, H. Dietary supplementation of β-glucan improves growth performance, the innate immune response and stress resistance of red sea bream, *Pagrus major*. *Aquac. Nutr.* **2017**, *23*, 148–159. [CrossRef]
17. Abu-Elala, N.M.; El-Sayed Ali, T.; Ragaa, N.M.; Ali, S.E.; Abd-Elsalam, R.M.; Younis, N.A.; Abdel-Moneam, D.A.; Hamdien, A.H.; Bonato, M.; Dawood, M.A.O. Analysis of the Productivity, Immunity, and Health Performance of Nile Tilapia (*Oreochromis niloticus*) Broodstock-fed Dietary Fermented Extracts Sourced from *Saccharomyces cerevisiae* (Hilyses): A Field Trial. *Animals* **2021**, *11*, 815. [CrossRef] [PubMed]
18. Mahdy, M.A.; Jamal, M.T.; Al-Harb, M.; Al-Mur, B.A.; Haque, M.F. Use of yeasts in aquaculture nutrition and immunostimulation: A review. *J. Appl. Biol. Biotechnol.* **2022**, *10*, 59–65. [CrossRef]
19. Zhou, X. Notes from the Aquaculture Statistician. *FAO Aquac. Newsl.* **2018**, *58*, 6–8.
20. Torrecillas, S.; Montero, D.; Izquierdo, M. Improved health and growth of fish fed mannan oligosaccharides: Potential mode of action. *Fish Shellfish Immunol.* **2014**, *36*, 525–544. [CrossRef]
21. Ahmadifar, E.; Moghadam, M.S.; Dawood, M.A.O.; Hoseinifar, S.H. Lactobacillus fermentum and/or ferulic acid improved the immune responses, antioxidative defence and resistance against *Aeromonas hydrophila* in common carp (*Cyprinus carpio*) fingerlings. *Fish Shellfish Immunol.* **2019**, *94*, 916–923. [CrossRef]
22. Dawood, M.A.O.; Magouz, F.I.; Mansour, M.; Saleh, A.A.; Asely, A.M.E.; Fadl, S.E.; Ahmed, H.A.; Al-Ghanim, K.A.; Mahboob, S.; Al-Misned, F. Evaluation of yeast fermented poultry by-product meal in Nile Tilapia (*Oreochromis niloticus*) feed: Effects on growth performance, digestive enzymes activity, innate immunity, and antioxidant capacity. *Front. Vet. Sci.* **2020**, *6*, 516. [CrossRef]
23. Navarrete, P.; Tovar-Ramírez, D. Use of yeasts as probiotics in fish aquaculture. *Sustain. Aquac. Tech.* **2014**, *1*, 135–172.
24. Nathanailides, C.; Kolygas, M.; Choremi, K.; Mavraganis, T.; Gouva, E.; Vidalis, K.; Athanassopoulou, F. Probiotics Have the Potential to Significantly Mitigate the Environmental Impact of Freshwater Fish Farms. *Fishes* **2021**, *6*, 76. [CrossRef]
25. Schiavone, M.; Vax, A.; Formosa, C.; Martin-Yken, H.; Dague, E.; François, J.M. A combined chemical and enzymatic method to determine quantitatively the polysaccharide components in the cell wall of yeasts. *FEMS Yeast Res.* **2014**, *14*, 933–947. [CrossRef] [PubMed]
26. Rawling, M.; Leclercq, E.; Foey, A.; Castex, M.; Merrifield, D. A novel dietary multi-strain yeast fraction modulates intestinal toll-like-receptor signalling and mucosal responses of rainbow trout (*Oncorhynchus mykiss*). *PLoS ONE* **2021**, *16*, e0245021. [CrossRef] [PubMed]
27. Rawling, M.; Schiavone, M.; Apper, E.; Merrifield, D.L.; Castex, M.; Leclercq, E.; Foey, A. Yeast cell wall extracts from *Saccharomyces cerevisiae* varying in structure and composition differentially shape the innate immunity and mucosal tissue responses of the intestine of zebrafish (*Danio rerio*). *Front. Immunol.* **2023**, *14*, 1158390. [CrossRef] [PubMed]
28. Bertolo, A.P.; Biz, A.P.; Kempka, A.P.; Rigo, E.; Cavalheiro, D. Yeast (*Saccharomyces cerevisiae*): Evaluation of cellular disruption processes, chemical composition, functional properties and digestibility. *J. Food Sci. Technol.* **2019**, *56*, 3697–3706. [CrossRef] [PubMed]
29. Hicks, T.M.; Verbeek, C.J.R. Meat Industry Protein by-Products: Sources and Characteristics. In *Protein Byproducts*; Dhillon, G.D., Ed.; Academic Press: Cambridge, MA, USA, 2016; pp. 37–61. [CrossRef]
30. Cuppett, S.L.; Soares, J.H. The metabolizable energy values and digestibilities of menhaden fish meal, fish solubles and fish oils. *Poult. Sci.* **1972**, *51*, 2078–2083. [CrossRef]
31. Li, P.; He, W.; Wu, G. Composition of amino acids in foodstuffs for humans and animals. In *Amino Acids in Nutrition and Health*; Wu, G., Ed.; Springer: Berlin/Heidelberg, Germany, 2021; pp. 189–210.
32. Deng, Y.; Borewicz, K.; van Loo, J.; Olabarrieta, M.Z.; Kokou, F.; Sipkema, D.; Verdegem, M.C.J. In-Situ Biofloc Affects the Core Prokaryotes Community Composition in Gut and Enhances Growth of Nile Tilapia (*Oreochromis niloticus*). *Microb. Ecol.* **2022**, *84*, 879–892. [CrossRef]
33. Bardhan, P.; Gupta, K.; Kishor, S.; Chattopadhyay, P.; Chaliha, C.; Kalita, E.; Goud, V.V.; Mandal, M. Oleaginous yeasts isolated from traditional fermented foods and beverages of Manipur and Mizoram, India, as a potent source of microbial lipids for biodiesel production. *Ann. Microbiol.* **2020**, *70*, 1–14. [CrossRef]
34. Parolini, G. Building Human and Industrial Capacity in European Biotechnology: The Yeast Genome. *Biotechnology* **1996**, *16*, 365–368.
35. Shurson, G.C. Yeast and yeast derivatives in feed additives and ingredients: Sources, characteristics, animal responses, and quantification methods. *Anim. Feed Sci. Technol.* **2018**, *235*, 60–76. [CrossRef]
36. Øverland, M.; Karlsson, A.; Mydland, L.T.; Romarheim, O.H.; Skrede, A. Evaluation of *Candida utilis*, *Kluyveromyces marxianus* and *Saccharomyces cerevisiae* yeasts as protein sources in diets for Atlantic salmon (*Salmo salar*). *Aquaculture* **2013**, *402*, 1–7. [CrossRef]
37. Lapeña, D.; Olsen, P.M.; Arntzen, M.Ø.; Kosa, G.; Eijsink, V.G.H.; Horn, S.J. Spruce sugars and poultry hydrolysate as growth medium in repeated fed-batch fermentation processes for production of yeast biomass. *Bioprocess Biosyst. Eng.* **2020**, *43*, 723–736. [CrossRef] [PubMed]

38. Abdel-Tawwab, M.; Adeshina, I.; Issa, Z.A. Antioxidants and immune responses, resistance to *Aspergilus flavus* infection, and growth performance of Nile tilapia, *Oreochromis niloticus*, fed diets supplemented with yeast, *Saccharomyces cerevisiae*. *Anim. Feed Sci. Technol.* **2020**, *263*, 114484. [CrossRef]
39. Vidakovic, A.; Huyben, D.; Sundh, H.; Nyman, A.; Vielma, J.; Passoth, V.; Kiessling, A.; Lundh, T. Growth performance, nutrient digestibility and intestinal morphology of rainbow trout (*Oncorhynchus mykiss*) fed graded levels of the yeasts *Saccharomyces cerevisiae* and *Wickerhamomyces anomalus*. *Aquac. Nutr.* **2020**, *26*, 275–286. [CrossRef]
40. Abass, D.A.; Obirikorang, K.A.; Campion, B.B.; Edziyie, R.E.; Skov, P.V. Dietary supplementation of yeast (*Saccharomyces cerevisiae*) improves growth, stress tolerance, and disease resistance in juvenile Nile tilapia (*Oreochromis niloticus*). *Aquac. Int.* **2018**, *26*, 843–855. [CrossRef]
41. Vidakovic, A.; Langeland, M.; Sundh, H.; Sundell, K.; Olstorpe, M.; Vielma, J.; Kiessling, A.; Lundh, T. Evaluation of growth performance and intestinal barrier function in Arctic Charr (*Salvelinus alpinus*) fed yeast (*Saccharomyces cerevisiae*), fungi (*Rhizopus oryzae*) and blue mussel (*Mytilus edulis*). *Aquac. Nutr.* **2016**, *22*, 1348–1360. [CrossRef]
42. Banu, M.R.; Akter, S.; Islam, M.R.; Mondol, M.N.; Hossain, M.A. Probiotic yeast enhanced growth performance and disease resistance in freshwater catfish gulsa tengra, *Mystus cavasius*. *Aquac. Rep.* **2020**, *16*, 100237. [CrossRef]
43. Ozório, R.O.A.; Portz, L.; Borghesi, R.; Cyrino, J.E.P. Effects of dietary yeast (*Saccharomyces cerevisia*) supplementation in practical diets of tilapia (*Oreochromis niloticus*). *Animals* **2012**, *2*, 16–24. [CrossRef]
44. Yones, A.-M.M.; Hussein, M.S.; Ali, M.W.; Abdel-Azem, A.-A.M. Effect of dietary Lacto cel-con probiotic on growth performance and hematology indices of fingerlings mono-sex Nile tilapia (*Oreochromis niloticus*). *Egypt. J. Aquat. Biol. Fish.* **2019**, *23*, 227–239. [CrossRef]
45. El-Bab, A.F.F.; Saghir, S.A.M.; El-Naser, I.A.A.; El-Kheir, S.M.M.A.; Abdel-Kader, M.F.; Alruhaimi, R.S.; Alqhtani, H.A.; Mahmoud, A.M.; Naiel, M.A.E.; El-Raghi, A.A. The effect of dietary *Saccharomyces cerevisiae* on growth performance, oxidative status, and immune response of sea bream (*Sparus aurata*). *Life* **2022**, *12*, 1013. [CrossRef]
46. Ferreira, I.M.P.L.V.O.; Pinho, O.; Vieira, E.; Tavarela, J.G. Brewer's *Saccharomyces* yeast biomass: Characteristics and potential applications. *Trends Food Sci. Technol.* **2010**, *21*, 77–84. [CrossRef]
47. Pradhan, D.; Swain, H.S.; Upadhyay, A.; Sahu, B.; Nanda, S.; Patra, S.K.; Kasturi, S.; Mohanta, K.N.; Giri, S.S. On Valorization of Brewer's Yeast as an Environmentally Sustainable Fishmeal Replacement in *Labeo rohita* Nutrition: Insight to Growth Attributes, Digestive Enzyme Activities and Haemato-biochemical Indices. *Waste Biomass Valorization* **2024**, *15*, 3503–3517. [CrossRef]
48. Pongpet, J.; Ponchunchoovong, S.; Payooha, K. Partial replacement of fishmeal by brewer's yeast (*Saccharomyces cerevisiae*) in the diets of Thai Panga (*Pangasianodon hypophthalmus* × *Pangasius bocourti*). *Aquac. Nutr.* **2016**, *22*, 575–585. [CrossRef]
49. Rosales, M.; Castillo, S.; Pohlenz, C.; Gatlin, D.M., III. Evaluation of dried yeast and threonine fermentation biomass as partial fish meal replacements in the diet of red drum *Sciaenops ocellatus*. *Anim. Feed Sci. Technol.* **2017**, *232*, 190–197. [CrossRef]
50. Hao, Q.; Xia, R.; Zhang, Q.; Xie, Y.; Ran, C.; Yang, Y.; Zhou, W.; Chu, F.; Zhang, X.; Wang, Y.; et al. Partially replacing dietary fish meal by *Saccharomyces cerevisiae* culture improve growth performance, immunity, disease resistance, composition and function of intestinal microbiota in channel catfish (*Ictalurus punctatus*). *Fish Shellfish Immunol.* **2022**, *125*, 220–229. [CrossRef]
51. Andriamialinirina, H.J.T.; Irm, M.; Taj, S.; Lou, J.H.; Jin, M.; Zhou, Q. The effects of dietary yeast hydrolysate on growth, hematology, antioxidant enzyme activities and non-specific immunity of juvenile Nile tilapia, *Oreochromis niloticus*. *Fish Shellfish Immunol.* **2020**, *101*, 168–175. [CrossRef]
52. Chen, M.; Chen, X.-Q.; Tian, L.-X.; Liu, Y.-J.; Niu, J. Enhanced intestinal health, immune responses and ammonia resistance in Pacific white shrimp (*Litopenaeus vannamei*) fed dietary hydrolyzed yeast (*Rhodotorula mucilaginosa*) and *Bacillus licheniformis*. *Aquac. Rep.* **2020**, *17*, 100385. [CrossRef]
53. Rimoldi, S.; Gini, E.; Koch, J.F.A.; Iannini, F.; Brambilla, F.; Terova, G. Erratum: Effects of hydrolyzed fish protein and autolyzed yeast as substitutes of fishmeal in the gilthead sea bream (*Sparus aurata*) diet, on fish intestinal microbiome. *BMC Vet. Res.* **2020**, *16*, 118, Erratum in *BMC Vet. Res.* **2020**, *16*, 219. [CrossRef]
54. Gong, Y.; Yang, F.; Hu, J.; Liu, C.; Liu, H.; Han, D.; Jin, J.; Yang, Y.; Zhu, X.; Yi, J.; et al. Effects of dietary yeast hydrolysate on the growth, antioxidant response, immune response and disease resistance of largemouth bass (*Micropterus salmoides*). *Fish Shellfish Immunol.* **2019**, *94*, 548–557. [CrossRef]
55. Ma, Y.; Li, L.-Y.; Li, M.; Chen, W.; Bao, P.-Y.; Yu, Z.-C.; Chang, Y.-Q. Effects of dietary probiotic yeast on growth parameters in juvenile sea cucumber, *Apostichopus japonicus*. *Aquaculture* **2019**, *499*, 203–211. [CrossRef]
56. Jarmołowicz, S.; Rożyński, M.; Kowalska, A.; Zakęś, Z. Growth in juvenile pikeperch (*Sander lucioperca* L.) stimulated with yeast, *Saccharomyces cerevisiae*, extract. *Aquac. Res.* **2018**, *49*, 614–620. [CrossRef]
57. Zhang, P.; Cao, S.; Zou, T.; Han, D.; Liu, H.; Jin, J.; Yang, Y.; Zou, X.; Xie, S.; Zhou, W. Effects of dietary yeast culture on growth performance, immune response and disease resistance of gibel carp (*Carassius auratus gibelio* CAS III). *Fish Shellfish Immunol.* **2018**, *82*, 400–407. [CrossRef] [PubMed]
58. Yuan, X.-Y.; Liu, W.-B.; Liang, C.; Sun, C.-X.; Xue, Y.-F.; Wan, Z.-D.; Jiang, G.-Z. Effects of partial replacement of fish meal by yeast hydrolysate on complement system and stress resistance in juvenile Jian carp (*Cyprinus carpio* var. *Jian*). *Fish Shellfish Immunol.* **2017**, *67*, 312–321. [CrossRef] [PubMed]
59. Gumus, E.; Aydin, B.; Kanyilmaz, M. Growth and feed utilization of goldfish (*Carassius auratus*) fed graded levels of brewer's yeast (*Saccharomyces cerevisiae*). *Iran. J. Fish. Sci.* **2016**, *15*, 1124–1133.

60. Yu, H.H.; Han, F.; Xue, M.; Wang, J.; Tacon, P.; Zheng, Y.H.; Wu, X.F.; Zhang, Y.J. Efficacy and tolerance of yeast cell wall as an immunostimulant in the diet of Japanese seabass (*Lateolabrax japonicus*). *Aquaculture* **2014**, *432*, 217–224. [CrossRef]
61. Peterson, B.C.; Booth, N.J.; Manning, B.B. Replacement of fish meal in juvenile channel catfish, *Ictalurus punctatus*, diets using a yeast-derived protein source: The effects on weight gain, food conversion ratio, body composition and survival of catfish challenged with *Edwardsiella ictaluri*. *Aquac. Nutr.* **2012**, *18*, 132–137. [CrossRef]
62. Goda, A.M.A.; Mabrouk, H.A.-H.H.; Wafa, M.A.E.-H.; El-Afifi, T.M. Effect of using baker's yeast and exogenous digestive enzymes as growth promoters on growth, feed utilization and hematological indices of Nile tilapia, *Oreochromis niloticus* fingerlings. *J. Agric. Sci. Technol. B* **2012**, *2*, 15–28.
63. Gause, B.; Trushenski, J. Replacement of fish meal with ethanol yeast in the diets of sunshine bass. *N. Am. J. Aquac.* **2011**, *73*, 97–103. [CrossRef]
64. Enes, P.; Panserat, S.; Kaushik, S.; Oliva-Teles, A. Dietary carbohydrate utilization by European sea bass (*Dicentrarchus labrax* L.) and gilthead sea bream (*Sparus aurata* L.) juveniles. *Rev. Fish. Sci.* **2011**, *19*, 201–215. [CrossRef]
65. Hoseinifar, S.H.; Mirvaghefi, A.; Merrifield, D.L. The effects of dietary inactive brewer's yeast *Saccharomyces cerevisiae* var. *ellipsoideus* on the growth, physiological responses and gut microbiota of juvenile beluga (*Huso huso*). *Aquaculture* **2011**, *318*, 90–94. [CrossRef]
66. Lunger, A.N.; Craig, S.R.; McLean, E. Replacement of fish meal in cobia (*Rachycentron canadum*) diets using an organically certified protein. *Aquaculture* **2006**, *257*, 393–399. [CrossRef]
67. Li, P.; Gatlin, D.M., III. Evaluation of brewer's yeast (*Saccharomyces cerevisiae*) as a feed supplement for hybrid striped bass (*Morone chrysops* × *M. saxatilis*). *Aquaculture* **2003**, *219*, 681–692. [CrossRef]
68. Long, M.; Lin, W.; Hou, J.; Gou, H.; Li, L.; Li, D.; Tang, R.; Yang, F. Dietary supplementation with selenium yeast and tea polyphenols improve growth performance and nitrite tolerance of Wuchang bream (*Megalobrama amblycephala*). *Fish Shellfish Immunol.* **2017**, *68*, 74–83. [CrossRef] [PubMed]
69. Cotter, P.A.; Craig, S.R.; McLean, E. Hyperaccumulation of selenium in hybrid striped bass: A functional food for aquaculture? *Aquac. Nutr.* **2008**, *14*, 215–222. [CrossRef]
70. Wang, L.; Zhang, X.; Wu, L.; Liu, Q.; Zhang, D.; Yin, J. Expression of selenoprotein genes in muscle is crucial for the growth of rainbow trout (*Oncorhynchus mykiss*) fed diets supplemented with selenium yeast. *Aquaculture* **2018**, *492*, 82–90. [CrossRef]
71. Wu, Y.; Wang, Y.; Ren, X.; Huang, D.; Si, G.; Chen, J. Replacement of fish meal with gamma-ray irradiated soybean meal in the diets of largemouth bass *Micropterus salmoides*. *Aquac. Nutr.* **2021**, *27*, 977–985. [CrossRef]
72. Wang, Y.; Ma, X.Z.; Wang, F.; Wu, Y.B.; Qin, J.G.; Li, P. Supplementations of poultry by-product meal and selenium yeast increase fish meal replacement by soybean meal in golden pompano (*Trachinotus ovatus*) diet. *Aquac. Res.* **2017**, *48*, 1904–1914. [CrossRef]
73. Wu, Y.; Fang, H.; Ma, H.; Wang, X. Supplementation of Selenium-Yeast Enhances Fishmeal Replacement by Soy Protein Concentrate in Diets for Golden Pompano (*Trachinotus ovatus*). *Aquac. Res.* **2023**, *2023*, 8953076. [CrossRef]
74. Dawood, M.A.O.; Magouz, F.I.; Essa, M.; Mansour, M. Impact of Yeast Fermented Poultry by-Product Meal on Growth, Digestive Enzyme Activities, Intestinal Morphometry and Immune Response Traits of Common Carp (*Cyprinus carpio*). *Ann. Anim. Sci.* **2020**, *20*, 939–959. [CrossRef]
75. Bob-Manuel, F.G. A comparative study of the effect of yeast single cell protein on growth, feed utilization and condition factor of the African catfish *Clarias gariepinus* (Burchell) and tilapia, *Oreochromis niloticus* (Linnaeus) fingerlings. *Afr. J. Agric. Res.* **2014**, *9*, 2005–2011.
76. Yousefi, S.; Shokri, M.M.; Navirian, H.A.; Hoseinifar, S.H. Effects of yeast cell membrane prebiotic (Immunowall®) on growth performance and hematological parameters in juvenile Persian sturgeon (*Acipenser persicus*). *J. Anim. Environ.* **2020**, *12*, 221–228.
77. Khalil, H.S.; Mansour, A.T.; Goda, A.M.A.; Omar, E.A. Effect of selenium yeast supplementation on growth performance, feed utilization, lipid profile, liver and intestine histological changes, and economic benefit in meagre, *Argyrosomus regius*, fingerlings. *Aquaculture* **2019**, *501*, 135–143. [CrossRef]
78. Hassaan, M.S.; Soltan, M.A.; Mohammady, E.Y.; Elashry, M.A.; El-Haroun, E.R.; Davies, S.J. Growth and physiological responses of Nile tilapia, *Oreochromis niloticus* fed dietary fermented sunflower meal inoculated with *Saccharomyces cerevisiae* and *Bacillus subtilis*. *Aquaculture* **2018**, *495*, 592–601. [CrossRef]
79. Tavana, B.G.Z.; Banaee, M.; Jourdehi, A.Y.; Haghi, B.N.; Hassani, M.H.S. Effects of dietary Sel-Plex supplement on growth performance, hematological and immunological parameters in Siberian sturgeon (*Acipenser baerii* Brandt, 1869). *Iran. J. Fish. Sci.* **2019**, *18*, 830–846. [CrossRef]
80. Ezenwaji, N.E.; Iluno, A.; Atama, C.; Nwaigwe, C.O.; Nwaigwe, C.U. Substitution of soyabean meal with bioactive yeast in the diet of *Clarias gariepinus*: Effect on growth rate, haematological and biochemical profile. *Afr. J. Biotechnol.* **2012**, *11*, 15802–15810.
81. Hauptman, B.S.; Barrows, F.T.; Block, S.S.; Gaylord, T.G.; Paterson, J.A. Evaluation of grain distillers dried yeast as a fish meal substitute in practical-type diets of juvenile rainbow trout, *Oncorhynchus mykiss*. *Aquaculture* **2014**, *432*, 7–14. [CrossRef]
82. Plaipetch, P.; Yakupitiyage, A. Effect of replacing soybean meal with yeast-fermented canola meal on growth and nutrient retention of Nile tilapia, *Oreochromis niloticus* (Linnaeus 1758). *Aquac. Res.* **2014**, *45*, 1744–1753. [CrossRef]
83. Ma, Y.; Liu, Z.; Yang, Z.; Li, M.; Liu, J.; Song, J. Effects of dietary live yeast *Hanseniaspora opuntiae* C21 on the immune and disease resistance against *Vibrio splendidus* infection in juvenile sea cucumber *Apostichopus japonicus*. *Fish Shellfish Immunol.* **2013**, *34*, 66–73. [CrossRef]

84. Plaipetch, P.; Yakupitiyage, A. Use of yeast-fermented canola meal to replace fishmeal in the diet of Asian sea bass *Lates calcarifer* (Bloch, 1790). *J. Aquac. Res. Dev.* **2012**, *3*, 1000125. [CrossRef]
85. Trosvik, K.A.; Rawles, S.D.; Thompson, K.R.; Metts, L.A.; Gannam, A.; Twibell, R.; Webster, C.D. Growth and body composition of Nile tilapia, *Oreochromis niloticus*, fry fed organic diets containing yeast extract and soybean meal as replacements for fish meal, with and without supplemental lysine and methionine. *J. World Aquac. Soc.* **2012**, *43*, 635–647. [CrossRef]
86. Buentello, J.A.; Neill, W.H.; Gatlin, D.M., III. Effects of dietary prebiotics on the growth, feed efficiency and non-specific immunity of juvenile red drum *Sciaenops ocellatus* fed soybean-based diets. *Aquac. Res.* **2010**, *41*, 411–418. [CrossRef]
87. Ebrahim, M.S.M.; Abou-Seif, R.A. Fish Meal Replacement by Yeast Protein (*Saccharomyces cerevisiae*) Supplemented with Biogenic L-Carintine as a Source of Methionine Plus Lysine Mixture in Feed for Nile Tilapia (*Oreochromis niloticus*) Fingerlinges. In Proceedings of the 8th International Symposium on Tilapia in Aquaculture 2008, Cairo, Egypt, 12–14 October 2008; pp. 999–1099. Available online: http://ag.arizona.edu/azaqua/ista/ISTA8/FinalPapers/11Nutrition/PDF/22Ramadan.pdf (accessed on 5 June 2024).
88. Ghosh, K.; Sen, S.K.; Ray, A.K. Feed Utilization Efficiency and Growth Performance in Rohu, Supplemented Diets. *Acta Ichthyol. Piscat.* **2005**, *35*, 111–117. [CrossRef]
89. Muzinic, L.A.; Thompson, K.R.; Morris, A.; Webster, C.D.; Rouse, D.B.; Manomaitis, L. Partial and total replacement of fish meal with soybean meal and brewer's grains with yeast in practical diets for Australian red claw crayfish *Cherax quadricarinatus*. *Aquaculture* **2004**, *230*, 359–376. [CrossRef]
90. Ziarati, M.; Zorriehzahra, M.J.; Hassantabar, F.; Mehrabi, Z.; Dhawan, M.; Shraun, K.; Emran, T.B.; Dhama, K.; Chaicumpa, W.; Shamsi, S. Zoonotic diseases of fish and their prevention and control. *Vet. Q.* **2022**, *42*, 95–118. [CrossRef] [PubMed]
91. Hill, C.; Guarner, F.; Reid, G.; Gibson, G.R.; Merenstein, D.J.; Pot, B.; Morelli, L.; Canani, R.B.; Flint, H.J.; Salminen, S.; et al. The International Scientific Association for Probiotics and Prebiotics consensus statement on the scope and appropriate use of the term probiotic. *Nat. Rev. Gastroenterol. Hepatol.* **2014**, *11*, 506–514. [CrossRef] [PubMed]
92. Mocanu, E.E.; Savin, V.; Popa, M.D.; Dima, F.M. The Effect of Probiotics on Growth Performance, Haematological and Biochemical Profiles in Siberian Sturgeon (*Acipenser baerii* Brandt, 1869). *Fishes* **2022**, *7*, 239. [CrossRef]
93. Kwasek, K.; Thorne-Lyman, A.L.; Phillips, M. Can human nutrition be improved through better fish feeding practices? a review paper. *Crit. Rev. Food Sci. Nutr.* **2020**, *60*, 3822–3835. [CrossRef]
94. Bunnoy, A.; Yanglang, A.; Tribamrung, N.; Keawthong, C.; Tumree, P.; Kumwan, B.; Meachasompop, P.; Saengrung, J.; Vanichvatin, K.; Muangrerk, C.; et al. Dietary administration of yeast (*Saccharomyces cerevisiae*) hydrolysate from sugar byproducts promotes the growth, survival, immunity, microbial community and disease resistance to VP (AHPND) in Pacific white shrimp (*Litopenaeus vannamei*). *Fish Shellfish Immunol.* **2024**, *145*, 109327. [CrossRef]
95. Da Silva Berto, R.; Pereira, G.D.V.; Mouriño, J.L.P.; Martins, M.L.; Fracalossi, D.M. Yeast extract on growth, nutrient utilization and haemato-immunological responses of Nile tilapia. *Aquac. Res.* **2016**, *47*, 2650–2660. [CrossRef]
96. Lazo-Vélez, M.A.; Serna-Saldívar, S.O.; Rosales-Medina, M.F.; Tinoco-Alvear, M.; Briones-García, M. Application of *Saccharomyces cerevisiae* var. *boulardii* in food processing: A review. *J. Appl. Microbiol.* **2018**, *125*, 943–951. [CrossRef]
97. Ansari, F.; Samakkhah, S.A.; Bahadori, A.; Jafari, S.M.; Ziaee, M.; Khodayari, M.T.; Pourjafar, H. Health-promoting properties of *Saccharomyces cerevisiae* var. *boulardii* as a probiotic; characteristics, isolation, and applications in dairy products. *Crit. Rev. Food Sci. Nutr.* **2023**, *63*, 457–485. [CrossRef]
98. Sharifuzzaman, S.M.; Austin, B. Probiotics for disease control in aquaculture. In *Diagnosis and Control of Diseases of Fish and Shellfish*; Austin, B., Newaj-Fyzul, A., Eds.; John Wiley and Sons: Hoboken, NJ, USA, 2017; pp. 189–222.
99. Darafsh, F.; Soltani, M.; Abdolhay, H.A.; Mehrejan, M.S. Improvement of growth performance, digestive enzymes and body composition of Persian sturgeon (*Acipenser persicus*) following feeding on probiotics: *Bacillus licheniformis*, *Bacillus subtilis* and *Saccharomyces cerevisiae*. *Aquac. Res.* **2020**, *51*, 957–964. [CrossRef]
100. Saba, A.O.; Yasin, I.S.M.; Azmai, M.N.A. Meta-analyses indicate that dietary probiotics significantly improve growth, immune response, and disease resistance in tilapia. *Aquac. Int.* **2024**, *32*, 4841–4867. [CrossRef]
101. Noman, M.; Kazmi, S.S.U.H.; Saqib, H.S.A.; Fiaz, U.; Pastorino, P.; Barcelò, D.; Tayyab, M.; Liu, W.; Wang, Z.; Yaseen, Z.M. Harnessing probiotics and prebiotics as eco-friendly solution for cleaner shrimp aquaculture production: A state of the art scientific consensus. *Sci. Total Environ.* **2024**, *915*, 169921. [CrossRef] [PubMed]
102. Wang, J.; Zhao, L.; Liu, J.; Wang, H.; Xiao, S. Effect of potential probiotic *Rhodotorula benthica* D30 on the growth performance, digestive enzyme activity and immunity in juvenile sea cucumber *Apostichopus japonicus*. *Fish Shellfish Immunol.* **2015**, *43*, 330–336. [CrossRef]
103. Pourgholam, M.A.; Khara, H.; Safari, R.; Sadati, M.A.Y.; Aramli, M.S. Influence of *Lactobacillus plantarum* inclusion in the diet of Siberian sturgeon (*Acipenser baerii*) on performance and hematological parameters. *Turkish J. Fish. Aquat. Sci.* **2017**, *17*, 1–5.
104. Hai, N.V. The use of probiotics in aquaculture. *J. Appl. Microbiol.* **2015**, *119*, 917–935. [CrossRef]
105. Rohani, M.F.; Islam, S.M.; Hossain, M.K.; Ferdous, Z.; Siddik, M.A.; Nuruzzaman, M.; Padeniya, U.; Brown, C.; Shahjahan, M. Probiotics, prebiotics and synbiotics improved the functionality of aquafeed: Upgrading growth, reproduction, immunity and disease resistance in fish. *Fish Shellfish Immunol.* **2022**, *120*, 569–589. [CrossRef]
106. Guluarte, C.; Reyes-Becerril, M.; Gonzalez-Silvera, D.; Cuesta, A.; Angulo, C.; Esteban, M.Á. Probiotic properties and fatty acid composition of the yeast *Kluyveromyces lactis* M3. In vivo immunomodulatory activities in gilthead seabream (*Sparus aurata*). *Fish Shellfish Immunol.* **2019**, *94*, 389–397. [CrossRef]

107. Pooramini, M.; Kamali, A.; Hajimoradloo, A.; Alizadeh, M.; Ghorbani, R. Effect of using yeast (*Saccharomyces cerevisiae*) as probiotic on growth parameters, survival and carcass quality in rainbow trout *Oncorhynchus mykiss* fry. *Int. Aquat. Res.* **2009**, *1*, 39.
108. Pratiwy, F.M.; Kharima, Z.P. The Use of Bread Yeast (*Saccharomyces cerevisiae*) in Increasing Growth, Disease Resistance, and as Immunostimulating for Various Kinds of Fish: A Review. *Asian J. Fish. Aquat. Res.* **2022**, *20*, 64–70. [CrossRef]
109. Tukmechi, A.; Bandboni, M. Effects of *Saccharomyces cerevisiae* supplementation on immune response, hematological parameters, body composition and disease resistance in rainbow trout, *Oncorhynchus mykiss* (Walbaum, 1792). *J. Appl. Ichthyol.* **2014**, *30*, 55–61. [CrossRef]
110. El-Boshy, M.E.; Ahmed, M.; AbdelHamid, F.M.; Gadalla, H.A. Immunomodulatory effect of dietary *Saccharomyces cerevisiae*, β-glucan and laminaran in mercuric chloride treated Nile tilapia (*Oreochromis niloticus*) and experimentally infected with *Aeromonas hydrophila*. *Fish Shellfish Immunol.* **2010**, *28*, 802–808. [CrossRef] [PubMed]
111. Sheikhzadeh, N.; Heidarieh, M.; Pashaki, A.K.; Nofouzi, K.; Farshbafi, M.A.; Akbari, M. Hilyses®, fermented *Saccharomyces cerevisiae*, enhances the growth performance and skin non-specific immune parameters in rainbow trout (*Oncorhynchus mykiss*). *Fish Shellfish Immunol.* **2012**, *32*, 1083–1087. [CrossRef] [PubMed]
112. Maccaferri, S.; Klinder, A.; Brigidi, P.; Cavina, P.; Costabile, A. Potential probiotic *Kluyveromyces marxianus* B0399 modulates the immune response in Caco-2 cells and peripheral blood mononuclear cells and impacts the human gut microbiota in an in vitro colonic model system. *Appl. Environ. Microbiol.* **2012**, *78*, 956–964. [CrossRef] [PubMed]
113. Ferreira, S.S.; Passos, C.P.; Madureira, P.; Vilanova, M.; Coimbra, M.A. Structure–function relationships of immunostimulatory polysaccharides: A review. *Carbohydr. Polym.* **2015**, *132*, 378–396. [CrossRef]
114. Rodrigues, M.V.; Zanuzzo, F.S.; Koch, J.F.A.; de Oliveira, C.A.F.; Sima, P.; Vetvicka, V. Development of fish immunity and the role of β-glucan in immune responses. *Molecules* **2020**, *25*, 5378. [CrossRef]
115. Voloski, A.P.D.S.; de Figueiredo Soveral, L.; Dazzi, C.C.; Sutili, F.; Frandoloso, R.; Kreutz, L.C. β-Glucan improves wound healing in silver catfish (*Rhamdia quelen*). *Fish Shellfish Immunol.* **2019**, *93*, 575–579. [CrossRef]
116. Divya, M.; Gopi, N.; Iswarya, A.; Govindarajan, M.; Alharbi, N.S.; Kadaikunnan, S.; Khaled, J.M.; Almanaa, T.N.; Vaseeharan, B. β-glucan extracted from eukaryotic single-celled microorganism *Saccharomyces cerevisiae*: Dietary supplementation and enhanced ammonia stress tolerance on *Oreochromis mossambicus*. *Microb. Pathog.* **2020**, *139*, 103917. [CrossRef]
117. Guzmán-Villanueva, L.T.; Ascencio-Valle, F.; Macías-Rodríguez, M.E.; Tovar-Ramírez, D. Effects of dietary β-1, 3/1, 6-glucan on the antioxidant and digestive enzyme activities of Pacific red snapper (*Lutjanus peru*) after exposure to lipopolysaccharides. *Fish Physiol. Biochem.* **2014**, *40*, 827–837. [CrossRef]
118. Reyes-Becerril, M.; Angulo, M.; Sanchez, V.; Guluarte, C.; Angulo, C. β-D-glucan from marine yeast *Debaryomyces hansenii* BCS004 enhanced intestinal health and glucan-expressed receptor genes in Pacific red snapper *Lutjanus peru*. *Microb. Pathog.* **2020**, *143*, 104141. [CrossRef]
119. Machuca, C.; Méndez-Martínez, Y.; Reyes-Becerril, M.; Angulo, C. Yeast β-Glucans as Fish Immunomodulators: A Review. *Animals* **2022**, *12*, 2154. [CrossRef] [PubMed]
120. Pionnier, N.; Falco, A.; Miest, J.J.; Shrive, A.K.; Hoole, D. Feeding common carp *Cyprinus carpio* with β-glucan supplemented diet stimulates C-reactive protein and complement immune acute phase responses following PAMPs injection. *Fish Shellfish Immunol.* **2014**, *39*, 285–295. [CrossRef] [PubMed]
121. Tungland, B. *Human Microbiota in Health and Disease: From Pathogenesis to Therapy*; Academic Press: Cambridge, MA, USA, 2018.
122. Rawling, M.D.; Pontefract, N.; Rodiles, A.; Anagnostara, I.; Leclercq, E.; Schiavone, M.; Castex, M.; Merrifield, D. The effect of feeding a novel multistrain yeast fraction on European seabass (*Dicentrachus labrax*) intestinal health and growth performance. *J. World Aquac. Soc.* **2019**, *50*, 1108–1122. [CrossRef]
123. Yousuf, S.; Tyagi, A.; Singh, R. Probiotic supplementation as an emerging alternative to chemical therapeutics in finfish aquaculture: A Review. *Probiotics Antimicrob. Proteins* **2023**, *15*, 1151–1168. [CrossRef]
124. Eissa, E.H.; Ahmed, N.H.; El-Badawi, A.A.; Munir, M.B.; Al-Kareem, O.M.A.; Eissa, M.E.H.; Hussien, E.H.M.; Sakr, S.E.-S. Assessing the influence of the inclusion of *Bacillus subtilis* AQUA-GROW® as feed additive on the growth performance, feed utilization, immunological responses and body composition of the Pacific white shrimp, *Litopenaeus vannamei*. *Aquac. Res.* **2022**, *53*, 6606–6615. [CrossRef]
125. Goda, A.M.; Omar, E.A.; Srour, T.M.; Kotiet, A.M.; El-Haroun, E.; Davies, S.J. Effect of diets supplemented with feed additives on growth, feed utilization, survival, body composition and intestinal bacterial load of early weaning European seabass, *Dicentrarchus labrax* post-larvae. *Aquac. Int.* **2018**, *26*, 169–183. [CrossRef]
126. Öztürk, F.; Esendal, Ö.M. Usage of *Lactobacillus rhamnosus* as a Probiotic in Sea Bass (*Dicentrarchus labrax*). *J. Anatol. Environ. Anim. Sci.* **2020**, *5*, 93–99. [CrossRef]
127. Hansen, J.Ø.; Lagos, L.; Lei, P.; Reveco-Urzua, F.E.; Morales-Lange, B.; Hansen, L.D.; Schiavone, M.; Mydland, L.T.; Arntzen, M.Ø.; Mercado, L. Down-stream processing of baker's yeast (*Saccharomyces cerevisiae*)—Effect on nutrient digestibility and immune response in Atlantic salmon (*Salmo salar*). *Aquaculture* **2021**, *530*, 735707. [CrossRef]
128. Langeland, M.; Lindberg, J.E.; Lundh, T. Digestive enzyme activity in Eurasian perch (*Perca fluviatilis*) and Arctic charr (*Salvelinus alpinus*). *J. Aquac. Res. Dev.* **2013**, *5*, 2.
129. Zhao, L.; Wang, W.; Huang, X.; Guo, T.; Wen, W.; Feng, L.; Wei, L. The effect of replacement of fish meal by yeast extract on the digestibility, growth and muscle composition of the shrimp *Litopenaeus vannamei*. *Aquac. Res.* **2017**, *48*, 311–320. [CrossRef]

130. Vidakovic, A.; Huyben, D.; Nyman, A.; Kiessling, A.; Lundh, T. Evaluation of growth performance, nutrient retention and apparent digestibility in rainbow trout (O. mykiss) fed graded levels of yeasts *Saccharomyces cerevisae* and *Wickerhamomyces anomalus*. In Proceedings of the Aquaculture Europe 2015 Conference, Rotterdam, The Netherlands, 20–23 October 2015.
131. Arup, T.; Patra, B.C. Oral administration of baker's yeast (*Saccharomyces cerevisiae*) acts as a growth promoter and immunomodulator in *Labeo rohita* (Ham.). *J. Aquac. Res. Dev.* **2011**, *2*, 109.
132. Das, J.; Hossain, M.S.; Hasan, J.; Siddique, M.A.M. Growth performance and egg ratio of a marine rotifer *Brachionus rotundiformis* fed different diets in captivity. *Thalass. An. Int. J. Mar. Sci.* **2021**, *37*, 113–118. [CrossRef]
133. Khatun, B.; Rahman, R.; Rahman, M.S. Evaluation of yeast *Saccharomyces cerevisiae* and algae *Chlorella vulgaris* as diet for Rotifer *Brachionus calyciflorus*. *Agriculturists* **2014**, *12*, 1–9. [CrossRef]
134. Talens-Perales, D.; Marín-Navarro, J.; Garrido, D.; Almansa, E.; Polaina, J. Fixation of bioactive compounds to the cuticle of Artemia. *Aquaculture* **2017**, *474*, 95–100. [CrossRef]
135. Ajah, P.O. Mass culture of Rotifera (*Brachionus quadridentatus* [Hermann, 1783]) using three different algal species. *Afr. J. Food Sci.* **2010**, *4*, 80–85.
136. Radhakrishnan, K.; Aanand, S.; Rameshkumar, S.; Divya, F. Effect of feeding rate and feeding frequency in mass culture of *Brachionus plicatilis* in semi-continuous method with a yeast-based diet. *J. Fish. Life Sci.* **2017**, *2*, 40–44.
137. Hu, J.; Wang, G.; Huang, Y.; Sun, Y.; Zhao, H.; Li, N. Effects of Substitution of Fish Meal with Black Soldier Fly (*Hermetia illucens*) Larvae Meal, in Yellow Catfish (*Pelteobagrus fulvidraco*) Diets. *Isr. J. Aquac.* **2017**, *69*, 9.
138. El-Khodary, G.M.; Mona, M.M.; El-Sayed, H.S.; Ghoneim, A.Z. Phylogenetic identification and assessment of the nutritional value of different diets for a copepod species isolated from Eastern Harbor coastal region. *Egypt. J. Aquat. Res.* **2020**, *46*, 173–180. [CrossRef]

animals

MDPI

Article

Effect of Substituting Soybean Meal in *Oreochromis niloticus* Diets with Pumpkin (*Cucurbita maxima*) Seed Cake on Water Quality, Growth, Antioxidant Capacity, Immunity, and Carcass Composition

Hossam A. M. Mounes [1], Zeinab M. A. Abd-El Azeem [2], Dena. A. Abd El-Bary [3], Adham A. Al-Sagheer [4,*], Yasmina M. Abd-Elhakim [5], Bayan A. Hassan [6], Sherif S. Sadek [7] and Kareem. M. Ahmed [2]

[1] Limnology Department, Central Laboratory for Aquaculture Research, Agricultural Research Center, Abassa, Abu Hammad 44662, Egypt; hmoans80@hotmail.com
[2] Fish Production Branch, Department of Animal Production, Faculty of Agriculture, Ain Shams University, Cairo 11241, Egypt
[3] Regional Center for Food and Feed, Agricultural Research Center, Giza 12619, Egypt
[4] Department of Animal Production, Faculty of Agriculture, Zagazig University, Zagazig 44511, Egypt
[5] Department of Forensic Medicine and Toxicology, Faculty of Veterinary Medicine, Zagazig University, Zagazig 44519, Egypt; yasmina.forensic@yahoo.com
[6] Pharmacology Department, Faculty of Pharmacy, Future University, Cairo 11835, Egypt; bayan.saffaf@fue.edu.eg
[7] Aquaculture Consultant Office, 9 Street No. 256, New-Maadi, Cairo 11435, Egypt; sadek_egypt35@hotmail.com
* Correspondence: adham_alsaht@zu.edu.eg

Simple Summary: Soybean meal is a widely utilized protein source in aquatic feeds; however, its application in fish diets may be restricted due to elevated prices; market fluctuations; and heightened human, poultry, and livestock consumption. Therefore, utilizing unused agro-industrial by-products as alternative protein sources can reduce dependence on soybean meal, promote economic viability in tilapia farming, and address sustainability issues in aquaculture. This research examined the effects of substituting soybean meal with pumpkin seed cake in Nile tilapia diets concerning growth, water quality, antioxidant capacity, immune status, and body composition. The findings indicate that replacing 40% of dietary soybean meal protein with pumpkin seed cake can lead to significant improvements in growth, antioxidant capacity, and immune status in Nile tilapia fish.

Abstract: A 10-week feeding experiment was performed to determine the impacts of partial substitution of soybean meal (SB) with pumpkin seed cake (PSC) in *Oreochromis niloticus* diets on water quality, growth rate, antioxidant capacity, immunity, and carcass composition. One hundred and fifty tilapia fish (average weight, 11.93 ± 0.17 g) were randomly allocated to five diets. The first diet (the basal diet) contained 420 g of SB per kg of feed. The remaining four diets, namely, D1, D2, D3, and D4, had SB partially replaced by PSC at 10%, 20%, 30%, and 40%, respectively. The results revealed that D4 and D1 significantly improved dissolved oxygen levels, while water temperature, pH, total ammonia, and nitrate levels were not significantly affected. Replacing SB with PSC significantly improved specific growth performance indicators and feed conversion compared to the control, with the D4 group showing the best values. Increasing PSC levels decreased serum glucose, aspartate aminotransferase, alanine aminotransferase, cholesterol, and triglyceride levels. In contrast, the D4 group had higher globulin, albumin, total protein, and lysozyme serum levels. Moreover, fish-fed PSC had significantly increased superoxide dismutase, glutathione peroxidase, and catalase activities and significantly decreased malondialdehyde levels. Increasing PSC substitution levels in fish diets increased the ash and crude lipid contents in the bodies of the fish, while crude protein and moisture decreased. In conclusion, replacing SB with PSC in fish diets significantly enhances growth performance, feed conversion, and fish health. Moreover, the findings suggest that PSC can be a promising alternative protein source for sustainable aquaculture practices.

Keywords: pumpkin seed cake; soybean meal; Nile tilapia; growth; antioxidant

1. Introduction

Aquaculture is an important industry that provides a significant source of protein for human consumption [1]. Nile tilapia (*Oreochromis niloticus*) is the world's third-most commonly farmed fish species. It played a significant role, contributing approximately 5% to the total production of fish aquaculture and surpassing a cumulative output of 4.41 million tons in 2020 [2]. It is considered one of the world's most extensively farmed fish species due to its rapid growth rate, adaptability to various environments, and high meat quality [3,4]. After China and Indonesia, Egypt is the third-largest producer of Nile tilapia globally [5]. However, the high cost of feed, especially protein sources such as soybean meal, is a major constraint on the sustainable development of tilapia farming [6]. Soybean meal (SM) is a beneficial source of protein in animal nutrition due to its superior amino acid profile and high crude protein content [7]. The increasing demand for SM and market price increases have prompted the aquaculture sector to search for environmentally friendly and cost-effective substitute feed ingredients [8–10]. Therefore, using unutilized agro-industrial by-products as alternative protein sources can reduce dependence on SM, ensure the economic viability of tilapia farming, and address sustainability challenges in aquaculture [6,11].

Pumpkin seed cake (PSC), produced when pumpkin seed oil is extracted, is rich in protein, fiber, and minerals [12]. Pumpkin seeds are pressed on a continuous screw press to produce PSC without affecting their nutrient contents [13]. Pumpkin seeds are known for various biological activities, including antioxidant, antimicrobial, anti-inflammatory, and antiparasitic effects [14]. In an earlier study, the nutritional composition of PSC was compared to that of SM [15]. The authors demonstrated that the gross energy, crude protein, and ether extract contents of PSC (59.8%, 12.46%, and 21.9 MJ/kg DM, respectively) were greater than those of SM (47.42%, 2.83%, and MJ/kg DM, respectively). Proteins isolated from PSC exhibited significantly higher concentrations of total sulfur amino acids than SM. Furthermore, PSC contains an abundant concentration of unsaturated fatty acids, including linoleic acid (50.9%), stearic acid (6.7%), palmitic acid (14.8%), and oleic acid (25.8%) [16,17]. Moreover, PSC has a low moisture content of about 7%, which makes it a suitable feed constituent for long-term storage and transport. In addition to being a source of protein, PSC is a prospective nutraceutical containing bioactive substances with beneficial health advantages, such as squalene, tocopherols, and phytosterols [12,16,18]. Utilizing by-products such as PSC in animal feeds can provide an economic benefit due to their reduced market price and provide feed producers with alternative feed ingredients. Nevertheless, limited studies have investigated the potential of PSC as a feed ingredient in practical fish feeds [17,19,20]. Furthermore, the results of these reports are inconsistent and require further investigation.

It has been hypothesized that replacing SB with PSC in tilapia diets would decrease SB usage without significantly impacting fish health or performance. The present study evaluated whether replacing SM with PSC would affect water quality, growth indicators, diet utilization, carcass composition, blood biochemical, immunity, or antioxidant status. Our research provides essential information on the prospective utilization of PSC as a sustainable protein alternative in tilapia feeds.

2. Materials and Methods

2.1. Diet Preparation

The PSC used in the present study consisted of 88.10% dry matter, 10.78% crude lipid, 55.20% crude protein, and 10.72% ash. PSC and the rest of the feed ingredients in this study were obtained from a local market in Sharkia Governorate, Egypt. Five iso-nitrogenous (crude protein, 315 g kg^{-1}) and isolipidic (crude lipid, 65 g kg^{-1}) diets were formulated.

The basal diet consisted of 420 g of SB per kg of diet, while in the remaining four diets, PSC replaced SB at 10%, 20%, 30%, and 40% (D1, D2, D3, and D4, respectively). All ingredients were ground into a fine powder, passed through a 200 μm mesh, and mixed thoroughly with vegetable and fish oils and water. The mixture was then pelletized with a feed mill (California Pellet Mill, San Francisco, CA, USA), dried for around a day in an oven until the dry matter levels reached about 90%, packed in sealed bags, and kept in a refrigerator at −20 °C until needed. All trial diets were prepared to meet the nutrient needs of tilapia, according to the recommendations of NRC [21]. Table 1 displays the nutrient composition and formulation of the diets.

Table 1. Ingredients and proximate composition of the experimental diets (g kg^{-1}) on a dry matter basis.

	Experimental Diets [1]				
	Control	D1	D2	D3	D4
Ingredients					
Soybean meal	420	378	336	294	252
Pumpkin seed cake [2]	0	33.5	67	100.4	133.9
Yellow corn	375	375	375	375	375
Fish meal	100	100	100	100	100
Wheat bran	70	70	70	70	70
Cellulose	0	11.2	22.4	33.7	44.9
Fish oil	10	10	10	10	10
Sunflower oil	20	17.3	14.6	11.9	9.2
Premix [3]	5	5	5	5	5
Proximate composition					
Dry matter	914.0	915.0	912.0	921.0	925.0
Crude protein (N × 6.25)	315.0	316.5	314.5	315.5	314.5
Crude lipid	65.5	65.7	65.4	65.1	65.9
Ash	75.5	79.5	76.5	74.5	76.0

[1] Control, D1, D2, D3, and D4 indicate that 0%, 10%, 20%, 30%, and 40% of soybean meal were replaced by pumpkin seed cake, respectively. [2] Pumpkin seed cake: 55.20% crude protein, 10.78% crude lipid, and 10.72% ash. [3] Provides the following per kg of diet: vitamin A 4000 IU, vitamin D_3 600 IU, vitamin E 20 mg, vitamin K_3 5 mg, vitamin B_1 3.6 mg, vitamin B_2 6 mg, vitamin B_5 12 mg, vitamin B_6 3.5 mg, vitamin B_{12} 0.02 mg, vitamin B_3 14.4 mg, biotin 0.07 mg, folic acid 0.9 mg, inositol 300 mg, vitamin C 50 mg, Mg 15 mg, Fe 30 mg, Zn 42 mg, Cu 4 mg, K 75 mg, Co 0.11 mg, Mn 1.6 mg, Se 0.04 mg, Mo 0.005 mg, and I 0.4 mg.

2.2. Fish and Rearing Conditions

Before the experiment's initiation, fiberglass tanks were cleaned and filled with dechlorinated tap water from the storage tank. Healthy Nile sex-reversed tilapia fingerlings were acquired from a commercial farm in Sharkia Governorate, Egypt. Fish were acclimated for ten days prior to the start of the experiment, during which they were fed the control diet three times a day, at 9:00, 13:30, and 17:00, until they reached apparent satiety. After acclimation, 150 juvenile (average weight, 11.93 ± 0.17 g) fish were randomly allocated to five groups, and each group had three fiberglass aquaria (water volume: 72 L; 10 fish per aquarium). The experimental diets (control, D1, D2, D3, and D4) were fed to the fish for 70 days, with the same schedule and procedures as during the acclimatization phase. The study was conducted with a photoperiod of 12 h light/12 h darkness. Every two days, 50% of the tank's water was siphoned to eliminate suspended and dissolved wastes in the tanks, and a 0.5 Hp air blower continuously aerated the experimental tanks. Throughout the experiment, fish were hand-fed thrice daily (9:00, 13:30, and 17:00) at a feeding rate of 3% body weight to apparent satiation. Feed intake was adjusted per the new body mass every two weeks throughout the trial.

2.3. Water Quality Parameters

Every two weeks, physicochemical parameters were evaluated. At a depth of 20 cm, water samples from each aquarium were collected. pH was measured on-site using a digital pH meter (Fisher Scientific, Waltham, MA, USA). Dissolved oxygen and water temperature were determined using an oxygen and temperature meter (Jenway, London, UK). The ammonia concentration in the water samples was estimated by a multi-parameter analyzer (HANNA Instruments, Smithfield, RI, USA). According to APHA [22], the nitrate concentration was measured with a spectrophotometer (model Milton Roy 21D) at wavelengths of 410 nm.

2.4. Growth, Feed Utilization, and Body Indices

On the 70th day, fish were harvested, counted, and weighed in each aquarium. Livers and viscera were removed (3 fish/aquarium; $n = 9$ fish/treatment) and weighed to compute the viscerosomatic index (VSI) and hepatosomatic index (HSI). Growth and body indices such as feed conversion ratio (FCR), body weight gain (BWG), specific growth rate (SGR), VSI, and HSI were calculated using the following standard formula:

$$\text{BWG}\left(\text{g fish}^{-1}\right) = \frac{\text{final wet weight} - \text{initial wet weight}}{\text{fish number per aquarium}} \quad (1)$$

$$\text{SGR}\ (\%/\text{day}) = \frac{\text{Ln of final wet weight} - \text{Ln of initial wet weight}}{\text{period(day)}} \times 100 \quad (2)$$

$$\text{FCR} = \frac{\text{total feed intake(g)}}{\text{total BWG(g)}} \quad (3)$$

$$\text{VSI}\ (\%) = \frac{\text{wet weight of viscera}}{\text{fish wet weight}} \times 100 \quad (4)$$

$$\text{HSI}\ (\%) = \frac{\text{wet weight of hepatopancreas}}{\text{fish wet weight}} \times 100 \quad (5)$$

2.5. Chemical Composition of the Diets and the Whole Carcass of Tilapia

Following the 10-week feeding experiment, three fish from each aquarium were sampled and frozen at −20 °C for chemical composition analysis of the carcass. The crude lipid, ash, moisture, and crude protein contents of the fish carcasses and experimental diets were estimated using AOAC [23] procedures. The moisture content was determined after three hours of dehydrating at 105 °C and a constant weight in an electric oven (JSON-100, Gongju, Republic of Korea). Five hours of combustion at 550 °C in a muffle furnace (Type 47900, Thermo Scientific, Waltham, MA, USA) were used to evaluate the ash content. The total lipid content was measured using a Soxhlet apparatus (El-Gomhouria Co., Zagazig, Egypt) and an ether extraction. After sulfuric acid digestion, the Kjeldal method was utilized to determine the crude protein level (N × 6.25).

2.6. Serum Biochemical, Immunity, and Antioxidant Parameters

At the end of the feeding trial, the fish were fasted for 24 h. Then, three fish per aquarium were anesthetized with MS-222 (100 mg/L) for blood sample collection from the caudal veins using non-heparinized syringes. The samples were centrifuged at 5000× *g* for 15 min at 4 °C to separate the serum, which was then stored at −20 °C for further examinations. The enzymatic activity of glutathione peroxidase (GPx), catalase (CAT), and superoxide dismutase (SOD), as well as the malondialdehyde (MDA) content, were measured by commercial kits from BioSource Inc., San Diego, CA, USA (Catalog number MBS480417, MBS2540413, MBS9718960, and MBS2540409, respectively). Lysozyme (LZM) activity was determined in line with the procedures of Ghareghanipoora et al. [24]. Serum total lipids, glucose, total protein (TP), albumin (ALB), creatinine, and uric acid concentrations were assessed colorimetrically using commercial kits (Bio-Diagnostics, Cairo, Egypt) with the catalog numbers TL 20 10, GL 13 20, TP 20 20, AB 10 10, CR 12 50, and UA 21 20, respectively. Globulin content was computed by subtracting ALB levels from TP levels. Serum alanine (ALT) and aspartate aminotransferase (AST) activities were quantified using Bio-diagnostics (Cairo, Egypt) colorimetric kits (AL 10 31 (45) and AS 10 61 (45), respectively).

2.7. Statistical Analysis

The significance of growth, feed utilization, body indices, water quality, carcass composition, and blood contents was discovered via one-way ANOVA using SAS software (Version 9.00) [25]. Then, Tukey's post hoc test was used. The optimal level of substitution using PSC was determined through the use of polynomial regression analysis, as explained by Yossa and Verdegem [26]. The third-order polynomial regression (cubic) provided a better fit to the observed patterns in our dataset compared to linear or quadratic models. Statistical significance was set at $p < 0.05$. In addition, correlations among growth performance indicators, the carcass composition, and serum biochemical parameters were examined by the Pearson coefficient of pairwise comparison between samples. They were recognized from the correlation matrix heatmap produced via GraphPad Prism version 8 (GraphPad Software, San Diego, CA, USA) [27].

3. Results

3.1. Water Quality

Figure 1A–E displays the estimated water quality parameters determined every second week throughout the experimental period. The values of temperature, pH, dissolved oxygen, ammonia, and nitrate were in the ranges of 28.67–29.13 °C, 8.29–8.42, 5.26–6.85 mg/L, 0.52–0.84 mg/L, and 0.08–0.16 mg/L, respectively, during the trial.

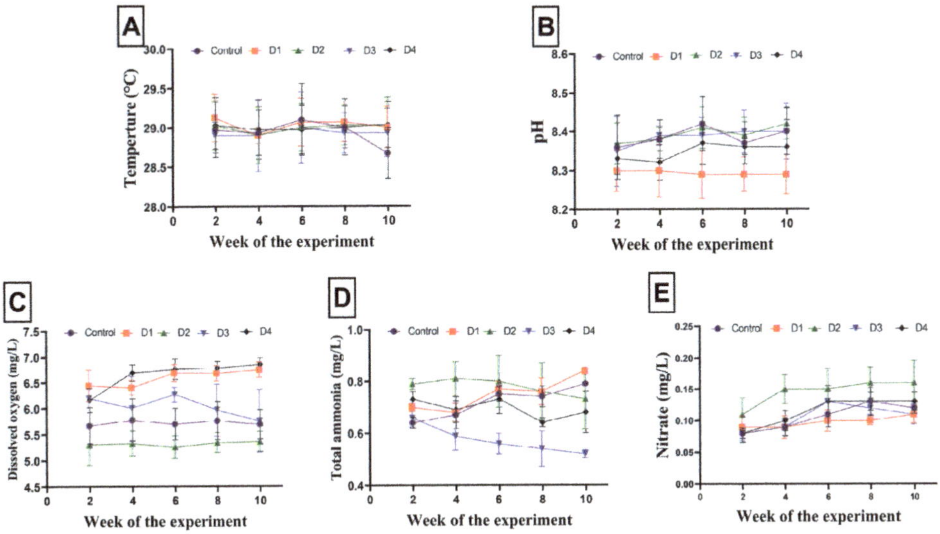

Figure 1. Average values of water quality parameters, determined every second week during the trial, including: (**A**) temperature, (**B**) pH, (**C**) dissolved oxygen, (**D**) total ammonia, and (**E**) nitrate. Control, D1, D2, D3, and D4 indicate that 0%, 10%, 20%, 30%, and 40% of soybean meal were replaced by pumpkin seed cake (PSC), respectively. Values are expressed as means ± standard error.

3.2. Growth Performance and Body Indices

Table 2 illustrates the growth indicators of *O. niloticus* fed the different experimental diets after 70 days of the experiment. The growth and feed efficiency parameters of Nile tilapia were enhanced by PSC-based diets compared to the control diet. The D4 group exhibited the greatest improvements in final weight, BWG, SGR, and FCR when PSC was substituted for SB. Feed intake was significantly ($p < 0.05$) increased in the fish fed on the D1 and D2 diets. The relationships between SGR, final weight, and FCR, as well as the different substitution levels using PSC, were best expressed by third-order polynomial regression (cubic) equations, where final weights were as follows: y = 17.35 + 25.46x −

$10.07x^2 + 1.183x^3$, $R^2 = 0.8132$ (Figure 2A); SGR: $y = 0.856 + 0.9868x - 0.3935x^2 + 0.04639x^3$, $R^2 = 0.5692$ (Figure 2B); and FCR: $y = 2.361 + 0.06212x + 1011x^2 - 0.02473x^3$, $R^2 = 0.7474$ (Figure 2C). Based on the polynomial regression analysis, the optimal substitution level using PSC for the maximum final weight and SGR and the best FCR were found in the fish fed on the D4 diet. As displayed in Figure 3, VSI and HSI showed significant differences in favor of PSC treatments compared to the control treatment. The lowest VSI value was detected in the D2 group, while the highest was found in the control treatment group. The highest HSI record was found in the control treatment, while the lowest was recorded in the D1 group.

Table 2. Growth performance and nutrient utilization as affected by substituting soybean meal (SBM) with pumpkin seed cake (PSC) in Nile tilapia diets.

Items	Experimental Diets [1]					SDM [2]
	Control	D1	D2	D3	D4	
Initial weight (g fish^{-1})	11.86 ^{NS}	11.87	12.08	11.92	11.94	0.17
Final weight (g fish^{-1})	33.83 ^{c}	37.74 ^{b}	34.51 ^{c}	34.02 ^{c}	40.60 ^{a}	3.01
Body weight gain (g fish^{-1})	21.97 ^{c}	25.87 ^{b}	22.43 ^{c}	22.10 ^{c}	28.66 ^{a}	2.98
Specific growth rate (%)	1.51 ^{b}	1.65 ^{ab}	1.51 ^{b}	1.51 ^{b}	1.75 ^{a}	0.20
Feed intake (g fish^{-1})	55.36 ^{b}	67.26 ^{a}	65.40 ^{a}	56.41 ^{b}	61.90 ^{ab}	5.38
Feed conversion ratio (g feed/g gain)	2.52 ^{ab}	2.60 ^{ab}	2.93 ^{a}	2.55 ^{ab}	2.16 ^{b}	0.29
Survival rate (%)	100 ^{NS}	100	100	100	100	0.00

[1] Control, D1, D2, D3, and D4 indicate that 0%, 10%, 20%, 30%, and 40% of SM were replaced by PSC, respectively. [2] SDM stands for the standard deviation of the mean. $^{a-c}$ Means in rows with different superscripts differ significantly ($p < 0.05$). NS = no significant differences between groups ($p > 0.05$).

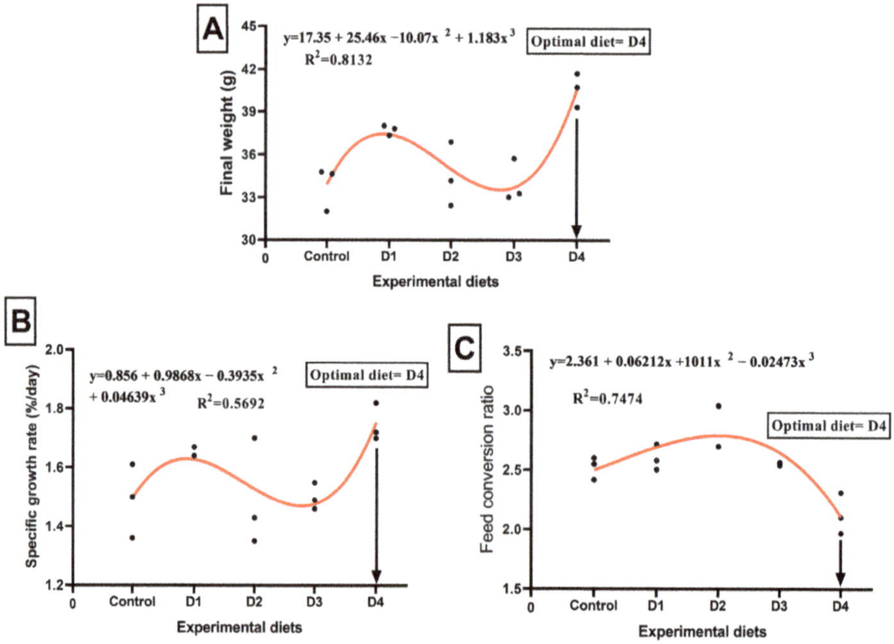

Figure 2. Relationship between final fish weight (**A**), specific growth rate (**B**), and feed conversion ratio (**C**) of Nile tilapia (*Oreochromis niloticus*) and different levels of pumpkin seed cake (PSC). Control, D1, D2, D3, and D4 indicate that 0%, 10%, 20%, 30%, and 40% of SM were replaced by PSC, respectively.

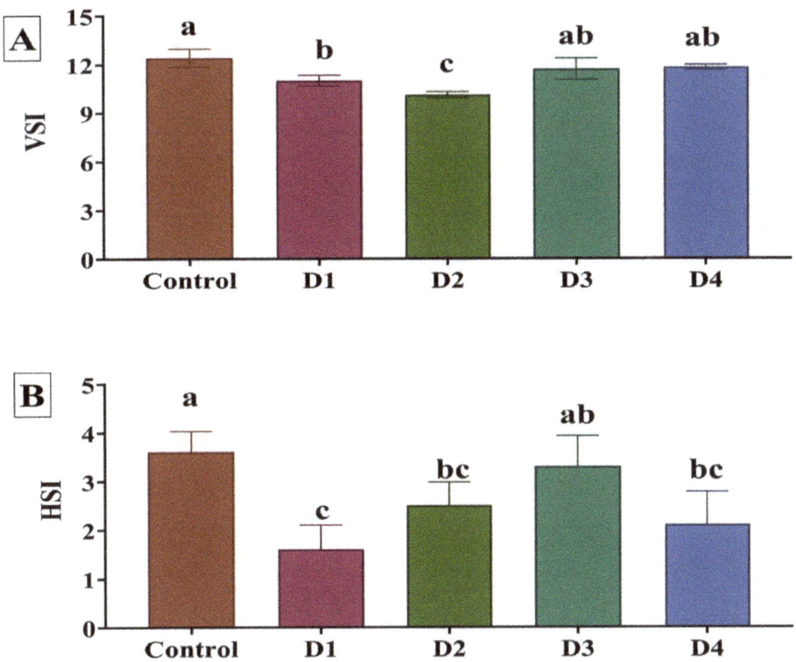

Figure 3. Viscerosomatic index (VSI; **A**) and hepatosomatic index (HSI; **B**) as affected by substituting soybean meal (SM) with pumpkin seed cake (PSC) in Nile tilapia diets. Control, D1, D2, D3, and D4 indicate that 0%, 10%, 20%, 30%, and 40% of SM were replaced by PSC, respectively. [a–c] Means in bars with different superscripts differ significantly. Different letters above the error bar indicate significant differences at $p < 0.05$, as determined by one-way ANOVA followed by Tukey's multiple comparisons test. Values are expressed as means ± standard deviation (3 fish/aquarium; $n = 9$ fish/treatment).

3.3. *Carcass Composition*

Table 3 displays the whole-body composition of *O. niloticus* fed the experimental diets which substituted SB with PSC for 70 days. Increasing PSC substitution levels in fish diets increased the ash (except the D4 group) and crude lipid contents in the bodies of the fish. In contrast, the crude protein and moisture contents in the fish decreased as PSC substitution increased in their diets.

Table 3. Carcass composition as affected by substituting soybean meal (SBM) with pumpkin seed cake (PSC) in Nile tilapia diets (% on a dry matter basis, except for moisture on a wet basis).

Items	Experimental Diets [1]					SDM [2]
	Control	D1	D2	D3	D4	
Moisture	76.9 [a]	72.4 [cd]	74.9 [b]	73.0 [c]	71.8 [d]	1.9
Crude lipids	14.4 [d]	18.0 [c]	16.8 [c]	20.3 [b]	23.7 [a]	3.4
Crude protein	64.8 [a]	58.7 [c]	60.0 [b]	57.4 [d]	57.5 [d]	2.8
Ash	20.0 [b]	22.5 [a]	22.5 [a]	21.8 [a]	18.3 [c]	1.9

[1] Control, D1, D2, D3, and D4 indicate that 0%, 10%, 20%, 30%, and 40% of SM were replaced by PSC, respectively. [2] SDM stands for the standard deviation of the mean. [a–d] Means in rows with different superscripts differ significantly ($p < 0.05$).

3.4. Blood Biochemical Biomarkers

The results concerning blood biochemical biomarkers, as affected by SB's replacement with PSC, are presented in Table 4. Reduced total cholesterol and triglyceride concentrations were observed as a dose–response with increasing PSC levels in the diets, and the D4 group had the lowest values ($p < 0.05$). Following the same trend, the serum contents of ALT, AST, creatinine, and urea were significantly decreased with all dietary inclusion levels of PSC. Significantly ($p < 0.05$) lower glucose was detected in the D4-fed fish compared to the control group. Total protein and globulin levels in the blood of fish fed the D1 and D4 diets were significantly ($p < 0.05$) higher than those of fish fed other experimental diets. In contrast, albumin levels in the blood of fish fed the D1, D3, and D4 diets were significantly ($p < 0.05$) higher than those of fish fed the control diet (Table 4).

Table 4. Blood biochemical parameters, as affected by substituting soybean meal (SBM) with pumpkin seed cake (PSC) in Nile tilapia diets.

Items	Experimental Diets [1]					SDM [2]
	Control	D1	D2	D3	D4	
Total cholesterol (mg/dL)	178.0 [a]	126.25 [c]	139.25 [b]	130.0 [bc]	123.05 [c]	6.39
Triglycerides (mg/dL)	280.81 [a]	188.25 [d]	235.0 [b]	208.92 [c]	173.25 [d]	6.37
Glucose (mg/dL)	122.83 [a]	112.05 [ab]	122.25 [a]	116.05 [ab]	105.36 [b]	7.48
ALT (U/L)	73.12 [a]	62.18 [c]	65.48 [c]	67.01 [bc]	69.09 [b]	3.24
AST (U/L)	111.25 [a]	87.5 [b]	76.25 [c]	78.5 [c]	85.91 [b]	3.78
Creatinine (mg/dL)	0.60 [a]	0.40 [bc]	0.47 [b]	0.41 [bc]	0.33 [c]	0.23
Urea (mg/dL)	16.2 [a]	10.15 [b]	12.1 [b]	11.75 [bc]	12.0 [b]	4.75
Total protein (g/dL)	1.38 [b]	3.29 [a]	1.33 [b]	1.54 [b]	3.09 [a]	0.31
Albumin (g/dL)	0.80 [c]	1.35 [ab]	0.84 [c]	1.08 [b]	1.54 [a]	0.28
Globulin (g/dL)	0.58 [b]	1.94 [a]	0.49 [b]	0.46 [b]	1.55 [a]	0.35

[1] Control, D1, D2, D3, and D4 indicate that 0%, 10%, 20%, 30%, and 40% of SM were replaced by PSC, respectively. [2] SDM stands for the standard deviation of the mean. [a–d] Means in rows with different superscripts differ significantly ($p < 0.05$). Alanine aminotransferase (ALT), aspartate aminotransferase (AST).

3.5. Antioxidant Activity and Immune Status

SOD, GPx, and CAT activities were markedly higher ($p < 0.05$) in all dietary incorporation levels of PSC relative to the control group (Figure 4). Notably, their highest values were detected in the D4 group. Conversely, reduced MDA levels were detected with all PSC substitution levels in diets, and the D1 and D4 groups had the lowest values ($p < 0.05$). Regarding lysozyme activity, fish fed PSC diets exhibited significantly ($p < 0.05$) higher activity than those fed the control diet (Figure 4). Fish fed the D1 diet had the most increased lysozyme activity in their serum compared with those fed the other diets (Figure 5).

3.6. Pearson Correlation between the Estimated Parameters

As revealed in the correlation matrix heatmap in Figure 6, a significant ($p < 0.01$) positive correlation was detected between the growth performance indices, including FBW and WG, and the serum levels of antioxidant enzymes (SOD, CAT, and GPx), proteins (TP, ALB, and GLOB), and the lysozyme activity. On the contrary, FBW and WG were negatively correlated with TG, GLU, HSI, and carcass moisture content. Furthermore, FCR was negatively correlated with SOD, CAT, GPx, and ALB.

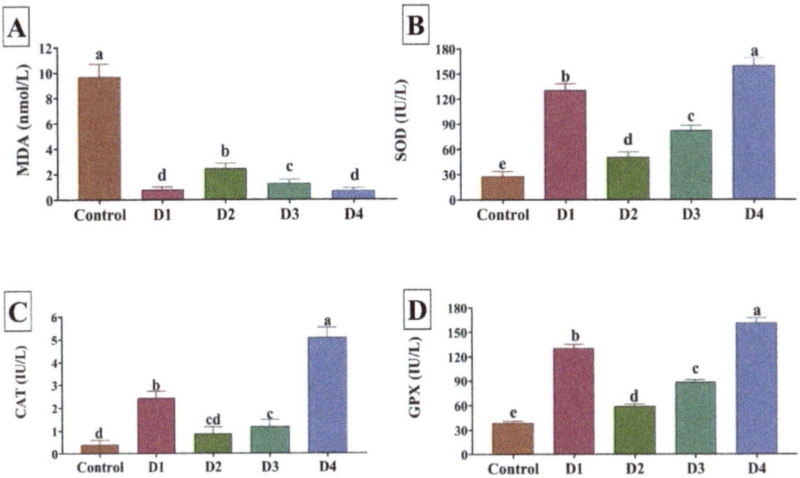

Figure 4. The serum levels of (**A**) malondialdehyde (MDA), (**B**) superoxide dismutase (SOD), (**C**) catalase (CAT), and (**D**) glutathione peroxidase (GPx) as affected by substituting soybean meal (SBM) with pumpkin seed cake (PSC) in Nile tilapia diets. Control, D1, D2, D3, and D4 indicate that 0%, 10%, 20%, 30%, and 40% of SM were replaced by PSC, respectively. Different letters above the error bar indicate significant differences at $p < 0.05$, as determined by one-way ANOVA followed by Tukey's multiple comparisons test. Values are expressed as means ± standard deviation ($n = 9$/treatment).

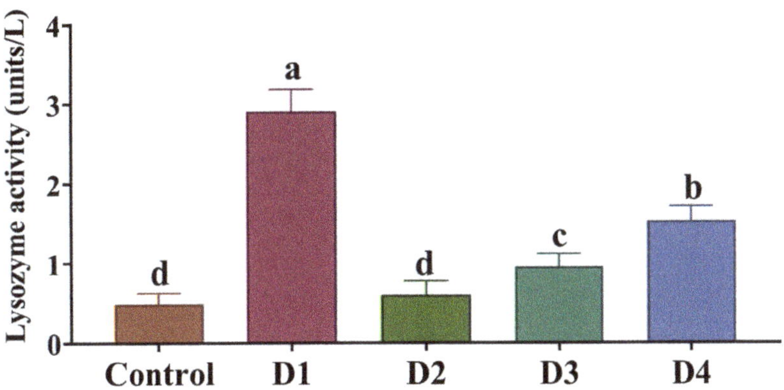

Figure 5. Lysozyme activity as affected by substituting soybean meal (SBM) with pumpkin seed cake (PSC) in Nile tilapia diets. Control, D1, D2, D3, and D4 indicate that 0%, 10%, 20%, 30%, and 40% of SM were replaced by PSC, respectively. Different letters above the error bar indicate significant differences at $p < 0.05$, as determined by one-way ANOVA followed by Tukey's multiple comparisons test. Values are expressed as means ± standard deviation ($n = 9$/treatment).

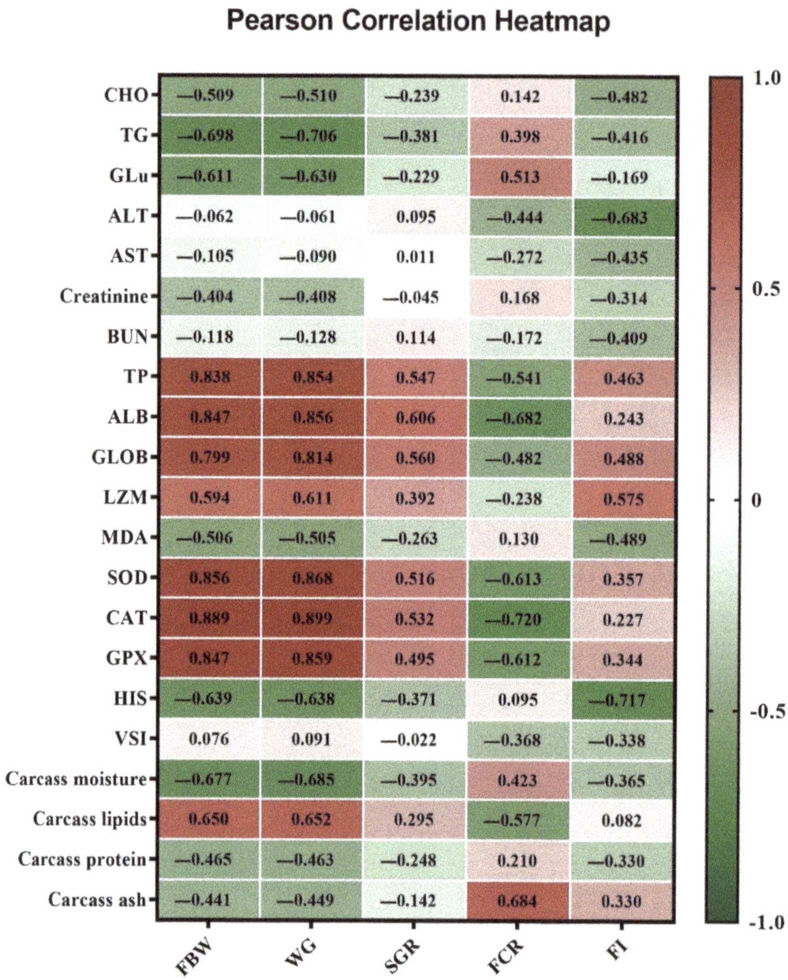

Figure 6. The correlation matrix heatmap shows the values of the Pearson correlation coefficient among growth performance indicators, carcass composition, and serum biochemical parameters. The positive values are in red, and the negative values are in green. The values range from −1 to 1, whereby 1 indicates a perfect negative linear relationship between variables, 1 indicates a perfect positive linear relationship between variables, and 0 indicates no relationship between studied variables. FBW: final body weight; WG: weight gain; SGR: specific growth rate; FCR: feed conversion ratio; and FI: feed intake; CHO: cholesterol; TG: triglycerides; GLU: glucose; ALT: alanine transaminase; AST: aspartate aminotransferase; BUN: blood urea nitrogen; TP: total protein; ALB: albumin; GLOB: globulin; LZM: lysozyme; MDA: malondialdehyde; SOD: superoxide dismutase; CAT: catalase; GPx: glutathione peroxidase, HIS: hepatosomatic index; and VSI: viscerosomatic index.

4. Discussion

Our findings showed that all water quality parameters, including water temperature, pH, total ammonia, dissolved oxygen, and nitrate, as estimated in all experimental groups, were within the recommended limits for warm-water fish [28,29].

Recent research has focused on the viability of incorporating agro-industrial and plant-based feed ingredients into fish feeds to gain financial advantages while enhancing the sustainability of aquaculture [30]. Several research investigations have found that replacing

SB with non-traditional feed constituents had no detrimental impact on the growth of rainbow trout [31,32], grass carp [33,34], Nile tilapia [6], or common carp [35]. In the current experiment, fish fed the D4 diet showed better FW, WG, SGR, and FCR. Recently, Musthafa et al. [36] stated that the *Mucuna pruriens* seed meal substantially enhanced the immune status and growth of the freshwater fish *O. niloticus*. Furthermore, Bai et al. [37] found that a diet containing 30% cottonseed meal improved the growth indicators of the red-bellied Pacu. The enhanced growth observed in fish fed with PSC could be attributed to this alternative feed ingredient's nutritional and bioactive properties. PSC is rich in protein, essential amino acids, and minerals [12], all of which are essential nutrients for the growth and development of fish. In addition, PSC contains bioactive compounds such as phytosterols and phenolic compounds [38] that have been shown to promote fish health and growth. Another important perspective is that the feed intake has significantly improved in most fish groups fed PSC-containing diets, which could be responsible for the recorded growth. Additionally, Greiling et al. [20] have demonstrated that increased intake of the PSC-containing diet by African catfish could be related to increased feed palatability.

VSI and HSI showed significant differences in favor of PSC inclusion compared to the control. Our findings contradicted those of Sezgin and Aydın [17], who concluded that PSC had no significant influence on HSI or VSI. However, our result could be explained by several studies, including those of Caili et al. [39], Nkosi et al. [40], Bardaa et al. [16], Musthafa et al. [41], and Salehi et al. [14], which suggest that the observed improvements in VSI and HSI in tilapia fed with diets containing PSC are likely due to a complex interplay of metabolic processes influenced by the nutritional composition of the feed. As a result, this could positively impact various metabolic processes, including those involved in growth.

Fish carcass composition can be influenced by changes in the composition of their diets [4]. In the present study, increasing PSC substitution levels in fish diets increased the ash and crude lipid contents, but reduced the crude protein and moisture contents. Likewise, substitution with other plant protein sources resulted in a comparable change in the carcass composition of Nile tilapia [4,6]. The increase in crude lipid content could be related to the increased feed intake, which has been shown to increase whole-body energy retention in the form of increased whole-body lipid deposition. In this regard, Greiling et al. [20] reported that the higher energy content of the PSC-containing diets could increase the weights of different sites of lipid deposition in fish carcasses. Meanwhile, the reduction in protein content could be related to the presence of some anti-nutritional factors or the deficiency of some amino acids, such as lysine [15].

Blood biochemical parameters are a conventional technique used regularly to evaluate the physiological, pathological, and immunological statuses of fish [42]. Herein, glucose, total cholesterol, and triglyceride concentrations were significantly reduced in *O. niloticus* fed with PSC diets, and the D4 group had the lowest values. The results of this study corroborate those of Caili et al. [39], who found that pumpkin seeds could efficiently reduce serum triglyceride levels and cholesterol. Similarly, Sezgin and Aydın [17] found that increasing the inclusion level of PSC in mirror carp (*Cyprinus carpio*) diets led to a decrease in both cholesterol and triglyceride levels, with a continued effect on cholesterol until reaching 100% inclusion. This notable reduction could be attributed to PSC's phytosterol bioactive components, which might affect both the expression levels of genes related to cholesterol and lipid metabolism and the activity of hepatic enzymes [17,43]. Our observation on serum glucose levels was similar to the findings of Sezgin and Aydın [17], who noted a decrease in serum glucose levels in groups fed on diets containing 66% and 100% PSC. In addition, the serum contents of ALT, AST, creatinine, and urea were significantly decreased with all dietary inclusion levels of PSC. The antioxidant activity of PSC may be responsible for its hepatoprotective and nephroprotective effects [44]. Major functions of serum proteins include sustaining pH, osmotic pressure, and transporting metabolites. TP is crucial to the innate immune response and humoral immunity of fish [45]. In this study, the TP and globulin levels in the blood of fish fed the D1 and D4 diets were significantly higher than in those fed other experimental diets. In contrast, the ALB levels of fish fed the D1, D3, and D4 diets were significantly higher than those fed the control diet. Our dataset is in contrast with the findings of Sezgin and Aydın [17], who found that the serum TP, ALB, ALT, and AST levels underwent only numerical changes in those parameters' values, with no statistically significant effect of PSC inclusion in the diets of mirror carp.

The antioxidant activity of enzymes, such as GPx, CAT, and SOD, is an essential biomarker for monitoring fish health and reactivity to external stimuli [30], and it can be used to assess fish antioxidant capability. Furthermore, Al-Sagheer et al. [46] stated that MDA, a marker of lipid peroxidation, can be utilized to assess the level of oxidative stress and antioxidant status. In this investigation, the activity of SOD, CAT, and GPx significantly increased while MDA decreased in fish fed formulated feeds with PSC. The addition of PSC, which contains active ingredients like proteins, minerals, phytosterols, triterpenes, carotenoids, lignans, polyunsaturated fatty acids, tocopherol, and phenolic and antioxidative compounds, may have triggered antioxidant defense and decreased MDA lipid oxidation in all treated groups [14,16,39,40]. Previous studies have found that PSC has potent antioxidative properties, as shown by its free radical scavenging activity and augmentation of antioxidase activity [47,48]. Furthermore, these effects could be attributed to the essential fatty acids found in pumpkin seeds, which play a critical role in maintaining cell membrane integrity and regulating gene expression, as well as the antioxidants found in pumpkin seeds, which could help to protect fish from oxidative stress by scavenging free radicals and reducing inflammation [20].

Lysozyme is a positively charged protein found in various bodily fluids, such as mucus, lymphoid tissue, and plasma, and is synthesized in multiple fish tissues [49]. It is essential to multiple defense mechanisms, including the immune response, opsonization, bacteriolysis, and antimicrobial activity [36]. Our dataset showed that lysozyme activity was boosted by incorporating PSC into feed compared to control groups. The findings of Musthafa et al. [41] on *O. mossambicus* fed diets including *C. mixta* seed meal corroborate these findings, since they showed an enhanced immune response to *A. hydrophila* and the Gram-negative bacterium. There is a strong link between oxidative stress and immune function in fish [50], and the liver plays an important role in synthesizing immune proteins [51]. The improved hepatic function and the antioxidant capacity of fish fed diets containing PSC in the current experiment could also have partly contributed to the recorded enhancement of the immune status. Notably, the positive relationship detected by the Pearson correlation analysis between growth indices and the antioxidant enzymes and lysozyme activity indicates that the enhancement of these parameters could be collectively responsible for the recorded enhanced growth performance and nutrient utilization of fish fed diets containing PSC.

5. Conclusions

Based on the study's findings, partial SB substitution (particularly 40% replacement) with PSC in the diets of Nile tilapia significantly enhanced growth performance, feed conversion, water quality, antioxidant capacity, and immunity. This study concludes that PSC could be a promising, sustainable, and cost-effective alternative protein source for Nile tilapia diets. Additional research is required in order to investigate the long-term effects of PSC on fish health and the environment.

Author Contributions: Conceptualization, H.A.M.M., A.A.A.-S., K.M.A. and S.S.S.; methodology, H.A.M.M., S.S.S. and D.A.A.E.-B.; software, H.A.M.M. and A.A.A.-S.; formal analysis, H.A.M.M., Z.M.A.A.-E.A. and K.M.A.; investigation, Z.M.A.A.-E.A., D.A.A.E.-B., A.A.A.-S. and K.M.A.; resources, D.A.A.E.-B., K.M.A., Y.M.A.-E. and B.A.H.; data curation, H.A.M.M., Z.M.A.A.-E.A., D.A.A.E.-B. and K.M.A.; writing—original draft preparation, Z.M.A.A.-E.A. and D.A.A.E.-B.; writing—review and editing, H.A.M.M., A.A.A.-S., Y.M.A.-E. and B.A.H.; visualization, H.A.M.M., Z.M.A.A.-E.A., D.A.A.E.-B. and K.M.A.; supervision, H.A.M.M. All authors have read and agreed to the published version of the manuscript.

Funding: This research received no external funding.

Institutional Review Board Statement: The Animal Experiments Local Ethics Committee of the Faculty of Agriculture at Ain Shams University approved the animal study protocol with approval number 4-202301 (date of approval; 12 February 2023).

Informed Consent Statement: Not applicable.

Data Availability Statement: The datasets used along with this research are available from the corresponding author upon reasonable request.

Conflicts of Interest: Author Sherif S. Sadek is the owner of the Aquaculture Consultant Office, Cairo, Egypt. The remaining authors declare that the research was conducted in the absence of any commercial or financial relationships that could be construed as a potential conflict of interest.

References

1. El-Houseiny, W.; Arisha, A.H.; Behairy, A.; Metwally, M.M.M.; Abdel-Warith, A.-W.A.; Younis, E.M.; Davies, S.J.; Hassan, B.A.; Abd-Elhakim, Y.M. The immunosuppressive, growth-hindering, hepatotoxic, and oxidative stress and immune related-gene expressions-altering effects of gibberellic acid in *Oreochromis niloticus*: A mitigation trial using alpha-lipoic acid. *Pestic. Biochem. Physiol.* **2024**, *198*, 105725. [CrossRef]
2. FAO. *The State of World Fisheries and Aquaculture 2022. Towards Blue Transformation*; Food and Agriculture Organization of the United Nations: Rome, Italy, 2022.
3. Ayyat, M.S.; Al-Sagheer, A.A.; Ayyat, A.M.; Abdelrhman, A.M.; Ahmed, N.H.; Naiel, M.A. Effects of a blend of herbal feed supplements on growth, associated blood indices and body chemical analysis in nile tilapia reared under high stocking density. *Aquac. Res.* **2022**, *53*, 5475–5485. [CrossRef]
4. Ayyat, M.S.; Ayyat, A.M.N.; Abdel-Rahman, M.S.; Al-Sagheer, A.A. Appraisal of leaf protein concentrate derived from sugar beet and carrot as a novel fish meal substitute for juvenile Nile tilapia (*Oreochromis niloticus*). *Anim. Feed Sci. Technol.* **2024**, *307*, 115833. [CrossRef]
5. El-Sayed, A.-F.M.; Fitzsimmons, K. From africa to the world—The journey of nile tilapia. *Rev. Aquac.* **2023**, *15*, 6–21. [CrossRef]
6. Ayyat, M.S.; A. Abdel-Rahman, G.; Ayyat, A.M.N.; Abdel-Rahman, M.S.; Al-Sagheer, A.A. Evaluation of leaf protein concentrate from *Beta vulgaris* and *Daucus carota* as a substitute for soybean meal in *Oreochromis niloticus* fingerlings diets. *Aquac. Res.* **2021**, *52*, 3256–3269. [CrossRef]
7. Gyan, W.R.; Ayiku, S.; Yang, Q. Effects of replacing fishmeal with soybean products in fish and crustaceans performance. *J. Aquat. Res. Dev.* **2019**, *10*, 573.
8. Li, H.; Hu, Z.; Liu, S.; Sun, J.; Ji, H. Influence of dietary soybean meal replacement with yellow mealworm (*Tenebrio molitor*) on growth performance, antioxidant capacity, skin color, and flesh quality of mirror carp (*Cyprinus carpio* var. Specularis). *Aquaculture* **2022**, *561*, 738686. [CrossRef]
9. Ayyat, M.S.; Ayyat, A.M.N.; Abd El-Latif, K.M.; Hessein, A.A.A.; Al-Sagheer, A.A. Inorganic mercury and dietary safe feed additives enriched diet impacts on growth, immunity, tissue bioaccumulation, and disease resistance in Nile tilapia (*Oreochromis niloticus*). *Aquat. Toxicol.* **2020**, *224*, 105494. [CrossRef]
10. El-Houseiny, W.; Arisha, A.H.; Metwally, M.M.M.; Abdel-Warith, A.-W.A.; Younis, E.M.; Davies, S.J.; Hassan, B.A.; Abd-Elhakim, Y.M. Alpha-lipoic acid suppresses gibberellic acid nephrotoxicity in Nile tilapia (*Oreochromis niloticus*) via modulating oxidative stress, inflammation, cytokine production, and apoptosis. *Pestic. Biochem. Physiol.* **2023**, *196*, 105598. [CrossRef]
11. Ajila, C.M.; Brar, S.K.; Verma, M.; Tyagi, R.D.; Godbout, S.; Valéro, J.R. Bio-processing of agro-byproducts to animal feed. *Crit. Rev. Biotechnol.* **2012**, *32*, 382–400. [CrossRef]
12. Ezzat, S.M.; Adel, R.; Abdel-Sattar, E. Pumpkin bio-wastes as source of functional ingredients. In *Mediterranean Fruits Bio-Wastes: Chemistry, Functionality and Technological Applications*; Ramadan, M.F., Farag, M.A., Eds.; Springer International Publishing: Cham, Switzerland, 2022; pp. 667–696.
13. Rabrenović, B.B.; Dimić, E.B.; Novaković, M.M.; Tešević, V.V.; Basić, Z.N. The most important bioactive components of cold pressed oil from different pumpkin (*Cucurbita pepo* L.) seeds. *LWT-Food Sci. Technol.* **2014**, *55*, 521–527. [CrossRef]
14. Salehi, B.; Capanoglu, E.; Adrar, N.; Catalkaya, G.; Shaheen, S.; Jaffer, M.; Giri, L.; Suyal, R.; Jugran, A.K.; Calina, D.; et al. Cucurbits plants: A key emphasis to its pharmacological potential. *Molecules* **2019**, *24*, 1854. [CrossRef] [PubMed]
15. Zdunczyk, Z.; Minakowski, D.; Frejnagel, S.; Flis, M. Comparative study of the chemical composition and nutritional value of pumpkin seed cake, soybean meal and casein. *Food/Nahr.* **1999**, *43*, 392–395. [CrossRef]
16. Bardaa, S.; Ben Halima, N.; Aloui, F.; Ben Mansour, R.; Jabeur, H.; Bouaziz, M.; Sahnoun, Z. Oil from pumpkin (*Cucurbita pepo* l.) seeds: Evaluation of its functional properties on wound healing in rats. *Lipids Health Dis.* **2016**, *15*, 73. [CrossRef] [PubMed]
17. Sezgin, A.; Aydın, B. Effect of replacing dietary soybean meal with pumpkin (*Cucurbita pepo*) seed cake on growth, feed utilization, haematological parameters and fatty acid composition of mirror carp (*Cyprinus carpio*). *Aquac. Res.* **2021**, *52*, 5870–5881. [CrossRef]
18. Akin, G.; Arslan, F.N.; Elmasa, S.K.; Yilmaz, I. Cold-pressed pumpkin seed (*Cucurbita pepo* L.) oils from the central anatolia region of Turkey: Characterization of phytosterols, squalene, tocols, phenolic acids, carotenoids and fatty acid bioactive compounds. *Grasas Aceites* **2018**, *69*, e232. [CrossRef]

19. Lovatto, N.D.M.; Goulart, F.R.; Loureiro, B.B.; Adorian, T.J.; de Freitas, S.T.; Pianesso, D.; Dalcin, M.O.; Athayde, M.L.; da Silva, L.P. Effects of phosphorylated protein concentrate of pumpkin seed meal on growth and digestive enzymes activity of silver catfish (*Rhamdia quelen*). *Aquac. Nutr.* **2017**, *23*, 201–209. [CrossRef]
20. Greiling, A.M.; Schwarz, C.; Gierus, M.; Rodehutscord, M. Pumpkin seed cake as a fishmeal substitute in fish nutrition: Effects on growth performance, morphological traits and fillet colour of two freshwater salmonids and two catfish species. *Arch. Anim. Nutr.* **2018**, *72*, 239–259. [CrossRef]
21. NRC. *Nutrient Requirements of Fish and Shrimp*; The National Academies Press: Washington, DC, USA, 2011.
22. APHA. *Standard Methods for the Examination of Water and Wastewater*, 23rd ed.; American Public Health Association: Washington DC, USA, 2017.
23. AOAC. *Official Methods of Analysis of the Association of Analytical Chemists International*, 18th ed.; Association of Official Analytical Chemists: Gaithersburg, MD, USA, 2005.
24. Ghareghanipoora, M.; Akbary, P.; Akhlaghi, M.; Fereidouni, M. Non-specific immune responses and immune related genes expression of rainbow trout (*Oncorhynchus mykiss*, walbaum) fed *Zataria multiflora* boiss extract. *Bull Environ. Pharmacol. Life Sci.* **2014**, *3*, 140–146.
25. SAS. *Note: SAS Proprietary Software Version 9.00 (TS m0)*; Sas Institute Inc.: Cary, NC, USA, 2002.
26. Yossa, R.; Verdegem, M. Misuse of multiple comparison tests and underuse of contrast procedures in aquaculture publications. *Aquaculture* **2015**, *437*, 344–350. [CrossRef]
27. Prism, G. *Prism 8 for Windows*; GraphPad Software Inc.: San Diego, CA, USA, 2019.
28. Boyd, C. *Water Quality in Ponds for Aquaculture*; Birmingham Publ. Co.: Birmingham, AL, USA, 1990; p. 482.
29. Boyd, C.E.; Tucker, C.S. Ecology of aquaculture ponds. In *Pond Aquaculture Water Quality Management*; Springer: Berlin/Heidelberg, Germany, 1998; pp. 8–86.
30. El-Houseiny, W.; Anter, R.G.A.; Arisha, A.H.; Mansour, A.T.; Safhi, F.A.; Alwutayd, K.M.; Elshopakey, G.E.; Abd El-Hakim, Y.M.; Mohamed, E.M.M. Growth retardation, oxidative stress, immunosuppression, and inflammatory disturbances induced by herbicide exposure of catfish, *Clarias gariepinus*, and the alleviation effect of dietary wormwood *Artemisia cina*. *Fishes* **2023**, *8*, 297. [CrossRef]
31. Dernekbaşı, S.; Karayücel, İ.; Akyüz, A.P. Evaluation of sesame (*Sesamum indicum*) seed meal as a replacer for soybean meal in the diets of rainbow trout (*Oncorhynchus mykiss* walbaum, 1792). *Ege J. Fish. Aquat. Sci.* **2017**, *34*, 31–39.
32. Ustaoglu Tiril, S.; Kerim, M. Evaluation of safflower meal as a protein source in diets of rainbow trout [*Oncorhynchus mykiss*, walbaum, 1792)]. *J. Appl. Ichthyol.* **2015**, *31*, 895–899. [CrossRef]
33. Li, X.; Chen, S.; Sun, J.; Huang, X.; Tang, H.; He, Y.; Pan, Q.; Gan, L. Partial substitution of soybean meal with faba bean meal in grass carp (*Ctenopharyngodon idella*) diets, and the effects on muscle fatty acid composition, flesh quality, and expression of myogenic regulatory factors. *J. World Aquac. Soc.* **2020**, *51*, 1145–1160. [CrossRef]
34. Lu, R.; Chen, Y.; Yu, W.; Lin, M.; Yang, G.; Qin, C.; Meng, X.; Zhang, Y.; Ji, H.; Nie, G. Defatted black soldier fly (*Hermetia illucens*) larvae meal can replace soybean meal in juvenile grass carp (*Ctenopharyngodon idellus*) diets. *Aquac. Rep.* **2020**, *18*, 100520. [CrossRef]
35. Wang, T.; Xu, M.; Wang, J.; Wan, W.; Guan, D.; Han, H.; Wang, Z.; Sun, H. A combination of rapeseed, cottonseed and peanut meal as a substitute of soybean meal in diets of yellow river carp *Cyprinus carpio* var. *Aquac. Nutr.* **2020**, *26*, 1520–1532. [CrossRef]
36. Musthafa, M.S.; Asgari, S.M.; Kurian, A.; Elumalai, P.; Ali, A.R.J.; Paray, B.A.; Al-Sadoon, M.K. Protective efficacy of *Mucuna pruriens* (L.) seed meal enriched diet on growth performance, innate immunity, and disease resistance in *Oreochromis mossambicus* against *Aeromonas hydrophila*. *Fish Shellfish Immunol.* **2018**, *75*, 374–380. [CrossRef]
37. Bai, T.L.; Ch, J.; Reddy, D.K. Comparison of effects of cotton seed meal with fish meal on growth, feed conversion ratio and survival of redbellied pacu (*Piaractus brachypomus*) int. *J. Sci. Environ. Technol.* **2016**, *5*, 1092–1099.
38. Dotto, J.M.; Chacha, J.S. The potential of pumpkin seeds as a functional food ingredient: A review. *Sci. Afr.* **2020**, *10*, e00575. [CrossRef]
39. Caili, F.; Huan, S.; Quanhong, L. A review on pharmacological activities and utilization technologies of pumpkin. *Plant Foods Hum. Nutr.* **2006**, *61*, 70–77. [CrossRef]
40. Nkosi, C.; Opoku, A.; Terblanche, S. In vitro antioxidative activity of pumpkin seed (*Cucurbita pepo*) protein isolate and its in vivo effect on alanine transaminase and aspartate transaminase in acetaminophen-induced liver injury in low protein fed rats. *Phytother. Res.* **2006**, *20*, 780–783. [CrossRef] [PubMed]
41. Musthafa, M.S.; Jawahar Ali, A.R.; Arun Kumar, M.S.; Paray, B.A.; Al-Sadoon, M.K.; Balasundaram, C.; Harikrishnan, R. Effect of *Cucurbita mixta* (L.) seed meal enrichment diet on growth, immune response and disease resistance in *Oreochromis mossambicus*. *Fish Shellfish Immunol.* **2017**, *68*, 509–515. [CrossRef] [PubMed]
42. El-Houseiny, W.; Algharib, S.A.; Mohamed, E.A.A.; Metwally, M.M.M.; Mahmoud, Y.K.; Alghamdi, Y.S.; Soliman, M.M.; Abd-Elhakim, Y.M.; El-Murr, A.E. Dietary parsley seed mitigates methomyl-induced impaired growth performance, hemato-immune suppression, oxidative stress, hepato-renal damage, and pseudomonas aeruginosa susceptibility in *Oreochromis niloticus*. *Antioxidants* **2022**, *11*, 1185. [CrossRef] [PubMed]
43. Niroomand, M.; Akbarzadeh, A.; Ebrahimi, E.; Sobhani, S.A.; Sheikhahmadi, A. Effects of dietary black cumin seed meal on growth performance, blood biochemistry and fatty acid composition of pacific white shrimp *Litopenaeus vannamei*. *Aquac. Nutr.* **2020**, *26*, 1072–1082. [CrossRef]

44. Latif, M.; Faheem, M.; Asmatullah; Hoseinifar, S.H.; Van Doan, H. Dietary black seed effects on growth performance, proximate composition, antioxidant and histo-biochemical parameters of a culturable fish, rohu (*Labeo rohita*). *Animals* **2020**, *11*, 48. [CrossRef]
45. Esmaeili, M. Blood performance: A new formula for fish growth and health. *Biology* **2021**, *10*, 1236. [CrossRef] [PubMed]
46. Al-Sagheer, A.A.; Mahmoud, H.K.; Reda, F.M.; Mahgoub, S.A.; Ayyat, M.S. Supplementation of diets for *Oreochromis niloticus* with essential oil extracts from lemongrass (*Cymbopogon citratus*) and geranium (*Pelargonium graveolens*) and effects on growth, intestinal microbiota, antioxidant and immune activities. *Aquac. Nutr.* **2018**, *24*, 1006–1014. [CrossRef]
47. Li, Y.; Zhang, G.; Fang, X.; Zhao, C.; Wu, H.; Lan, Y.; Che, L.; Sun, Y.; Lv, J.; Zhang, Y. Effects of replacing soybean meal with pumpkin seed cake and dried distillers grains with solubles on milk performance and antioxidant functions in dairy cows. *Animal* **2021**, *15*, 100004. [CrossRef]
48. Krimer-Malešević, V. Pumpkin seeds: Phenolic acids in pumpkin seed (*Cucurbita pepo* L.). In *Nuts and Seeds in Health and Disease Prevention*; Elsevier: Amsterdam, The Netherlands, 2020; pp. 533–542.
49. Zhang, G.; Gong, S.; Yu, D.; Yuan, H. Propolis and herba epimedii extracts enhance the non-specific immune response and disease resistance of *Chinese sucker, Myxocyprinus asiaticus*. *Fish Shellfish Immunol.* **2009**, *26*, 467–472. [CrossRef]
50. Biller, J.D.; Takahashi, L.S. Oxidative stress and fish immune system: Phagocytosis and leukocyte respiratory burst activity. *An. Acad. Bras. Ciênc.* **2018**, *90*, 3403–3414. [CrossRef]
51. Reynaud, S.; Raveton, M.; Ravanel, P. Interactions between immune and biotransformation systems in fish: A review. *Aquat. Toxicol.* **2008**, *87*, 139–145. [CrossRef] [PubMed]

animals

MDPI

Article

Inclusion of Sorghum in *Cyprinus carpio* L. Diet: Effects on Growth, Flesh Quality, Microbiota, and Oxidative Status

Cristian-Alin Barbacariu [1], Gabriela Dumitru [2,*], Cristina Mihaela Rimbu [3], Cristina Elena Horhogea [3], Lenuța Dîrvariu [1], Elena Todirașcu-Ciornea [2], Dana Andreea Șerban [1,4] and Marian Burducea [1,*]

[1] Research and Development Station for Aquaculture and Aquatic Ecology, "Alexandru Ioan Cuza" University, Carol I, 20A, 700505 Iasi, Romania; alin.barbacariu@uaic.ro (C.-A.B.); lus22grigorica@yahoo.com (L.D.); dana.serban@uaic.ro (D.A.Ș.)

[2] Faculty of Biology, "Alexandru Ioan Cuza" University, Carol I, 20A, 700505 Iasi, Romania; ciornea@uaic.ro

[3] Department of Public Health, Faculty of Veterinary Medicine, University of Life Sciences "Ion Ionescu de la Brad" Iași, Mihail Sadoveanu Alley 6-8, 700490 Iasi, Romania; cristina.rimbu@iuls.ro (C.M.R.); cristina.horhogea@iuls.ro (C.E.H.)

[4] Faculty of Food and Animal Sciences, University of Life Sciences "Ion Ionescu de la Brad" Iași, Mihail Sadoveanu Alley 6-8, 700490 Iasi, Romania

* Correspondence: gabriela.dumitru@uaic.ro (G.D.); marian.burducea@uaic.ro (M.B.)

Simple Summary: Sorghum is a nutritious crop that thrives in diverse climates and can withstand harsh growing conditions, making it an attractive option for feeding farmed fish in various regions around the world. In this study, the effects of different sorghum-containing diets on carp were tested, revealing a reduction in growth rates, notable changes in flesh protein content, and significant effects on intestinal microbiota and oxidative stress. This indicates sorghum's potential health and metabolic impacts on *Cyprinus carpio*.

Abstract: This study investigates the impact of including sorghum in the diet of the common carp (*Cyprinus carpio*) on its growth, blood parameters, meat composition, intestinal microbiota, and oxidative stress. Experimental diets with varying sorghum content (0%—V0 or control, 10%—V1, 20%—V2, and 30%—V3) were administered to carp weighing 43 g initially. Notably, in the 30% variant, sorghum entirely replaced corn and barley in the diet. Chemical analysis of sorghum unveiled a protein content of 14% and a fat content of 3.9%. Sorghum inclusion led to a decline in final body weight and weight gain, particularly notable in the V3 group with 30% sorghum. However, other physiological parameters, such as feed conversion ratio, specific growth rate, and organ indices, remained unaffected. Protein and salt content in carp flesh increased with higher sorghum inclusion levels, while hematological parameters showed minimal variations. Analysis of the intestinal microbiota revealed increases in both aerobic and anaerobic bacterial populations with sorghum inclusion. Furthermore, sorghum concentration inversely correlated with glutathione levels and positively correlated with malondialdehyde content, indicating a disruption of antioxidant defense mechanisms and elevated oxidative stress.

Keywords: sorghum; common carp; aquaculture; growth performance; hematology; intestinal microbiota; oxidative stress

1. Introduction

Aquaculture plays a pivotal role in meeting the global demand for food and nutrition, being the fastest growing food industry sector. According to the latest report from the Food and Agriculture Organization (FAO) released in 2022 [1], aquaculture contributed 122.6 million tons out of the total 214 million tons of global fishery and aquaculture production in 2020, valued at USD 281.5 billion. Looking ahead, the FAO projects a 14 percent increase in fish production by 2030, with aquaculture playing a crucial role in this growth trajectory [1]. The sustained expansion of aquaculture will contribute significantly to achieving Sustainable Development Goals (SDGs), particularly SDG 2 (Zero Hunger), SDG 14 (Life Below Water), and SDG 8 (Decent Work and Economic Growth). However, the sector faces significant challenges, particularly related to feed, which represents over 50% of production costs [1]. The challenge regarding feeding stems from the rising costs and scarcity of traditional ingredients like fish meal and fish oil, necessitating the exploration of sustainable alternatives. In this context, identifying and integrating alternative ingredients into aquafeed formulations becomes increasingly crucial. This not only addresses economic concerns, but also aligns with sustainability goals by reducing reliance on finite marine resources.

The chemical composition of sorghum (*Sorghum bicolor*) is similar to that of other cereals used in aquaculture (such as corn, wheat, barley, etc.), with approximately 11.4% protein, 3.3% fat, and 1.9% fiber [2]. One of the beneficial features of sorghum is its ability to be cultivated in semi-arid areas, with the United States being the world's leading producer, contributing to around 15% of total production [3]. The main advantages of using sorghum in fish feed include high antioxidant content, nutrient source, non GMO status, sustainability, and significant global production [4–7]. However, challenges in using sorghum include its lower protein content (similar to other cereals), reduced starch digestibility, the presence of antinutritional factors, lysine and threonine deficiencies, and limited studies on its application in fish feeding [4–7]. Key antinutritional factors in sorghum include trypsin inhibitors, amylase inhibitors, tannins, and phytic acid [8].

In addition to its various advantages and challenges in aquaculture, sorghum's cultivation in semi-arid areas faces increasing pressures due to climate change. These regions are experiencing heightened aridity, exacerbated by factors such as deforestation and desertification, affecting the yield and quality of sorghum crops [9]. Furthermore, amidst the global challenges posed by the COVID-19 pandemic and geopolitical tensions, including the Ukraine conflict, contributing to a surge in cereal prices, the exploration of alternatives gains significance [10]. Although Europe, including Romania, is not among the top sorghum-producing regions globally, its yields are quite impressive. According to Popescu et al. [11], in 2017, the EU produced 1.18% of the world sorghum output on only 0.12% of the global sorghum cultivation area. Romania contributed 7.18% to the EU's total production. With a yield per surface unit of 3879 kg/ha, Romania surpassed the world average by 2.71 times. Recognizing the potential benefits of sorghum cultivation and the urgent need to address climate change, EU policies are geared towards encouraging the expansion of cultivated areas and production.

The common carp (*Cyprinus carpio* L.) is the most important species cultivated in Romania and is considered a traditional species in many European countries such as Hungary, Poland, the Czech Republic, Bulgaria, Slovenia, and Lithuania, with its production reaching 36,403 tons in 2021 in the mentioned countries, according to Eurostat [3]. Carp-farming systems in Romania are typically semi-intensive, utilizing earthen ponds and incorporating cereals and oilseeds in feed formulations. The main issues related to the use of plant-based ingredients stem from the presence of antinutritional factors and their low digestibility. In this context, D'mello [12] demonstrated that sorghum digestibility in carp is roughly similar to that of rye in terms of lipids and slightly lower in terms of protein. Degani highlighted in his study that the protein digestibility of sorghum in carp is 71.86%, while lipid digestibility is 76.71% [13].

Considering the traditional aquaculture nature of carp farming, farms engaging in this activity also play a role in maintaining these practices, contributing to the creation of local ecosystem services [14]. The relevance of sustainable feeds for carp farming lies in their crucial role in fostering sustainable aquaculture practices, which are essential for preserving local ecosystem services. In animal farming (pigs, poultry, ruminants), sorghum is commonly used and usually replaces corn [15]. A study by Anderolu [16] demonstrated the possibility of replacing corn with sorghum in ratios of up to 50% without negative effects on growth or health in the African catfish (*Clarias gariepinus*). Another study showed the potential replacement of cassava with sorghum in ratios of up to 20% in *Pangasius pangasius* [17].

In light of these considerations, this study aims to assess the viability of sorghum as a determinant ingredient in the common carp's diet, with a specific focus on determining the optimal inclusion levels and investigating its effects on growth performance, blood parameters, meat quality, intestinal microbiota, and oxidative stress.

2. Materials and Methods

2.1. Fish Farming and Management

In the spring of 2023, at the Research and Development Station for Aquaculture and Aquatic Ecology, directed natural reproduction was conducted in an earthen pond using carp broodstock from the station. A total of 120 carp fingerlings were selected from a pool of 8000 individuals, with an initial weight of 43 g. The fish were transferred to the station's recirculating aquaculture system (RAS) and acclimated for two weeks before the start of the experiment. The RAS was equipped with water tanks, drainage pipes, feeding systems, a mechanical filter, a biological filter, a UV filter, recirculation pumps, automation, O_2 sensors, temperature sensors, and pH sensors. The feeding trial had a duration of six weeks, beginning on 7 August and ending on 15 September. Throughout the trial, the water temperature started at 25.6 °C and decreased to 21.7 °C by the end (Table 1). Daily measurements included dissolved oxygen and temperature, while pH, ammonia, nitrites, nitrates, and phosphorus were measured weekly. Water parameters (ammonia, nitrites, nitrates, phosphates) were measured using the Hanna Iris HI801 Spectrophotometer and Hanna reagent kits (Hanna Instruments, Salaj, Romania), Hach HQ30d (Hach Company, Loveland, CO, USA) (dissolved oxygen), and Hach HQ11d (Hach Company) (pH and conductivity). Water parameters are presented in Table 1.

Table 1. Water quality in the RAS during the fish trial with experimental sorghum diets.

Parameter	Units	Week 1	Week 2	Week 3	Week 4	Week 5	Week 6
Temperature	°C	25.60	22.90	25.00	25.10	23.00	21.70
pH	units	8.10	7.79	7.88	7.77	7.42	7.86
Dissolved oxygen	mg/L	8.00	8.10	8.00	8.00	8.10	8.00
Nitrate (NO_3^-)	mg/L	0.22	0.16	0.10	0.02	0.07	0.04
Nitrite (NO_2^-)	mg/L	0.28	0.14	0.10	0.02	0.10	0.03
Ammonia (NH_3^+)	mg/L	0.05	0.12	0.06	0.02	0.01	0.20
Ammonium (NH_4^+)	mg/L	12.60	14.40	11.00	10.00	12.00	10.00
Phosphorus total (PO_4^{3-})	mg/L	7.80	8.00	8.10	7.80	7.70	8.00
Hardness total	°G	98.00	95.00	100.00	84.00	88.00	96.00
Calcium (Ca_2^+)	mg/l	54.00	66.00	64.00	61.00	57.00	59.00
Magnesium (Mg_2^+)	mg/l	0.00	0.00	0.00	0.00	0.00	0.00
Iron (Fe)	mg/l	116.00	128.00	122.00	100.00	112.00	124.00
Sulfate (SO_4^{2-})	mg/l	0.10	0.00	0.10	0.06	0.00	0.10

2.2. Experimental Diets

Four feed diets were created, with the proportion of sorghum varying as follows: control—0%, V1—10%, V2—20%, V3—30%. The control diet consisted of sunflower meal 20%, green peas 20%, corn 15%, barley 12%, corn DDGS 11%, fish meal 20%, and sunflower oil 2%, without sorghum. The ingredients used (sunflower meal, green peas, corn, barley, corn DDGS, fish meal, sunflower oil, and sorghum) were specific to carp diets and originated from Romania, except for the fishmeal. The diets were produced through extrusion, grinding, mixing, and pelleting. A low percentage of fish meal (20%) was used, resulting in a diet with 22% protein, close to the diets used by carp farmers in earthen ponds for the lowest possible cost.

Analysis of the proximate composition of ingredients and diets was performed with a DA 7250 NIR Analyser (Perten Instruments, Hagersten, Sweden) (Tables 2 and 3). The fish were distributed in 12 cylindrical fiberglass tanks with a volume of 0.75 m^3. Each water tank housed 10 fish, resulting in 30 fish per experimental variant.

Table 2. Proximate composition of the ingredients used in experimental sorghum diets for carp.

Ingredient (%)	Moisture	Protein	Fat	Fiber	Starch	Ash	Sugar	Calcium	Phosphorus
Sorghum	13.17	14.31	3.97	4.55	50.50	12.30	2.59	1.06	0.68
Sunflower meal	10.46	41.10	1.63	22.10	-	7.54	-	-	-
Green peas	11.2	22.35	-	-	-	-	-	-	-
Corn	12.34	7.51	3.60	-	63.11	-	-	-	-
Barley	11.29	11.78	-	4.97	53.20	-	-	-	-
Corn DDGS	10.47	26.67	-	9.48	3.72	3.96	-	-	0.69
Fish meal	10.09	36.50	8.40	-	-	24.40	-	-	-

Table 3. Ingredients and proximate composition of experimental diets for carp.

Ingredient (%)	V0 (0% Sorghum)	V1 (10% Sorghum)	V2 (20% Sorghum)	V3 (30% Sorghum)
Sunflower meal	20	20	20	23
Green peas	20	20	20	17
Corn	15	13	8	-
Barley	12	5	-	-
Corn DDGS	11	11	11	10
Fish meal	20	20	20	20
Sunflower oil	2	1	1	-
Sorghum	-	10	20	30
Proximate composition (%)				
Moisture	9.71 ± 0.35	10.16 ± 0.10	10.42 ± 0.18	10.34 ± 0.26
Protein	22.34 ± 0.24	22.17 ± 0.12	22.07 ± 0.15	22.48 ± 0.19
Fat	5.30 ± 0.08	5.20 ± 0.04	5.74 ± 0.03	5.05 ± 0.02
Fiber	6.66 ± 0.07	6.33 ± 0.01	5.99 ± 0.05	6.43 ± 0.01
Starch	60.10 ± 0.26	58.87 ± 0.07	58.47 ± 0.28	57.43 ± 0.23
Ash	8.63 ± 0.12	8.17 ± 0.07	8.63 ± 0.09	8.97 ± 0.07
Sugar	2.90 ± 0.15	2.90 ± 0.04	2.89 ± 0.09	2.22 ± 0.07
Phosphorus	0.06 ± 0.02	0.07 ± 0.01	0.10 ± 0.03	0.08 ± 0.01

In the beginning of the experiment, the fish were weighed and measured. Manual feeding with the experimental diets was performed three times a day (8:30 AM, 12:00 PM, 3:00 PM). The daily feed quantity was 3%, determined based on water temperature and fish weight.

2.3. Evaluation of Growth Performance

To determine the growth parameters, the fish were measured (in cm) and weighed (in g) three times during the indicated growth period. Based on these measurements, the main growth indices were determined and calculated, and were as follows:

- IBW (initial body weight, g);
- FBW (final body weight, g);
- WG (weight gain, g) = FBW − IBW;
- SGR (specific growth rate, % day^{-1}) = (ln FBW − ln IBW)/days of experiment × 100;
- FCR (feed conversion ratio, g/g) = Feed intake (g)/WG;
- CF (condition factor) = FBW/body $length^3$ × 100;
- PER (protein efficiency ratio) = WG/Protein intake (g);
- LER (lipid efficiency ratio) = WG/Lipid intake (g);
- HSI (hepatosomatic index, %) = 100 × [final liver weight (g)/final body weight (g)];
- VSI (viscerosomatic index, %) = 100 × [final visceral weight (g)/final body weight (g)].

2.4. Proximate Composition of Fish Meat

For the sampling procedure of fish flesh (muscle), the fish were euthanized using clove oil (2%). Nine samples from five fish were taken per variant to ensure accuracy and reliability in the analysis. The proximate composition analysis of the flesh was carried out using a DA 7250 NIR Analyzer manufactured by Perten Instruments in Hagersten, Sweden.

2.5. Blood Hematological Profile

Blood samples were collected from five fish per experimental variant after anesthetization in a 0.2% clove oil concentration for five minutes, using EDTA (ethylenediaminetetraacetic acid)-coated tubes. The evaluations were made using the hematology analyzer Abacus Vet 5, Budapest, Hungary (courtesy of Nova Group).

2.6. Oxidative Status

After the feeding trial period ended, liver, muscle, and intestine tissue samples were precisely dissected and collected from four specimens for each experimental variant for oxidative status assessment. The tissue samples were homogenized in an ice-cold potassium phosphate buffer solution 0.1 M, KCl 1.15%, pH 7.4, in a ratio of 1:10 (*w*/*v*). The homogenates were centrifuged (20 min at 3000 rpm and 4 °C) and the supernatants were further used to measure the activities of superoxide dismutase (SOD), catalase (CAT), and glutathione peroxidase (GPX) and to determine the content of reduced glutathione (GSH) and malondialdehyde (MDA) according to the methods described in Barbacariu et al. [18]. SOD, CAT, and GPX activities and the levels of GSH and MDA were normalized to the total content of soluble proteins measured by the Bradford method [19].

2.7. Intestinal Microbiota

Intestine samples from five fish per variant were taken under laboratory conditions. After dissection, the intestinal contents were aseptically aspirated and 1 g of it was added to 9 mL of sterile saline solution using a sterile Pasteur pipette.

The quantitative determination of some groups of representative microorganisms in the gut microbiota of *Cyprinus carpio* fish was performed according to the methods described by Barbacariu et al. [20]. The microbiological indicators used in the quantitative assessment of the microbiota were as follows: the total number of aerobic bacteria (TNG-A), the total number of anaerobic bacteria (TNG-AN), the total number of sulfite-reducing *Clostridia* (TN-SRC), and the total number of *Enterobacteriaceae* (TN-E). The quantitative results of the microbiological tests were expressed in colony-forming units present in one gram of intestinal contents (CFU/g). To facilitate statistical analysis, the CFU/g values determined for each indicator were converted to decimal logarithms.

2.8. Statistical Analysis

Growth performance, flesh composition, blood biochemical profile, and intestinal microflora data were statistically processed by analysis of variance (ANOVA) followed by the Tukey test ($p < 0.05$) using SPSS software version 21 (IBM Corp, Armonk, NY, USA). Results are reported as means ± standard error of the mean (S.E.M.). Biochemical results (oxidative status) were analyzed by a two-way ANOVA followed by Tukey's multiple comparisons tests using GraphPad Prism software v9.3.1 (La Jolla, CA, USA). Differences were considered significant when $p < 0.05$ and the values are expressed as means ± S.E.M.

3. Results

3.1. Evaluation of Growth Performance

The growth performance of carp is summarized in Table 4. The results demonstrate a clear trend in the growth performance of common carp in response to varying levels of sorghum inclusion in their diets. As the proportion of sorghum increased from V0 to V3, there was a noticeable decline in both final body weight (FBW) and weight gain (WG). Specifically, carp fed with the V3 diet containing 30% sorghum exhibited the lowest

FBW and WG compared to those on other diets. However, no significant differences were observed in parameters such as feed conversion ratio (FCR), specific growth rate (SGR), condition factor (CF), protein efficiency ratio (PER), lipid efficiency ratio (LER), hepatosomatic index (HSI), and viscerosomatic index (VSI) across the experimental groups. These results indicate that while sorghum inclusion negatively impacted the growth performance of common carp in terms of final weight and weight gain, it did not significantly affect the other production traits measured in this study. Moreover, between the control group (V0) and the group with the lowest sorghum inclusion (V1), there were no significant differences observed in either FBW or WG. This suggests that the introduction of a small amount of sorghum (10%) into the diet did not markedly alter the growth performance of the common carp compared to the control group.

Table 4. Growth performance parameters of common carp fed sorghum experimental diets.

Parameter	V0 (0% Sorghum)	V1 (10% Sorghum)	V2 (20% Sorghum)	V3 (30% Sorghum)	ANOVA
IBW (g)	44.39 ± 1.33 ns	43.6 ± 0.92 ns	43.01 ± 1.26 ns	41.72 ± 1.44 ns	0.49
FBW (g)	76.54 ± 1.81 a	70.71 ± 1.52 ab	66.48 ± 1.74 b	65.46 ± 2.49 b	0.00
WG (g)	32.05 ± 2.26 a	26.8 ± 0.47 ab	23.07 ± 1.01 b	24.47 ± 1.36 b	0.01
FCR (g/g)	2.91 ± 0.49 ns	2.99 ± 0.45 ns	3.13 ± 0.39 ns	3.18 ± 0.37 ns	0.969
SGR (% day^{-1})	1.95 ± 0.19 ns	1.71 ± 0.03 ns	1.53 ± 0.02 ns	1.65 ± 0.12 ns	0.133
CF	2.29 ± 0.27 ns	1.94 ± 0.03 ns	1.96 ± 0.03 ns	1.95 ± 0.03 ns	0.214
PER (g/g)	1.94 ± 0.33 ns	1.78 ± 0.27 ns	1.65 ± 0.21 ns	1.51 ± 0.22 ns	0.699
LER (g/)	7.56 ± 1.3 ns	8.97 ± 1.36 ns	7.34 ± 0.95 ns	8.41 ± 1.22 ns	0.763
HSI (%)	0.69 ± 0.08 ns	0.8 ± 0.05 ns	0.75 ± 0.03 ns	0.77 ± 0.03 ns	0.525
VSI (%)	15.99 ± 2.51 ns	17.14 ± 3.03 ns	20.11 ± 4.3 ns	22.28 ± 3.11 ns	0.555

IBW—initial body weight, FBW—final body weight, WG—weight gain, FCR—feed conversion ratio, CF—condition factor, PER—protein efficiency ratio, LER—lipid efficiency ratio, HSI—hepatosomatic index, and VSI—viscerosomatic index. Different lowercase letters represent statistically significant differences according to Tukey's test at $p < 0.05$. 'ns' denotes non-significant differences.

3.2. Proximate Composition of Fish Meat

The biochemical composition of common carp flesh fed with sorghum experimental diets is summarized in Table 5. Notable variations were observed in protein and salt content across the experimental groups. Protein content increased with higher levels of sorghum inclusion, reaching the highest value in the V2 group (17.00%). Conversely, salt content exhibited significant differences among the groups, peaking in the V2 group (3.74%). Moisture, fat, ash, and collagen content showed no significant variations among the experimental groups ($p > 0.05$). These results suggest that sorghum inclusion primarily influences protein and salt content in common carp flesh, potentially impacting its nutritional profile.

Table 5. Biochemical composition of common carp flesh fed sorghum experimental diets.

Parameter (%)	V0 (0% Sorghum)	V1 (10% Sorghum)	V2 (20% Sorghum)	V3 (30% Sorghum)	ANOVA
Moisture	69.12 ± 0.51 ns	69.86 ± 0.33 ns	68.74 ± 0.88 ns	69.12 ± 0.59 ns	0.630
Protein	15.64 ± 0.31 b	16.22 ± 0.24 ab	17.00 ± 0.25 a	15.98 ± 0.16 b	0.008
Fat	6.28 ± 0.53 ns	5.32 ± 0.58 ns	6.56 ± 1.24 ns	5.60 ± 1.34 ns	0.798
Ash	2.76 ± 0.17 ns	3.14 ± 0.2 ns	3.14 ± 0.16 ns	3.04 ± 0.16 ns	0.392
Collagen	1.06 ± 0.09 ns	0.78 ± 0.15 ns	0.70 ± 0.28 ns	0.84 ± 0.13 ns	0.540
Salt	2.34 ± 0.12 b	3.46 ± 0.43 ab	3.74 ± 0.37 a	2.82 ± 0.34 ab	0.041

Different lowercase letters represent statistically significant differences according to Tukey's test at $p < 0.05$. 'ns' denotes non-significant differences.

3.3. Results of Blood Hematological Examination

The blood parameters of common carp are presented in Table 6. White blood cell count (WBC) remained consistent across all groups, with no significant differences observed

(p = 0.885). A slight lymphopenia associated with marked monocytosis and neutrophilia (doubling of the number of neutrophiles), suggestive for an inflammatory process due to sorghum inclusion, can be observed, however, these variations were not statistically significant. Eosinophil count (EOS) also exhibited slight variations across the groups, but these differences were not statistically significant (p = 0.297). Basophil count (BAS) and platelet count (PLT) remained stable across all groups, suggesting minimal impact on clotting function due to sorghum inclusion. Mean platelet volume (MPV) and platelets (PCT) did not exhibit significant differences between the groups, indicating consistent platelet size and volume across all dietary treatments.

Table 6. Effect of experimental diets with sorghum on hematological parameters in common carp.

Parameter	V0 (0% Sorghum)	V1 (10% Sorghum)	V2 (20% Sorghum)	V3 (30% Sorghum)	ANOVA
WBC ($X10^9/l$)	49.3 ± 2.36 ns	50.92 ± 2.56 ns	48.76 ± 4.42 ns	47.47 ± 2.64 ns	0.885
LYM ($X10^9/l$)	44.41 ± 2.02 ns	38.84 ± 1.50 ns	40.91 ± 6.08 ns	41.34 ± 5.69 ns	0.839
MON ($X10^9/l$)	0.37 ± 0.10 ns	3.96 ± 0.27 ns	3.74 ± 1.29 ns	1.61 ± 1.36 ns	0.077
NEU ($X10^9/l$)	6.25 ± 1.48 ns	8.12 ± 0.80 ns	4.12 ± 2.26 ns	4.52 ± 1.84 ns	0.376
EOS ($X10^9/l$)	0.81 ± 0.21 ns	0.61 ± 0.12 ns	0.48 ± 0.04 ns	0.49 ± 0.08 ns	0.297
BAS ($X10^9/l$)	0.04 ± 0.01 ns	0.03 ± 0.01 ns	0.02 ± 0.01 ns	0.02 ± 0.01 ns	0.524
PLT ($X10^9/l$)	42.33 ± 1.33 ns	45.33 ± 5.46 ns	45.33 ± 5.70 ns	34.00 ± 7.37 ns	0.453
MPV (fl)	9.33 ± 0.07 ns	9.4 ± 0.12 ns	9.67 ± 0.52 ns	9.40 ± 0.32 ns	0.878
PCT (%)	0.04 ± 0.00 ns	0.04 ± 0.00 ns	0.05 ± 0.01 ns	0.03 ± 0.01 ns	0.181

WBC (white blood cells), LYM (lymphocytes), MON (monocytes), NEU (neutrophils), EOS (eosinophils), BAS (basophils), PLT (platelets), MPV (mean platelet volume), PCT (plateletcrit). 'ns' denotes non-significant differences.

3.4. Intestinal Microbiota

The influence of sorghum experimental diets on the gut microbiota of carp is illustrated in Figure 1. Total aerobic bacteria (TNG-A) showed significant increases in V1, V2, and V3 compared to V0. Similarly, total anaerobic bacteria (TNG-AN) were notably higher in V2 and V3 compared to V0 and V1. No significant differences were observed in total sulfite-reducing clostridia (TN-SRC) and total enterobacteria (TN-E) across the groups ($p > 0.05$). The observed increases in both aerobic and anaerobic bacteria populations suggest a positive effect of sorghum inclusion on the gut microbiota of carp. This indicates that sorghum experimental diets may contribute to promoting a more diverse and potentially beneficial microbial community in the carp intestine.

3.5. Oxidative Stress

The activity of the antioxidant enzymes SOD, CAT, and GPX is presented in Figure 2. The variation in enzyme activity in common carp is influenced by both tissue type and the concentration of sorghum in the experimental diets. As sorghum concentration increased, there were notable alterations in enzyme activities, particularly with decreasing trends observed in SOD activity, especially in the muscle (2.7 USOD/mg protein) and intestine (1.4 USOD/mg protein) tissue of V2 ($p < 0.0001$) compared to the control (3.6 USOD/mg protein in muscle and 2.2 USOD/mg protein in intestine). SOD activity decreased in liver tissue as well, from 4.2 USOD/mg protein in the control to 3.4 USOD/mg protein in V2; however, the differences were statistically insignificant. We observed that the lowest activity was detected in intestinal tissue, followed by muscle and liver tissues.

CAT activity in the fish liver decreased in V1 in comparison with the control group. We registered an increase in CAT activity in muscle tissue in V3 (44.8 UCAT/mg protein) compared with the control (39.7 UCAT/mg protein). In intestinal tissue, the variations were insignificant.

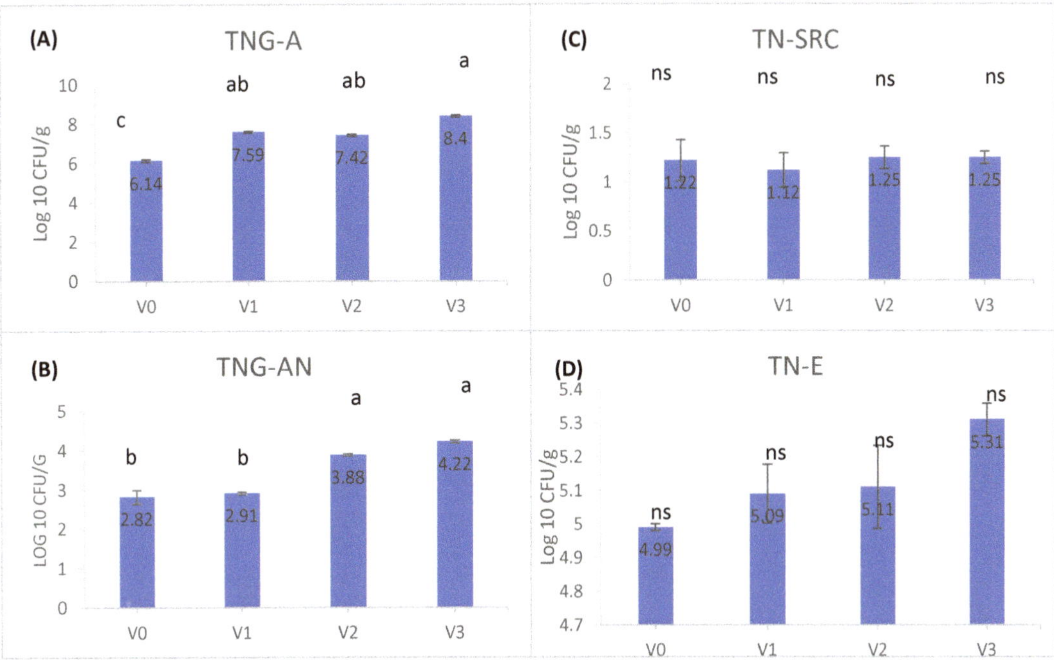

Figure 1. Graphical representation of the logarithm of the average number of CFU/g intestinal contents, for the indicators: (**A**) TNG-A (total number of aerobic bacteria), (**B**) TNG-AN (total number of anaerobic bacteria), (**C**) TN-SRC (total number of sulfite-reducing *Clostridia*), (**D**) TNE (total number of *Enterobacteriaceae*). Different lowercase letters represent statistically significant differences according to Tukey's test at $p < 0.05$. 'ns' denotes non-significant differences.

GPx in the liver of the control group was 4.5 UGPx/mg protein; the activity decreased in an inversely proportional relationship with the administered sorghum concentration (4.1 UGPx/mg protein in the case of the V1 variant, 3.2 UGPx/mg protein in the case of the V2 variant, and 2.3 UGPx/mg protein in the case of the group with 30% sorghum). The enzyme activity was almost identical in the samples taken from the intestinal tissue, the average value of GPx being 2.8 UGPx/mg protein in carp specimens from the control group and 2.9 UGPx/mg protein in V1. In the experimental variants where sorghum was administered in a concentration of 20 and 30%, the activity level reached thresholds of 1.3 UGPx/mg protein and 0.7 UGPx/mg protein. A completely different situation was observed at the muscle tissue level, where the highest average activity was recorded in V1 (3.3 UGPx/mg protein), while in variants V2 and V3, the activity of the enzyme was clearly inferior to that in both the control group (0.9 UGPx/mg protein in V2 and 0.7 UGPx/mg protein in V3, compared to 1.5 UGPx/mg protein in the control) and in V1.

The concentration of GSH in carp decreased in correlation with sorghum concentration in the diet (Figure 3). In the control variant, GSH values were 28.5 µg GSH/µg protein in the muscle, 41.7 µg GSH/µg protein in the liver, and 23.6 µg GSH/µg protein in the intestine. However, in variant V2 with 20% sorghum, average values dropped to 20.0 µg GSH/µg protein in the muscle, 24.3 µg GSH/µg protein in the liver, and 14.8 µg GSH/µg protein in the intestine. Notably, the group treated with 30% sorghum exhibited significantly lower GSH concentrations across all tissue types, with average values halved compared to the control group: 12.6 µg GSH/µg protein in the muscle, 21.5 µg GSH/µg protein in the liver, and 13.0 µg GSH/µg protein in the intestine.

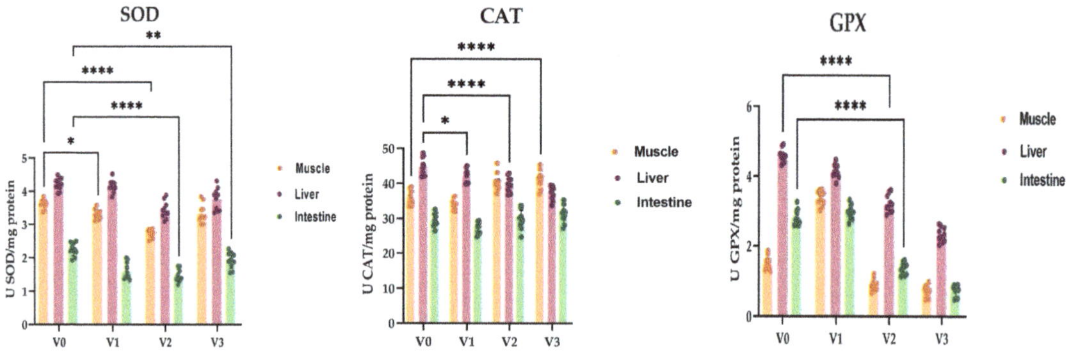

Figure 2. Sorghum influence on the oxidative status determined in muscle, liver, and intestine tissues of *Cyprinus carpio*. The enzymatic parameters consisted of measuring SOD, CAT, and GPX-specific activities. The values are expressed as means ± S.E.M. (n = 12). Two-way ANOVA analysis revealed overall significant differences between the experimental groups. For Tukey multiple comparisons analysis, **** $p < 0.0001$, ** $p < 0.01$, * $p < 0.05$.

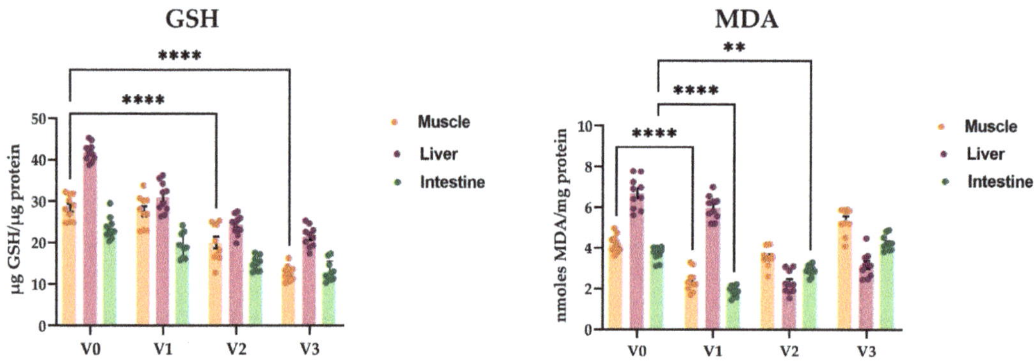

Figure 3. Sorghum influence on the oxidative status determined in muscle, liver, and intestine tissues of *Cyprinus carpio*. The non-enzymatic parameters consisted of estimating the levels of GSH and MDA. The values are expressed as means ± S.E.M. (n = 12). Two-way ANOVA analysis revealed overall significant differences between the experimental groups. For Tukey multiple comparisons analysis, **** $p < 0.0001$, ** $p < 0.01$.

Sorghum administration in carp diets had varied effects on MDA content in the analyzed tissues (Figure 3). In muscle tissue, the average MDA concentration in the reference group was 4.2 nmoles MDA/mg protein, decreasing in experimental variants V1 (2.3 nmoles MDA/mg protein) and V2 (3.5 nmoles MDA/mg protein). However, in carp fed with 30% sorghum, MDA levels were 1.27 times higher than in the control (5.3 nmoles MDA/mg protein). A similar trend was observed in intestinal tissue, with maximum MDA recorded in the control group (3.7 nmoles MDA/mg protein) and decreasing in the 10% sorghum concentration group (1.8 nmoles MDA/mg protein). Liver tissue exhibited the highest MDA concentrations, with the control and V1 variants showing maximum average values of 6.6 nmoles MDA/mg protein and 5.9 nmoles MDA/mg protein, respectively. Notably, the V2 variant displayed the lowest value, at 2.2 nmoles MDA/mg protein.

The experimental findings demonstrate a clear association between sorghum concentration in carp diets and alterations in both glutathione (GSH) concentration and malondialdehyde (MDA) content across different tissue types. As sorghum concentration increased, there was a significant decrease in GSH levels, indicating a potential disruption in antioxidant defense mechanisms, while MDA levels concurrently increased, suggesting elevated oxidative stress levels.

4. Discussion

The inclusion of sorghum in carp feed formulations at concentrations of 20% and 30% resulted in significant reductions in final body weight and weight gain. However, at 10% sorghum inclusion, these parameters were not significantly affected. Furthermore, weight gain, feed conversion ratio, and condition factor were not affected in any of the variants, even when corn and barley were completely replaced by sorghum at 30%. These findings are supported by the proximate composition of sorghum, which revealed a protein content of 14.31%, fat content of 3.97%, and starch content of 50.5%. This suggests that sorghum can be a viable and sustainable alternative in enhancing the efficiency and cost-effectiveness of carp farming operations. Based on the current study, sorghum is effective when comprising a total of 10% of the diet. Sorghum protein content varies based on environmental conditions, growing region, and soil type, ranging from 8 to 15% on a dry matter basis [2]. It can be classified into various soluble fractions, such as albumins, globulins, prolamins, kafirins, cross-linked kafirins, cross-linked glutelins, and unextracted protein residue [2]. Despite the diversity in amino acid profiles among sorghum varieties, sorghum proteins, like other plant proteins, tend to be lower in essential amino acids such as lysine, tryptophan, and threonine compared to animal-based ingredients [2]. While limited data are available, they suggest that carp, tilapia, and silver catfish can digest and metabolize sorghum-based diets [2,16,21]. In carp, the protein digestibility of sorghum was 71.86%, while fat digestibility was 76.71% [16]. In silver catfish (*Rhamdia quelen*), the inclusion of sorghum in diet led to increased FCR, likely linked to its adverse effects on nutrient absorption in the fish. The predominant protein in sorghum is kafirin, which exhibits a rigid structural configuration within the starch protein matrix, low solubility, and an imbalanced amino acid profile, contributing to reduced food digestibility [21]. The most common carp farming technology is in earthen ponds, in a semi-intensive system. In these systems, feeding is conducted using either pelleted feeds or a ground mixture of various ingredients, primarily cereals, but also animal- and plant-origin flours and meals. As feeding represents over 50% of production costs, identifying more sustainable ingredients is timely. While for super-intensive systems, the FCR (feed conversion ratio) can reach a value of 1, in semi-intensive systems such as those for carp, the FCR can reach values of 3.5 [22]; thus, creating cheap recipes with new ingredients and lower FCRs is of interest.

The proximate composition of fish meat, including moisture, protein, fat, ash, collagen, and salt content, reflects its nutritional quality and consumer appeal. While moisture content showed minimal variation among treatment groups (68.74% to 69.86%), significant differences were observed in protein levels, with the highest content in V2 (17%) and the lowest in the control group (15.64%), indicating the potential of dietary interventions to influence protein deposition in fish. These findings highlight the potential of dietary interventions to influence fish meat composition, impacting its sensory and nutritional properties. Additionally, Hussein et al. (2016) [23] found an increase in protein content in the meat of Nile tilapia (*Oreochromis niloticus*), while Obe (2014) [24] observed a similar increase in the meat of a catfish hybrid (*Heterobranchus bidorsalis* × *Clarias gariepinus*). Sorghum was the studied ingredient used in the diets in the previously mentioned studies.

Various factors, including diet composition, environmental conditions, and physiological status, can significantly influence hematological parameters in fish [25,26]. Leukocyte count (WBC) is an important parameter in the assessment of immune status in vertebrates, encompassing various cell types such as lymphocytes, monocytes, neutrophils, eosinophils, and basophils. In our study, leukocyte count notably decreased in the sorghum-treated variants, particularly in V2 and V3. Interestingly, monocyte levels increased in V2, indicating a potential inflammatory response. The cause of these variations may be represented by the effect of microbiota changes due to the action of feed compounds, favoring the multiplication of some species to the detriment of others, causing a dysmicrobism, with effects on cell populations. Other factors can be related to environmental parameters.

The gut microbiome of fish is extremely complex, consisting of aerobic, facultative anaerobic, and obligate anaerobic microorganisms varying between 107 and 1011 cells/gram of gut contents, according to some authors [27,28]. The microbiological indicators determined in this study, used as a quantitative benchmark in the evaluation of the effect induced by sorghum nutrients on the gut microbiota, are the main colonizers of the gastrointestinal tract of fish. Of course, the gut microflora varies depending on the complexity of the fish's digestive system, aquatic habitat, aquatic microflora, and feeding technology [29,30]. More and more studies show that the diet used in fish nutrition is related to the composition of the gut microbiome [31], which is closely related to the immune system, and together they are responsible for the overall health status. In the experimental conditions of our study, the results of the microbiological indicators in the experimental groups fed diets supplemented with 10% (V1), 20% (V2), and 30% (V3) sorghum were found to exceed the results of the control group fed the basal diet. The analysis of the four microbiological indicators clearly showed the dominance of aerobic/facultative anaerobic microbiota compared to strictly anaerobic microbiota in the intestine of *Cyprinus carpio*. A comparison of these results with those of the control group shows that the diet supplemented with different concentrations of sorghum provides favorable nutrients for the maintenance and even proliferation of commensal bacteria in the digestive tract. The presence of anaerobic bacteria in the digestive tract of fish is crucial for maintaining a balanced microbial ecosystem, and they have been frequently detected in the guts of carp and tilapia [29,31]. This anaerobic microbiota also include sulfite-reducing Clostridia, which are representative of monitoring water hygiene [32] and are of major public health concern when they contaminate food [33]. However, we can consider a balanced diet that favors a moderate proliferation of bacterial populations in the digestive tube of carp to be beneficial, as it increases protection against digestive diseases, given the known role of commensal microorganisms in the host's defense against transient pathogenic and nutrient-hungry microorganisms in the digestive tube [34].

Oxidative stress, the imbalance between the production of reactive oxygen species (ROS) and antioxidant defenses, can alter the structure and the function of mitochondrial and cellular components [35]. In fact, the term oxidative stress relates to an excess of pro-oxidative factors (ROS) over antioxidants. Due to having an unpaired electron, free radicals are unstable and highly reactive. A free radical reacts with susceptible compounds, including lipids, proteins, and/or DNA. All these reactions result in the formation of a new free radical, excessive concentrations of ROS causing major oxidative damage [36]. This so-called oxidative damage can have a pronounced impact on fitness by reducing fertility and accelerating ageing. Oxidative stress has been linked to many pathologies; these associations emphasize that a balance must be found between the relative abundance of ROS and antioxidants. Cells possess complex biochemical and genetic mechanisms to maintain such a balance, and it is clear that their disruption can have profound pathophysiological consequences [37]. In order to avoid excessive levels of oxidative damage following an increase in ROS production, organisms can increase their synthesis of endogenous antioxidant enzymes (superoxide dismutase, catalase, glutathione peroxidase) to counteract the action of ROS or replace molecular constituents susceptible to the action of ROS (e.g., polyunsaturated fatty acids) with less sensitive ones (e.g., monounsaturated fatty acids) [38,39].

Data in the literature show that sorghum has comparative nutritional value to other cereals in terms of chemical composition, protein, fat, carbohydrates, and non-starch polysaccharides, as well as bioactive components like vitamin B and fat-soluble vitamins (D, E, and K), dietary fibers, policosanols, iron, zinc, copper and manganese, tocols, phytosterols, γ-oryzanol, phytic acid, micronutrients, macronutrients, essential minerals, and non-nutrients like carotenoids and polyphenols [40,41]. Sorghum varieties are rich in bioactive phenolic compounds [41] such as ferulic acid, gallic acid, vanillic acid, caffeic acid, cinnamic acid, salicylic acid, p-coumarinic acid, luteolin, apigenin, and 3-deoxyanthocyanidins. These constituents contribute to the numerous health benefits of this grain, including anti-oxidative, anti-inflammatory, anti-proliferative, anti-diabetic, anti-atherogenic activities, antimicrobial properties, and anti-cancer effects, as well as free radical scavenging and high antioxidant activity [42–46]. Tannins in sorghum impact the availability of minerals, proteins, and starch. Due to their significant interaction with starch and particularly amylose, polymeric tannins are key in producing resistant starch. The tannin oligomers present human health benefits through their immunomodulatory, antiradical, vasodilatory, cardioprotective, antithrombotic, and anti-UV actions [47–49]. Others potential health benefits are risk reduction for obesity, dyslipidemia, hypertension, and promoting urination and blood circulation [50]. Due to these properties, but also to its agronomic advantages (high yield, drought tolerance, and low production costs), interest in sorghum for human and animal consumption has increased considerably worldwide [51]. In addition, sorghum stands out compared to other grains due to its high antioxidant activity [52].

Our experimental results showed wide variations in superoxide dismutase (SOD) activity depending on the investigated tissue and the experimental variant. While the control group exhibited SOD activity levels of 3.6 USOD/mg protein in muscle, 4.2 USOD/mg protein in liver tissue, and 2.2 USOD/mg protein in intestinal samples, a decrease in enzyme activity was observed in direct correlation with the added sorghum concentration in experimental variants, particularly in variant V2 with 20% sorghum. For variant V2, activity thresholds of 1.4 USOD/mg protein in the intestine, 2.7 USOD/mg protein in muscle, and 3.4USOD/mg protein in liver tissue were recorded. Superoxide dismutase is a critical enzyme in the body's first line of defense, an important and indispensable marker in the entire antioxidant defense strategy, especially in reference to the superoxide anion radical (*O2), which is perpetually generated in normal body metabolism, especially through the mitochondrial energy production pathway [53].

Our results also highlighted a decrease in CAT activity in fish liver in all experimental variants compared to the reference group, while a slight increase in activity was observed in muscle and intestinal tissue in variants with sorghum. In another series of experiments, we determined the activity of catalase, an enzyme that plays a crucial role in the adaptive response to H_2O_2 [54], suggesting that its role is somehow auxiliary to that of glutathione peroxidase [55]. Using radio-labeling and aminotriazole inhibition, it was shown that CAT takes care of 50% of the hydrogen peroxide, and glutathione peroxidase/reductases take care of the other half [56]. Moreover, the literature data indicate increased levels of the enzyme, obtained by overexpression in different organelles, which extended the lifespan of transgenic laboratory mice by 20%, the greatest increase being related to the mitochondria, where the highest levels of reactive oxygen species are generated [56–59].

The situation is different in the case of glutathione-peroxidase, an important cellular antioxidant enzyme found in the cytoplasm and mitochondria of eukaryotic cells that modulates the balance between necessary and harmful levels of reactive oxygen species, using glutathione (GSH) as a reducing agent. The GPx enzyme exhibited a completely different behavior in carp specimens from experimental variants, showing a close dependence on the type of tissue analyzed and the concentration of sorghum added to the diet. GSH concentration in *Cyprinus carpio* L. specimens decreased in correlation with the concentration of sorghum in the diet, with the lowest concentrations observed in the group treated with 30% sorghum across all tissue types analyzed. The literature data emphasizes that muscular exertion can cause an imbalance between prooxidants and antioxidants in skeletal muscle, with oxygen flow being able to increase considerably during powerful exercise, while in fish antioxidant enzyme activities appear to depend on mainly by their swimming ability and lifestyle rather than evolutionary group [60]. An important component of the antioxidant defense system in fish is glutathione (GSH), a tripeptide that directly or indirectly regulates the elimination of ROS and their reaction products [61,62], through a cumulative action with several other associated enzymes, such as GSH-peroxidase (EC 1.11.1.9), GSH-reductase (EC 1.6.4.2), and Glutathione-S-transferase (EC 2.5.1.18). The literature in the field highlights the importance of fish nutrition, as the enrichment of food with antioxidants is able to improve resistance to various stress conditions, increasing antioxidant capacity and ensuring superior qualities of aquaculture products [63].

In the final stage of our experiment, we determined the concentration of MDA, a small and reactive organic molecule that occurs ubiquitously among eukaryotes, formed by three carbon molecules with two aldehyde groups at the carbon 1 and carbon 3 positions. The measurement of malondialdehyde content has long been used as a lipid peroxidation marker in studies related to oxidative stress and redox signaling [64]. The administration of sorghum had varied influences on the content of MDA in the analyzed biological tissues, with significant changes observed in muscle, liver, and intestinal tissue samples. Scientific data indicate the significant influence that dietary supplementation with different agents has on the concentration of MDA in different tissues and organs, the highest values being found, as a rule, in the liver [65,66]. It should also be noted that sorghum is a grain rich in protein, lipids, and, above all, carbohydrates; the literature data emphasize that a diet rich in carbohydrates and lipids can cause a significant decrease in the concentration of MDA in fish and, implicitly, increase their physiological condition [41,67].

5. Conclusions

This study highlights the significant impact of including sorghum in the common carp's diets on various physiological parameters. While sorghum positively influenced protein and salt content in fish flesh and promoted intestinal microbiota diversity, it adversely affected growth performance and oxidative stress indicators. Decreased glutathione levels and increased malondialdehyde content suggest compromised antioxidant defense mechanisms and elevated oxidative stress levels with higher sorghum concentrations. These findings underscore the importance of carefully considering dietary composition in fish nutrition to optimize growth performance and overall health. Notably, at a 10% sorghum inclusion level, no significant effects on final body weight and weight gain were observed, suggesting a potential threshold where the detrimental impacts on growth performance may be mitigated.

Author Contributions: Conceptualization, C.-A.B., L.D. and M.B., methodology, C.-A.B., G.D., C.M.R., C.E.H., L.D., E.T.-C., D.A.Ș. and M.B.; software, M.B.; writing—original draft preparation, C.-A.B., G.D., C.M.R., C.E.H., L.D., E.T.-C., D.A.Ș. and M.B.; writing—review and editing, C.-A.B., G.D., C.M.R., C.E.H., L.D., E.T.-C., D.A.Ș. and M.B. All authors have read and agreed to the published version of the manuscript.

Funding: This research received no external funding.

Institutional Review Board Statement: The animal study protocol was approved by the Ethics Committee of the Unit Using Animals for Scientific Purposes, approval no. 60, 1 March 2023.

Informed Consent Statement: Not applicable.

Data Availability Statement: All data generated by this study are present in this article.

Conflicts of Interest: The authors declare no conflicts of interest.

References

1. FAO. *Brief to The State of World Fisheries and Aquaculture 2022—Towards Blue Transformation*; FAO: Rome, Italy, 2022.
2. Zarei, M.; Amirkolaei, A.K.; Trushenski, J.T.; Sealey, W.M.; Schwarz, M.H.; Ovissipour, R. Sorghum as a Potential Valuable Aquafeed Ingredient: Nutritional Quality and Digestibility. *Agriculture* **2022**, *12*, 669. [CrossRef]
3. Production from Aquaculture Excluding Hatcheries and Nurseries (from 2008 Onwards). Available online: https://ec.europa.eu/eurostat/databrowser/view/fish_aq2a/default/table?lang=en&category=fish.fish_aq (accessed on 1 February 2024).
4. Nyachoti, C.; Atkinson, J.; Leeson, S. Sorghum tannins: A review. *World's Poult. Sci. J.* **1997**, *53*, 5–21. [CrossRef]
5. Godfray, H.C.J.; Beddington, J.R.; Crute, I.R.; Haddad, L.; Lawrence, D.; Muir, J.F.; Pretty, J.; Robinson, S.; Thomas, S.M.; Toulmin, C. Food security: The challenge of feeding 9 billion people. *Science* **2010**, *327*, 812–818. [CrossRef]
6. Mohammed, N.A.; Ahmed, I.A.M.; Babiker, E.E. Nutritional evaluation of sorghum flour (*Sorghum bicolor* L. Moench) during processing of injera. *World Acad. Sci. Eng. Technol.* **2011**, *51*, 99–103.
7. Al-Ogaily, S.; Al-Asgah, N.; Ali, A. Effect of feeding different grain sources on the growth performance and body composition of tilapia, *Oreochromis niloticus* (L.). *Aquac. Res.* **1996**, *27*, 523–529. [CrossRef]

8. Gassem, M.A.; Osman, M.A. Proximate composition and the content of sugars, amino acids and anti-nutritional factors of three sorghum varieties. *Agric. Res. Cent. King Saud. Univ. Res. Bull.* **2003**, *125*, 5–19.
9. Spinoni, J.; Barbosa, P.; Cherlet, M.; Forzieri, G.; McCormick, N.; Naumann, G.; Vogt, J.V.; Dosio, A. How will the progressive global increase of arid areas affect population and land-use in the 21st century? *Glob. Planet. Change* **2021**, *205*, 103597. [CrossRef]
10. Leal Filho, W.; Fedoruk, M.; Paulino Pires Eustachio, J.H.; Barbir, J.; Lisovska, T.; Lingos, A.; Baars, C. How the War in Ukraine Affects Food Security. *Foods* **2023**, *12*, 3996. [CrossRef] [PubMed]
11. Popescu, A.; Dinu, T.A.; Stoian, E. Sorghum—An Important Cereal in the World, in the European Union and Romania. *Sci. Pap. Ser. Manag. Econ. Eng. Agric. Rural. Dev.* **2018**, *18*, 271.
12. D'mello, J. Antinutritional factors and mycotoxins. In *Farm Animal Metabolism and Nutrition*; CABI Publishing: Wallingford, UK, 2000; pp. 383–403.
13. Degani, G. Digestible energy in dietary sorghum, wheat bran, and rye in the common carp (*Cyprinus carpio* L.). *Isr. J. Aquac.* **2006**, *58*, 71–77. [CrossRef]
14. Frélichová, A.; Vačkář, D.; Pártl, A.; Loučková, B.; Harmáčková, Z.V.; Lorencová, E. Integrated assessment of ecosystem services in the Czech Republic. *Ecosystem Services* **2014**, *8*, 110–117. [CrossRef]
15. Dowling, L.; Arndt, C.; Hamaker, B. Economic viability of high digestibility sorghum as feed for market broilers. *Agron. J.* **2002**, *94*, 1050–1058. [CrossRef]
16. Aderolu, A.Z.; Kuton, M.P.; Odu-Onikosi, S.G. Substitution effect of sorghum meal for maize meal in the diet of catfish (*Clarias gariepinus*, Burchell, 1822) juvenile. *Res. J. Fis. Hyd* **2009**, *4*, 41–45.
17. Dec, S.; Thap, D. *Sorghum as Source of Ingredient to Replace Cassava Forfeeding Pangasius in Vietnam*; US Grains Council: Kuala Lumpur, Malaysia, 2017.
18. Babacariu, C.A.; Rîmbu, C.M.; Burducea, M.; Dîrvariu, L.; Miron, L.D.; Boiangiu, R.S.; Dumitru, G.; Todirașcu-Ciornea, E. Comparative Study of Flesh Quality, Blood Profile, Antioxidant Status, and Intestinal Microbiota of European Catfish (*Silurus glanis*) Cultivated in a Recirculating Aquaculture System (RAS) and Earthen Pond System. *Life* **2023**, *13*, 1282. [CrossRef]
19. Bradford, M.M. A rapid and sensitive for the quantitation of microgram quantitites of protein utilizing the principle of proteindye binding. *Anal. Biochem.* **1976**, *72*, 248–254. [CrossRef] [PubMed]
20. Barbacariu, C.-A.; Rimbu, C.M.; Dirvariu, L.; Burducea, M.; Boiangiu, R.S.; Todirascu-Ciornea, E.; Dumitru, G. Evaluation of DDGS as a Low-Cost Feed Ingredient for Common Carp (*Cyprinus carpio* Linneus) Cultivated in a Semi-Intensive System. *Life* **2022**, *12*, 1609. [CrossRef]
21. Rodrigues, M.L.; Sanchez, M.S.S.; Pessini, J.E.; Weiler, K.A.; Deparis, A.; Boscolo, W.R.; Bittencourt, F.; Signor, A. Replacement of corn by sorghum and phytase supplementation in silver catfish (*Rhamdia quelen*) diets: Growth performance, physiological variables and bone mineralization. *J. Appl. Anim. Res.* **2020**, *48*, 142–150. [CrossRef]
22. Woynarovich, A.; Moth-Poulsen, T.; Péteri, A. *Carp polyculture in Central and Eastern Europe, the Caucasus and Central Asia: A manual*; FAO Fisheries and Aquaculture Technical Paper No. 554; FAO: Rome, Italy, 2010; 73p.
23. Hussein, M.S.; Younes, A.E.A.; Ali, M.W.; Abd El-Azim, A.M.A. Effect of sorghum replacement and probiotic on growth performance and feed utilization of the Nile Tilapia (*Oreochromis niloticus*). *Egypt. J. Aquat. Biol. Fish.* **2016**, *20*, 113–121. [CrossRef]
24. Obe, B.W. Growth Performance and Nutrient Utilization of Catfish Hybrid (*Heterobranchus bidorsalis* X *Clarias gariepinus*) Fed Fermented Sorghum (*Sorghum bicolor*) Waste Meal Diets. *Int. J. Appl. Sci. Technol.* **2014**, *4*, 130.
25. Paul, T.; Shukla, S.P.; Kumar, K.; Poojary, N.; Kumar, S. Effect of temperature on triclosan toxicity in *Pangasianodon hypophthalmus* (Sauvage, 1878): Hematology, biochemistry and genotoxicity evaluation. *Sci. Total Environ.* **2019**, *668*, 104–114. [CrossRef]
26. Witeska, M.; Kondera, E.; Ługowska, K.; Bojarski, B. Hematological methods in fish—Not only for beginners. *Aquaculture* **2022**, *547*, 737498. [CrossRef]
27. Nayak, S.K. Role of gastrointestinal microbiota in fish. *Aquac. Res.* **2010**, *41*, 1553–1573. [CrossRef]
28. Llewellyn, M.S.; Boutin, S.; Hoseinifar, S.H.; Derome, N. Teleost microbiotas: The state of the art in their characterization, manipulation and importance in aquaculture and fisheries. *Front. Microbiol.* **2014**, *5*, 207. [CrossRef]
29. Cahill, M.M. Bacterial Flora of Fishes: A Review. *Microb. Ecol.* **1990**, *19*, 21–41. Available online: https://www.jstor.org/stable/4251098 (accessed on 6 February 2024). [CrossRef]
30. Butt, R.L.; Volkoff, H. Gut microbiota and energy homeostasis in fish. *Front. Endocrinol.* **2019**, *10*, 9. [CrossRef] [PubMed]
31. Luan, Y.; Li, M.; Zhou, W.; Yao, Y.; Yang, Y.; Zhang, Z.; Ringø, E.; Erik Olsen, R.; Liu Clarke, J.; Xie, S.; Research Progress and Potential Applications. *Engineering* **2023**, *29*, 137–146. [CrossRef]
32. Stelma, G.N., Jr. Use of bacterial spores in monitoring water quality and treatment. *J. Water Health* **2018**, *16*, 491–500. [CrossRef] [PubMed]
33. Beres, C.; Colobatiu, L.; Tabaran, A.; Mihaiu, R.; Mihaiu, M. Prevalence and Characterisation of *Clostridium perfringens* Isolates in Food-Producing Animals in Romania. *Microorganisms* **2023**, *11*, 1373. [CrossRef]
34. de Bruijn, I.; Liu, Y.; Wiegertjes, G.F.; Raaijmakers, J.M. Exploring fish microbial communities to mitigate emerging diseases in aquaculture. *FEMS Microbiol. Ecol.* **2017**, *94*, fix161. [CrossRef]
35. Beaulieu, M.; Haas, A.; Schaefer, M. Self-supplementation and effects of dietary antioxidants during acute thermal stress. *J. Exp. Biol.* **2014**, *217*, 370–375. [CrossRef]
36. Lichtenberg, D.; Pinchuk, I. Oxidative stress, the term and the concept. *Biochem. Biophys. Res. Commun.* **2015**, *461*, 441–444. [CrossRef]

37. Hayes, J.D.; Dinkova-Kostova, A.T.; Tew, K.D. Oxidative stress in cancer. *Cancer Cell* **2020**, *38*, 167–197. [CrossRef]
38. Pamplona, R.; Costantini, D. Molecular and structural antioxidant defenses against oxidative stress in animals. *Am. J. Physiol. Regul. Integr. Comp. Physiol.* **2011**, *301*, R843–R863. [CrossRef] [PubMed]
39. García-Caparrós, P.; De Filippis, L.; Gul, A.; Hasanuzzaman, M.; Ozturk, M.; Altay, V.; Lao, M.T. Oxidative Stress and Antioxidant Metabolism under Adverse Environmental Conditions: A Review. *Bot. Rev.* **2021**, *87*, 421–466. [CrossRef]
40. Xiong, Y.; Zhang, P.; Warner, R.D.; Shen, S.; Fang, Z. Cereal grain-based functional beverages: From cereal grain bioactive phytochemicals to beverage processing technologies, health benefits and product features. *Crit. Rev. Food Sci. Nutr.* **2020**, *62*, 2404–2431. [CrossRef]
41. Tanwar, R.; Panghal, A.; Chaudhary, G.; Kumari, A.; Chhikara, N. Nutritional, Phytochemical and Functional Potential of Sorghum: A Review. *Food Chem. Adv.* **2023**, *3*, 100501. [CrossRef]
42. Xu, J.; Wang, W.; Zhao, Y. Phenolic Compounds in Whole Grain Sorghum and Their Health Benefits. *Foods* **2021**, *10*, 1921. [CrossRef]
43. Chhikara, N.; Abdulahi, B.; Munezero, C.; Kaur, R.; Singh, G.; Panghal, A. Exploring the nutritional and phytochemical potential of sorghum in food processing for food security. *Nutr. Food Sci.* **2018**, *49*, 318–332. [CrossRef]
44. Xu, S.; Shen, Y.; Xu, J.; Qi, G.; Chen, G.; Wang, W.; Sun, X.; Li, Y. Antioxidant and anticancer effects in human hepatocarcinoma (HepG2) cells of papain-hydrolyzed sorghum kafirin hydrolysates. *J. Funct. Foods* **2019**, *58*, 374–382. [CrossRef]
45. Serna-Loaiza, S.; Miltner, A.; Miltner, M.; Friedl, A. A review on the feedstocks for the sustainable production of bioactive compounds in biorefineries. *Sustainability* **2019**, *11*, 6765. [CrossRef]
46. Xiong, Y.; Damasceno Teixeira, T.V.; Zhang, P.; Warner, R.D.; Shen, S.; Fang, Z. Cellular Antioxidant Activities of Phenolic Extracts from Five Sorghum Grain Genotypes. *Food Biosci.* **2021**, *41*, 101068. [CrossRef]
47. Abdelhalim, T.S.; Abdalla, A.A.; Sir Elkhatim, K.A.; Othman, M.H.; Alkhair, T.M.M.; Almaiman, S.A.; Almaiman, S.A.; Osman, M.A.; Hassan, A.B. Indigenous Sudanese sorghum-based food: Secondary metabolites and antioxidant activities of traditional Sudanese nonalcoholic beverage Hulu-mur from two sorghum landraces. *Food Sci. Nutr.* **2023**, *11*, 2654–2662. [CrossRef] [PubMed]
48. Xu, W.T.; An, Q.; Ruan, L.H.; Zhou, F.; Zhou, L.; Peng, M.; Li, L.I.; Yang, X.S.; Li, Q.J.; Yang, J. Potential antiplatelet aggregation metabolites from the discarded sorghum (*Sorghum bicolor* L.) root. *Nat. Prod. Res.* **2023**, *37*, 967–973. [CrossRef] [PubMed]
49. Seo, J.W.; Lee, J.G.; Ghimire, B.K.; Kim, M.J.; Yu, C.Y.; Seong, E.S. Characterization of antioxidant and anti-inflammatory activities in the grains of collected Sorghum bicolor varieties. *J. Appl. Biol. Chem.* **2023**, *66*, 227–235. [CrossRef]
50. Aluko, O.O.; Ajayi, O.B. Investigation of antilipidemic efficacy of condensed tannins from three varieties of *Sorghum bicolor* seeds on high-fat diet-induced obese rats. *Comp. Clin. Pathol.* **2023**, *32*, 847–858. [CrossRef]
51. Pinheiro, S.S.; Anunciação, P.C.; de Morais Cardoso, L.; Della Lucia, C.M.; de Carvalho, C.W.P.; Queiroz, V.A.V.; Sant'Ana, H.M.P. Stability of B vitamins, vitamin E, xanthophylls and flavonoids during germination and maceration of sorghum (*Sorghum bicolor* L.). *Food Chem.* **2021**, *345*, 128775. [CrossRef]
52. Ghimire, B.K.; Seo, J.W.; Yu, C.Y.; Kim, S.H.; Chung, I.M. Comparative Study on Seed Characteristics, Antioxidant Activity, and Total Phenolic and Flavonoid Contents in Accessions of *Sorghum bicolor* (L.) Moench. *Molecules* **2021**, *26*, 3964. [CrossRef] [PubMed]
53. Ighodaro, O.M.; Akinloye, O.A. First line defence antioxidants-superoxide dismutase (SOD), catalase (CAT) and glutathione peroxidase (GPX): Their fundamental role in the entire antioxidant defence grid. *Alex. J. Med.* **2018**, *54*, 287–293. [CrossRef]
54. Ransy, C.; Vaz, C.; Lombès, A.; Bouilland, F. Use of H_2O_2 to Cause Oxidative Stress, the Catalase Issue. *Int. J. Mol. Sci.* **2020**, *21*, 9149. [CrossRef]
55. Jemec, A.; Drobne, D.; Tišler, T.; Sepčić, K. Biochemical biomarkers in environmental studies-lessons learnt from enzymes catalase, glutathione S-transferase and cholinesterase in two crustacean species. *Environ. Sci. Pollut. Res.* **2010**, *17*, 571–581. [CrossRef]
56. Tehrani, H.S.; Moosavi-Movahedi, A.A. Catalase and its mysteries. *Prog. Biophys. Mol. Biol.* **2018**, *140*, 5–12. [CrossRef] [PubMed]
57. Barbacariu, C.-A.; Burducea, M.; Dîrvariu, L.; Oprea, E.; Lupu, A.-C.; Teliban, G.-C.; Agapie, A.L.; Stoleru, V.; Lobiuc, A. Evaluation of diet supplementation with wheat grass juice on growth performance, body composition and blood biochemical profile of carp (Cyprinus carpio L.). *Animals* **2021**, *11*, 2589. [CrossRef] [PubMed]
58. Burducea, M.; Dincheva, I.; Dirvariu, L.; Oprea, E.; Zheljazkov, V.D.; Barbacariu, C.-A. Wheat and Barley Grass Juice Addition to a Plant-Based Feed Improved Growth and Flesh Quality of Common Carp (*Cyprinus carpio*). *Animals* **2022**, *12*, 1046. [CrossRef]
59. Burducea, M.; Lobiuc, A.; Dirvariu, L.; Oprea, E.; Olaru, S.M.; Teliban, G.-C.; Stoleru, V.; Poghirc, V.A.; Cara, I.G.; Filip, M.A.; et al. Assessment of the fertilization capacity of the aquaculture sediment for wheat grass as sustainable alternative use. *Plants* **2022**, *11*, 634. [CrossRef] [PubMed]
60. Vélez-Alavez, M.; De Anda-Montañez, J.A.; Galván-Magaña, F.; Zenteno-Savín, T. Comparative study of enzymatic antioxidants in muscle of elasmobranch and teleost fishes. *Comp. Biochem. Physiol. Part. A Mol. Integr. Physiol.* **2015**, *187*, 61–65. [CrossRef] [PubMed]
61. Srikanth, K.; Pereira, E.; Duarte, A.C.; Ahmad, I. Glutathione and its dependent enzymes' modulatory responses to toxic metals and metalloids in fish-a review. *Environ. Sci. Pollut. Res.* **2013**, *20*, 2133–2149. [CrossRef] [PubMed]
62. Sukhovskaya, I.V.; Borvinskaya, E.V.; Smirnov, L.P.; Kochneva, A.A. Role of Glutathione in Functioning of the System of Antioxidant Protection in Fish (Review). *Inland. Water Biol.* **2017**, *10*, 97–102. [CrossRef]

63. Ming, J.; Ye, J.; Zhang, Y.; Yang, X.; Shao, X.; Qiang, J.; Xu, P. Dietary optimal reduced glutathione improves innate immunity, oxidative stress resistance and detoxification function of grass carp (*Ctenopharyngodon idella*) against microcystin-LR. *Aquaculture* **2019**, *498*, 594–605. [CrossRef]
64. Morales, M.; Munné-Bosch, S. Malondialdehyde: Facts and artifacts. *Plant Physiol.* **2019**, *180*, 1246–1250. [CrossRef]
65. Karanth, J. Oxidative stress and antioxidant status in rat blood, liver and muscle: Effect of dietary lipid, carnitine and exercise. *Int. J. Vitam. Nutr. Res.* **2005**, *75*, 333–339. [CrossRef]
66. Hafez, A.; Nassef, E.; Fahmy, M.; Elsabagh, M.; Bakr, A.; Hegazi, E. Impact of dietary nano-zinc oxide on immune response and antioxidant defense of broiler chickens. *Environ. Sci. Pollut. Res.* **2020**, *27*, 19108–19114. [CrossRef] [PubMed]
67. Wang, L.N.; Liu, W.B.; Lu, K.L.; Xu, W.N.; Cai, D.S.; Zhang, C.N.; Qian, Y. Effects of dietary carbohydrate/lipid ratios on non-specific immune responses, oxidative status and liver histology of juvenile yellow catfish Pelteobagrus fulvidraco. *Aquaculture* **2014**, *426*, 41–48. [CrossRef]

animals

Article

Effects of Dietary Inclusion of Enzymatically Hydrolyzed Compound Soy Protein on the Growth Performance and Intestinal Health of Juvenile American Eels (*Anguilla rostrata*)

Yichuang Xu [1,2], Chengyao Wu [1,2], Pan Wang [1,2], Xiaozhao Han [1,2], Jinyue Yang [3] and Shaowei Zhai [1,2,*]

1 Fisheries College, Jimei University, Xiamen 361021, China; xuyichuang@jmu.edu.cn (Y.X.); 54chengyaowu@sina.com (C.W.); 202311908028@jmu.edu.cn (P.W.); 202414908011@jmu.edu.cn (X.H.)
2 Engineering Research Center of the Modern Technology for Eel Industry, Ministry of Education of China, Xiamen 361021, China
3 College of Ocean Food and Biological Engineering, Jimei University, Xiamen 361021, China; 202461000113@jmu.edu.cn
* Correspondence: zhaisw@jmu.edu.cn; Tel.: +86-592-6181420

Simple Summary: The drive to diminish reliance on fishmeal and reduce feed costs has pushed the development of fishmeal alternatives. In recent years, the use of enzymatically hydrolyzed soybean as a fishmeal substitute in aquafeeds has garnered growing attention and application. This study demonstrated that appropriate levels of dietary enzymatically hydrolyzed compound soy protein (EHCS), as a substitute for white fishmeal, significantly enhanced growth performance and improved the intestinal health of American eels. Based on the quadratic regression analysis of weight gain rate and feed efficiency against the dietary EHCS inclusion levels, the optimal inclusion levels for American eels ranged from 7.78% to 8.33% with maximum levels spanning 17.59% to 17.77%, which corresponded to white fishmeal replacement levels of 24.31–26.03% and 54.97–55.53%, respectively. To ensure normal growth performance, it is recommended that the inclusion level of EHCS should not exceed 17.77%. Our findings offer a valuable reference for the application of EHCS as a white fishmeal substitution in the diets of American eels.

Abstract: The enzymatic hydrolysis of soybeans could enhance their application as an ingredient and alternative to fishmeal in aquafeeds. Here, a 10-week feeding trial was conducted to evaluate the impacts of different dietary inclusion levels of enzymatically hydrolyzed compound soy protein (EHCS) on the growth performance and intestinal health of juvenile American eels (*Anguilla rostrata*). Five experimental diets were formulated with graded EHCS inclusion levels at 0% (EHCS0), 8% (EHCS8), 16% (EHCS16), 24% (EHCS24), and 32% (EHCS32). Each diet was randomly assigned to four replicate tanks. The results showed that eels fed the EHCS8 diet exhibited superior growth performance, decreased serum lipid content, and increased immunity compared to those fed the EHCS0 diet. Eels fed the EHCS8 diet also displayed improved intestinal histology, enhanced antioxidant capacity and balance of intestinal microbiota as well as an enhanced proliferation of probiotics compared to those receiving the EHCS0 diet. Compared with eels fed the EHCS0 diet, those fed the EHCS16 diet exhibited comparable growth performance and values for the aforementioned markers. The quadratic regression analysis of weight gain rate and feed efficiency against the dietary EHCS inclusion levels determined the maximum levels of dietary EHCS inclusion for American eels range from 17.59% to 17.77%.

Keywords: enzymatically hydrolyzed soy protein; growth performance; intestinal health; American eels

1. Introduction

Fishmeal is an optimal protein source in aquafeeds due to its high protein content, well-balanced amino acid composition, and excellent palatability [1]. Nevertheless, the escalating scarcity of marine fishery resources coupled with the rapid development of aquaculture have led to a supply shortage of fishmeal and a consequent rise in fishmeal prices, which raised the cost of aquafeeds [2,3]. Therefore, the quest to reduce reliance on fishmeal and lower feed costs have propelled the development of cost-effective fishmeal substitutions to the forefront of research in the aquaculture [4–6].

Among the numerous protein sources, soybean is commonly utilized as a fishmeal substitution in aquafeeds owing to its ideal protein content, rich amino acid profile, abundant availability, and low cost [7,8]. However, untreated soybean contains various antinutritional factors that not only impair protein digestibility for aquatic animals but also have adverse effects on the health of aquatic animals, limiting their use in aquafeeds [7,8]. Therefore, the elimination of antinutritional factors in soybeans is advantageous for their application in aquafeeds. Enzymatically hydrolysis is particularly noteworthy as a technique employed to reduce antinutritional factors due to its mild conditions and high efficiency, which result in the effective removal of antinutritional factors and minimal nutrient loss in soybean [6,9]. Additionally, the large molecular proteins, starch, cellulose, and mannans present in soybeans undergo enzymatic hydrolysis into smaller peptides, amino acids, glucose, maltose, and other nutrients that are readily absorbable [10]. Therefore, compared with untreated soybean, enzymatically hydrolyzed soybean exhibits a significant reduction in antinutritional factor content and an enhanced nutritional value.

In recent years, the utilization of enzymatically hydrolyzed soybean as a fishmeal substitute in aquafeeds has seen increasing attention and application [4–6]. Huang et al. [6] demonstrated that replacing up to 45% of dietary fishmeal with enzymatically hydrolyzed soybean did not adversely impact the growth performance of juvenile Chinese mitten crabs (*Eriocheir sinensis*). Tibaldi et al. [4] also revealed that substituting 50% of dietary fishmeal with enzymatically hydrolyzed soybean meal has no adverse effects on the growth performance and whole-body composition of European sea bass (*Dicentrarchus labrax*). Moreover, hydrolyzing soybean using a multi-enzyme strategy exhibits a higher efficacy in substituting fishmeal in aquafeeds compared to its single-enzyme counterpart [5]. It was reported that protease-treated soybean meal could substitute for 20% of fishmeal in the diets of largemouth bass (*Micropterus salmoides*), whereas a combination treatment with protease and non-starch polysaccharide enzymes could facilitate soybean meal replacing up to 47.27% of dietary fishmeal for largemouth bass [5]. Therefore, soybean protein hydrolyzed through multiple enzymes represents a highly efficacious substitute for fishmeal in aquafeeds.

Eels (*Anguilla* spp.) are widely distributed throughout the world, and American eels (*Anguilla rostrata*) represent one of the primary cultured eel species in southern China. Eels have a high protein requirement, and the highest-grade white fishmeal traditionally serves as the primary protein source in their diets. Due to the escalating costs of fishmeal, there has been a significant focus on investigating substitutes for fishmeal in eel diets. Considering that the potential of soybean hydrolyzed by a combination of multiple enzymes to replace fishmeal in feed, this study was carried out to assess the effects of substituting dietary white fishmeal with enzymatically hydrolyzed compound soy protein (EHCS), treated with a combination of multiple enzymes, on the growth performance and intestinal health of juvenile American eels. Our findings are anticipated to offer valuable insights into the application of EHCS in the diets for American eels, as well as potentially for other fish species, to reduce the feed cost and promote the development of low/non-fishmeal diets in the near future.

2. Materials and Methods

2.1. Ethics Statement

This study on American eels culture and management adhered to the Management Rule of Laboratory Animals (Chinese Order No. 676 of the State Council, revised 1 March 2017). All necessary measures were taken to minimize animal suffering. The Jimei University Ethics Committee approved the research protocols (identification code: 2021-11 and approved in 11 March 2021).

2.2. Diet Preparation and Fish Culture

Five isonitrogenous and isolipidic experimental diets were prepared, incorporating enzymatically hydrolyzed compound soy protein (EHCS) at graded levels of 0% (EHCS0), 8% (EHCS8), 16% (EHCS16), 24% (EHCS24), and 32% (EHCS32), corresponding to progressive replacement levels of white fishmeal at 0%, 25%, 50%, 75%, and 100%, respectively (Table 1). The EHCS was sourced from Xiamen Longli Industry and Trade Co., Ltd., Xiamen, China. The EHCS comprised a mixture of soybean meal, soy protein concentrate, and soy protein isolate in a 4:5:1 ratio, which was enzymatically hydrolyzed with alkaline protease (750 IU/g substrate), neutral protease (250 IU/g substrate), and acid protease (250 IU/g substrate) at a controlled temperature of 4 °C for a duration of 24 h. For the preparation of experimental diets, the ingredients were first crushed using a feed grinder (ZFJ-300, Jiangyin Ruizong Machinery Manufacturing Co., Ltd., Wuxi, China); then, they were sieved through an 80 μm mesh, thoroughly mixed, and stored at −20 °C until feeding. Prior to feeding, the powdered diet was hydrated with water at a ratio of 1:1.2 to attain a dough-like consistency. The amino acid composition of the experimental diets and the levels of antinutritional factors in EHCS are detailed in Tables 2 and 3, respectively.

After a 4-week acclimatization period, a total of 500 health American eels (mean initial weight: 26.00 ± 0.02 g per fish, mean ± S.D.) were randomly assigned to the EHCS0 group, EHCS8 group, EHCS16 group, EHCS24 group, and EHCS32 group, respectively. Each tank contained 25 fish with four replicates per group. The feeding trial lasted for 10 weeks. An automatic temperature control system with flow through was utilized throughout the trial. All fish were fed to satiation twice daily at 07:00 and 17:00. The uneaten feed was collected to assess the feed consumption. The water quality parameters were monitored manually twice a week and are detailed as follows: temperature 25.3–27.5 °C, pH 7.2–8.2, dissolved oxygen 6.5–8.5 mg/L, total ammonia nitrogen ≤ 0.12 mg/L, nitrite ≤ 0.02 mg/L.

Table 1. Formulation and proximate composition analysis (% of dry weight basis) of the basal diet.

Ingredients (%)	**EHCS0**	**EHCS8**	**EHCS16**	**EHCS24**	**EHCS32**
EHCS	0.00	8.00	16.00	24.00	32.00
White fishmeal	32.00	24.00	16.00	8.00	0.00
Brown fishmeal	40.00	40.00	40.00	40.00	40.00
α-starch	26.00	25.50	25.00	24.50	24.00
Fish oil	0.50	1.00	1.50	2.00	2.50
Choline chloride	0.50	0.50	0.50	0.50	0.50
$Ca(H_2PO_4)_2$	0.50	0.50	0.50	0.50	0.50
Mineral premix [a]	0.10	0.10	0.10	0.10	0.10
Vitamin premix [b]	0.40	0.40	0.40	0.40	0.40
Total	100.00	100.00	100.00	100.00	100.00
Proximate analysis %, dry weight					
Moisture	7.24	7.47	7.68	7.95	7.84
Crude protein	48.93	48.64	48.19	48.21	47.77
Crude lipid	6.72	6.58	6.60	6.45	6.54
Ash	11.80	10.80	9.87	8.63	7.61
Calcium	3.08	2.70	2.30	1.86	1.39
Phosphorus	2.04	1.91	1.71	1.50	1.28
Total energy	18.56	18.65	18.75	18.88	19.07

[a] Minerals premix (mg kg^{-1}diet): Na_2SeO_3 0.89, $Cu_2(OH)_3Cl$ 12.05, $Ca(IO_3)_2$ 2.59, $FeSO_4 \cdot H_2O$ 666.67, $MnSO_4 \cdot H_2O$ 94.34, $CoSO_4$ 3.64, $ZnSO_4 \cdot H_2O$ 202.89. [b] Vitamin premix (IU or mg/kg diet): Vitamin A 3 333.33 IU, VC 33.33 mg, VD_3 416.67 IU, VE 15.00 mg, VK_3 1.00 mg, VB_1 2.67 mg, VB_2 5.00 mg, VB_6 2.00 mg, VB_{12} 0.01 mg, nicotinic acid 23.33 mg, folate acid 0.33 mg, biotin 0.05 mg, calcium pantothenate 10.00 mg, inositol 33.33 mg.

Table 2. Amino acid levels of trial diets (% of dry weight basis).

	EHCS0	EHCS8	EHCS16	EHCS24	EHCS32
Essential amino acid (EAA)					
Lysine	3.71	3.63	3.55	3.47	3.38
Methionine	1.29	1.26	1.23	1.20	1.17
Threonine	2.13	2.06	1.99	1.92	1.85
Tryptophan	0.53	0.54	0.54	0.55	0.56
Leucine	3.55	3.50	3.44	3.39	3.34
Arginine	2.72	2.72	2.71	2.71	2.71
Histidine	1.60	1.58	1.56	1.54	1.52
Valine	2.38	2.33	2.28	2.22	2.17
Isoleucine	2.02	2.00	1.98	1.97	1.95
Phenylalanine	1.73	1.75	1.78	1.81	1.84
Total EAA	21.65	21.36	21.07	20.77	20.48
Non-essential amino acid (NEAA)					
Aspartic acid	4.31	4.31	4.31	4.31	4.31
Cysteine	0.47	0.48	0.48	0.49	0.50
Tyrosine	1.72	1.66	1.59	1.52	1.46
Serine	1.93	1.92	1.90	1.89	1.87
Glutamate	6.07	6.18	6.29	6.39	6.50
Proline	1.91	1.90	1.89	1.87	1.86
Glycine	2.96	2.81	2.66	2.52	2.37
Alanine	2.92	2.80	2.69	2.57	2.45
Total NEAA	22.30	22.05	21.81	21.56	21.31

Table 3. The contents of antinutritional factors in compound soy protein treated with or without enzymatic hydrolysis (mg g^{-1}).

	Before Enzymatic Hydrolysis	After Enzymatic Hydrolysis
Trypsin inhibitor factor	1.75	1.30
Glycinin	23.60	<2.00
β-conglycinin	31.90	3.00

2.3. *Sample Collection*

At the end of the feeding trial, the eels were fasted for 24 h. The number and weight of eels in each tank was recorded. Eels were anesthetized using eugenol 50 mg/L in 1:4 dilution with rearing water for about 5 min. The collected serum samples were stored at −80 °C for subsequent biochemical analysis. Anterior intestine samples from four eels per tank were sampled and rapidly frozen in liquid nitrogen and stored at −80 °C for determining intestinal digestive enzyme activity and antioxidant-related indices. Anterior intestine samples from two eels per tank were fixed in Bouin's solution and stored at 4 °C for morphological observation. The anterior intestines of six eels per tank were rapidly frozen in liquid nitrogen and then stored at −80 °C for intestinal microbiota analysis. Ten eels from each tank were randomly selected for whole-body composition analysis.

2.4. *Growth Performance*

The final body weight (FBW), weight gain rate (WGR), specific growth rate (SGR), feed intake (FI), feed efficiency (FE), feed conversion ratio (FCR), protein efficiency ratio (PER), and survival rate (SR) were calculated in accordance with methodologies described in previous studies [6,11].

2.5. *Proximate Composition, Amino Acid Composition, and Antinutritional Factors*

The moisture content of whole-body fish and experimental diets samples were measured by a freeze-dryer (Labogene, Lillerød, Denmark). The contents of crude protein, crude lipid, ash, calcium (Ca), and phosphorus (P) were evaluated based on the description of the AOAC [12]. The levels of trypsin inhibitor factor (TIF, #KTI-EA), glycinin (#11S-EA), and β-conglycinin (#7S-EA) were analyzed with the utilization of kits (Tianjin Longke Xinyu Biotechnology Co., Ltd., Tianjin, China). The amino acid composition was measured according to the "Determination of amino acids in feeds" (GB/T 18246-2019) published by the Standardization Administration of China in 10 December 2019 [13].

2.6. Serum Biochemical Indexes

Alkaline phosphatase (AKP, #A059-2-2), complement 3 (C3, #H186-1-2), diamine oxidase (DAO, #A088-2-1), D-lactic acid (D-lac, #A019-3-1), glutamic-oxalacetic transaminase (GOT, #C010-1-1), glutamic-pyruvic transaminase (GPT, #C009-1-1), high-density lipoprotein cholesterol (HDL-C, #A112-1-1), immunoglobulin M (IgM, #A109-1-2), low-density lipoprotein cholesterol (LDL-C, #A113-1-1), lysozyme (LZM, #A050-1-1), total cholesterol (TC, #A111-1-1), and triglyceride (TG, #A110-1-1) were determined with the utilization of kits (Nanjing Jiancheng Bioengineering Co., Ltd., Nanjing, China).

2.7. Activities of Intestinal Digestive Enzymes

The intestinal protease activity was evaluated using the Folin phenol reagent. The intestinal activities of amylase (#C016-1-1) and lipase (#A0542-1) were measured using commercial kits (Nanjing Jiancheng Bioengineering Co., Ltd.).

2.8. Intestinal Antioxidant Parameters

The intestinal activities of superoxide dismutase (SOD, #A001-1-1), catalase (CAT, #A007-1-1), and glutathione peroxidase (GSH-PX, #A005-1-2), as well as the total antioxidant capacity (T-AOC, #A015-2-1) and the level of malondialdehyde (MDA, #A003-1-2), were quantified using commercial assay kits (Nanjing Jiancheng Biotechnology Co., Ltd., Nanjing, China).

2.9. Histological Observation

The preparation of intestinal tissue sections was carried out based on the methodology described by Xu et al. [14]. The tissue sections were stained with hematoxylin and eosin (H&E) and subsequently observed under a microscope (BX80-JPA, Olympus, Tokyo, Japan) for capturing representative micrographs. Morphometric analysis was conducted using Image-Pro Plus 6.0 software (Media Cybernetics, Silver Spring, MD, USA).

2.10. Intestinal Microbiota Profiling

The intestinal microbiota analysis was based on the description from Lu et al. [15]. In summary, total bacterial DNA was extracted from American eel intestinal samples. The forward (338F, 5′-ACTCCTACGGGAGGCAGCAG-3′) and reverse (806R, 5′-GACTACHVGG GTWTCTAAT-3′) primers were employed for PCR amplification of the V3-V4 hypervariable region of the bacterial 16S rRNA gene. High-throughput sequencing was then conducted using the Illumina Miseq PE300 platform (Beijing Allwegene Tech. Co., Ltd., Beijing, China).

2.11. Statistical Analysis

The data were presented as mean ± standard deviation (S.D., $n = 4$). One-way ANOVA was performed using SPSS 26.0 (SPSS, Chicago, IL, USA), and Duncan's method was used for multiple comparisons. Differences were considered significant at $p < 0.05$. When the analyzed data were expressed as percentages, and the overall rate was less than 30% or greater than 70%, a square root arcsine transformation was applied before statistical analysis. QIIME (v1.8.0) software was used to analyze the alpha diversity indices of the intestinal microbiota. PLS-DA was used for beta diversity analysis, and Lefse analysis was conducted using Python v2.7 software with an LDA score threshold of 3 and $p < 0.05$.

3. Results

3.1. Growth Performance

As shown in Table 4, the WGR, SGR, FE, and PER values were higher for American eels fed the EHCS8 diet than those fed the EHCS0 diet. There were no significant differences in the WGR, SGR, FE, FCR, and PER values between American eels fed the EHCS16 diet and the EHCS0 diet (Table 4). American eels fed the EHCS24 and EHCS32 diets showed lower WGR, SGR, FE, FCR, and PER values compared to those fed the EHCS0 diet (Table 4). American eels fed the EHCS32 diet also exhibited lower SRs than those fed other four diets (Table 4). Based on the quadratic regression analysis of WGR and FE against the dietary EHCS inclusion levels, the optimal levels of dietary EHCS inclusion were 7.78% and 8.33%, respectively, while the maximum levels that exhibited comparable effect of the diet without EHCS inclusion were 17.77% and 17.59%, respectively, for American eels (Figure 1A,B).

Table 4. Effects of dietary EHCS inclusion on the growth performance of juvenile American eel.

	EHCS0	EHCS8	EHCS16	EHCS24	EHCS32
IBW (g)	26.00 ± 0.01	25.99 ± 0.02	26.01 ± 0.04	26.00 ± 0.01	26.01 ± 0.02
FBW (g)	49.90 ± 0.73 ^{c}	52.77 ± 0.83 ^{d}	49.96 ± 0.52 ^{c}	46.61 ± 0.66 ^{b}	42.86 ± 0.81 ^{a}
WGR (%)	91.91 ± 2.78 ^{c}	103.06 ± 3.29 ^{d}	92.06 ± 2.04 ^{c}	79.32 ± 2.63 ^{b}	64.76 ± 3.04 ^{a}
SGR (%/d)	0.93 ± 0.02 ^{c}	1.01 ± 0.02 ^{d}	0.93 ± 0.02 ^{c}	0.84 ± 0.02 ^{b}	0.71 ± 0.03 ^{a}
FI (g)	32.15 ± 0.78 ^{c}	34.32 ± 1.48 ^{d}	31.87 ± 0.63 ^{c}	30.16 ± 0.26 ^{b}	28.08 ± 0.12 ^{a}
FE (%)	74.34 ± 0.84 ^{c}	78.07 ± 1.29 ^{d}	75.14 ± 0.42 ^{c}	68.37 ± 2.44 ^{b}	60.01 ± 2.87 ^{a}
FCR	1.35 ± 0.02 ^{a}	1.28 ± 0.02 ^{a}	1.33 ± 0.01 ^{a}	1.46 ± 0.05 ^{b}	1.67 ± 0.08 ^{c}
PER (%)	154.42 ± 1.74 ^{c}	162.91 ± 2.71 ^{d}	157.49 ± 0.87 ^{cd}	143.94 ± 5.14 ^{b}	126.91 ± 6.08 ^{a}
SR (%)	95.00 ± 2.00 ^{b}	96.00 ± 3.27 ^{b}	96.00 ± 3.27 ^{b}	93.00 ± 8.25 ^{ab}	87.00 ± 3.83 ^{a}

All values are shown as mean ± S.D. (n = 4). Letters (a–d) denote significance at $p < 0.05$. IBW, initial body weight; FBW, final body weight; WGR, weight gain rate; SGR, specific growth rate; FI, feed intake; FE, feed efficiency; FCR, feed conversion ratio; PER, protein efficiency ratio; SR, survival rate.

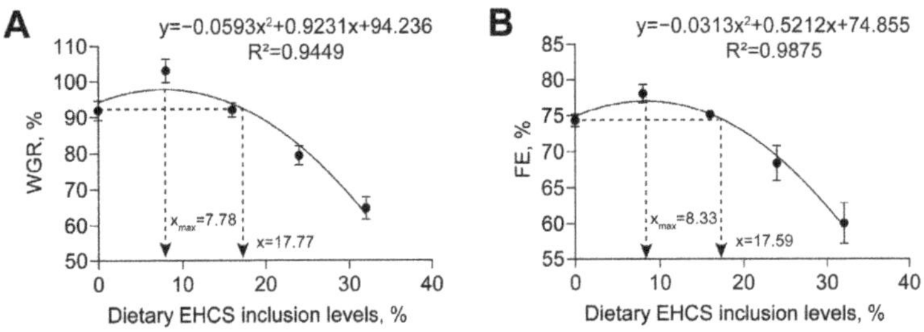

Figure 1. The relationship between dietary inclusion EHCS levels and the WGR (**A**) and FE (**B**) of juvenile American eels. Values are shown as mean ± S.D. (n = 4).

3.2. Body Composition

As shown in Table 5, the different dietary inclusion levels of EHCS did not affect the levels of moisture, crude protein, crude lipid, ash, calcium, and phosphorus in the whole body of juvenile American eels.

Table 5. Effects of dietary EHCS inclusion on the body composition of juvenile American eels (wet weight basis).

	EHCS0	EHCS8	EHCS16	EHCS24	EHCS32
Moisture (%)	68.43 ± 1.05	67.99 ± 1.10	69.09 ± 0.61	70.15 ± 2.40	69.24 ± 1.20
Crude protein (%)	17.95 ± 0.64	18.49 ± 0.50	18.06 ± 0.60	17.71 ± 0.25	18.07 ± 0.65
Crude lipid (%)	9.13 ± 0.34	9.04 ± 0.52	9.01 ± 0.30	8.64 ± 0.36	8.68 ± 0.23
Ash (%)	2.06 ± 0.25	1.95 ± 0.13	2.03 ± 0.22	2.12 ± 0.23	2.26 ± 0.07
Calcium (g kg^{-1})	4.35 ± 0.15	4.33 ± 0.17	4.24 ± 0.13	4.18 ± 0.11	4.36 ± 0.19
Phosphorus (g kg^{-1})	3.56 ± 0.50	3.49 ± 0.49	3.64 ± 0.26	3.74 ± 0.43	4.05 ± 0.28

Values are shown as mean ± S.D. (n = 4).

3.3. Serum Biochemical Parameters

The effects of dietary EHCS inclusion on the serum biochemical parameters of American eels are presented in Figure 2. The serum activities of GOT, GPT, and DAO, along with the serum levels of TG and D-lac, initially decreased and then increased as the dietary inclusion levels of EHCS increased (Figure 2). American eels fed the EHCS8 diets exhibited the lowest serum activities of GOT, GPT, and DAO, and the lowest serum contents of TG and D-lac, compared to those fed the other four diets (Figure 2). The serum activities of AKP and LZM, alongside the serum levels of lgM and C3, initially increased and subsequently decreased as dietary inclusion levels of EHCS increased (Figure 2). American eels fed the EHCS8 diet exhibited the highest serum activities of AKP and LZM, along with the serum levels of lgM and C3, compared to those fed the other four diets (Figure 2). The serum TC content was higher, and the serum HDL-C content was lower, for eels fed the EHCS24 and EHCS32 diets with no significant differences among eels fed the other three diets (Figure 2). The serum LDL-C content was the highest for eels fed the EHCS32 diets and showed no significant differences among the eels that were fed the other four diets (Figure 2).

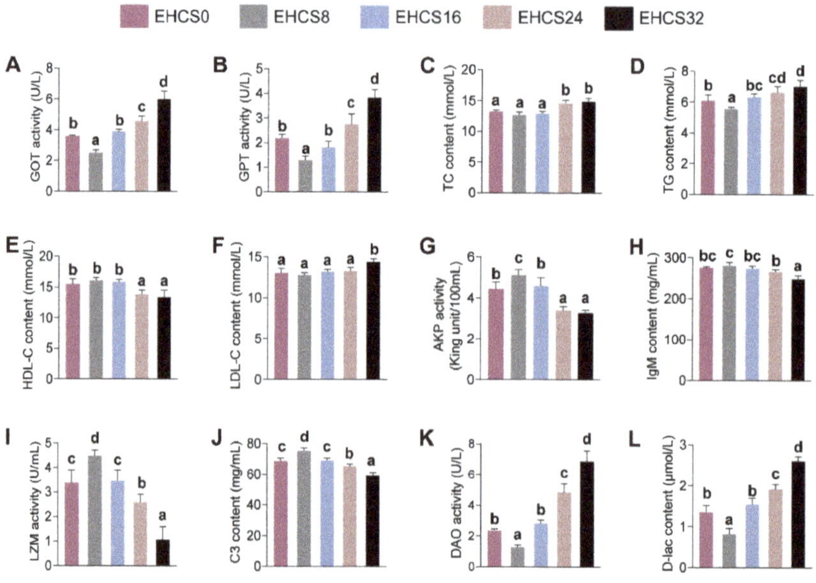

Figure 2. Effects of dietary EHCS inclusion on the serum biochemical parameters of juvenile American eels. (**A**) GOT activity. (**B**) GPT activity. (**C**) TC content. (**D**) TG content. (**E**) HDL-C content. (**F**) LDL-C content. (**G**) AKP activity. (**H**) lgM content. (**I**) LZM activity. (**J**) C3 content. (**K**) DAO activity. (**L**) D-lac content. Values are shown as mean ± S.D. (n = 4). Letters (a–d) denote significance at $p < 0.05$.

AKP, acid phosphatase; C3, complement 3; D-lac, D-lactate; DAO, diamine oxide; GOT, glutamic-oxalacetic transaminase; GPT, glutamic-pyruvic transaminase; HDL-C, high-density lipoprotein cholesterol; IgM, immunoglobulin M; LDL-C, low-density lipoprotein cholesterol; LZM, lysozyme; TC, total cholesterol; TG, triglyceride.

3.4. Intestinal Histology

The effects of dietary EHCS inclusion on the intestinal histology are illustrated in Figure 3. American eels fed the EHCS8 diet exhibited increased muscular thickness (MT), whereas those fed the EHCS24 and EHCS32 diets exhibited decreased MT and villus height (VH) compared to those fed the EHCS0 diet (Figure 3). There was no significant difference in MT and VH between American eels fed the EHCS0 and EHCS16 diets (Figure 3).

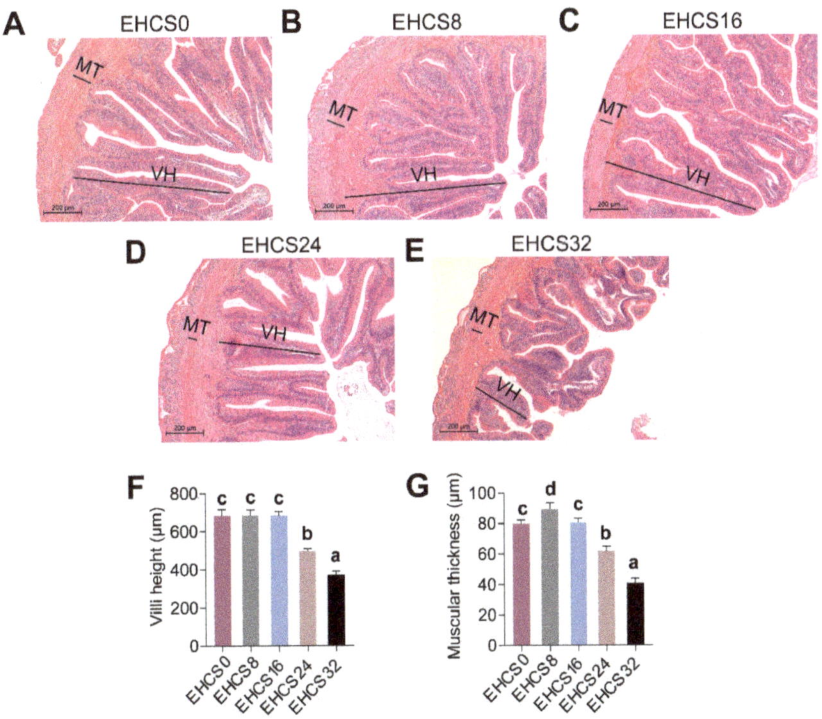

Figure 3. Effects of dietary EHCS inclusion on intestinal histology of juvenile American eels. (**A–E**) The representative photomicrographs of intestine sections of hematoxylin and eosin (H&E) staining, scale bar, 200 μm. VH, villus height. MT, muscular thickness. (**F**,**G**) The analysis of villus height and muscular thickness. Values are shown as mean ± S.D. (n = 4). Letters (a–d) denote significance at $p < 0.05$.

3.5. Intestinal Digestive Enzymes Activities

The effects of dietary EHCS inclusion on the intestinal digestive enzyme activities are illustrated in Figure 4. American eels fed the EHCS8 and EHCS16 diets showed no significant differences in lipase and protease activities, whereas those fed the EHCS24 and EHCS32 diets exhibited lower lipase and protease activities compared to those fed the EHCS0 diet (Figure 4A,B). Different dietary inclusion levels of EHCS had no effects on amylase activity (Figure 4C).

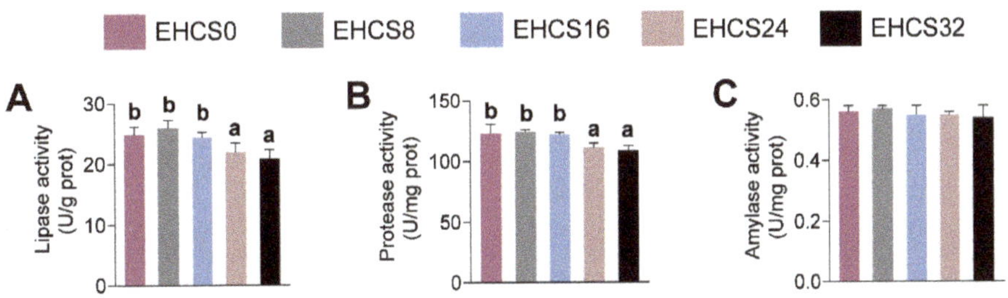

Figure 4. Effects of dietary EHCS inclusion on the activities of intestinal amylase (**A**), lipase (**B**), and proteases (**C**) activities of juvenile American eels. Values are shown as mean ± S.D. (n = 4). Letters (a–b) denote significance at $p < 0.05$.

3.6. Intestinal Antioxidant Capacity

The impacts of dietary EHCS inclusion on the intestinal antioxidant capacity are shown in Figure 5. American eels fed the EHCS8 diet exhibited elevated T-AOC levels and increased activities of CAT, T-SOD and GSH-PX, while those fed the EHCS24 and EHCS32 diets exhibited reduced T-AOC levels and diminished activities of CAT, T-SOD and GSH-PX compared to those fed the EHCS0 diet (Figure 5A–D). American eels fed the EHCS8 and EHCS16 diets showed no significant differences in MDA content, whereas those fed the EHCS24 and EHCS32 diets exhibited elevated MDA content in comparison to those fed the EHCS0 diet (Figure 5E).

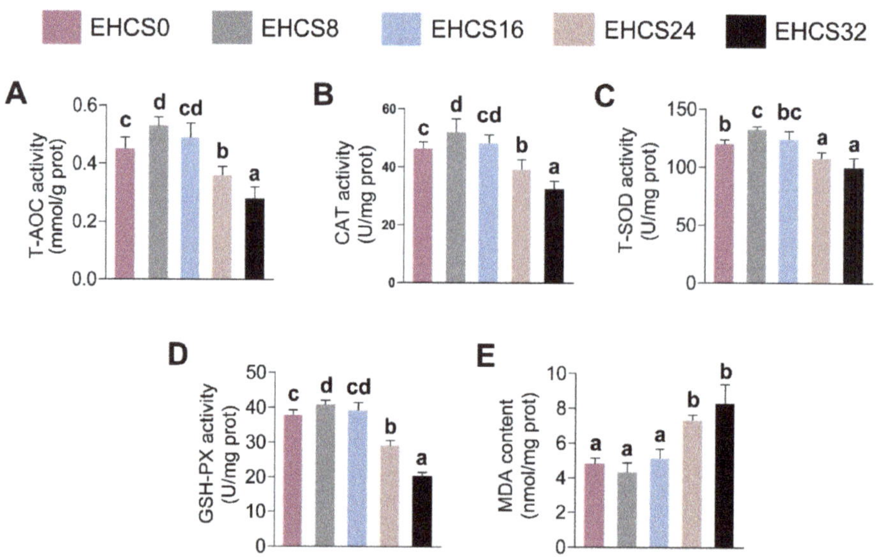

Figure 5. Effects of dietary EHCS inclusion on the intestinal antioxidant capacity of the juvenile American eels. (**A**) Total antioxidant capacity (T-AOC); (**B**) catalase (CAT) activity; (**C**) total superoxide dismutase (SOD) activity; (**D**) glutathione peroxidase (GSH-Px) activity; (**E**) malondialdehyde (MDA) level. Values are shown as mean ± S.D. (n = 4). Letters (a–d) denote significance at $p < 0.05$.

3.7. Intestinal Microbiota

Considering that American eels fed the EHCS8 diet exhibited superior growth performance and intestinal health, while those fed the EHCS32 diet experienced growth arrest and intestinal injury compared to those fed the EHCS0 diet, an analysis of intestinal microbiota was carried out among American eels fed these three diets to examine whether different dietary EHCS inclusion levels affect the microbiota composition of American eels. As shown in Figure 6A, there was no significant difference in the alpha diversity of the intestinal flora among eels fed the EHCS0, EHCS8, and EHCS32 diets. PLS-DA analysis revealed that the beta diversity of the intestinal microbiota among eels fed the EHCS0, EHCS8, and EHCS32 diets was dispersed, indicating differences in intestinal microbiota composition among the three groups (Figure 6B). Compared to American eels fed the EHCS0 diet, the relative abundance of Cyanobacteria was increased in those fed the EHCS8 diet, and no difference was observed between eels fed the EHCS0 and EHCS32 diets (Figure 6C,D). Compared to American eels fed the EHCS0 diet, the relative abundances of *Paraclostridium*, *Serratia*, and *Ralstonia* were significantly decreased in those fed the EHCS8 and EHCS32 diets; the relative abundances of *Chryseobacterium*, *Acidithiobacillaceae_KCM_B_112*, and *Pseudomonas* were significantly increased in those fed the EHCS8 diet; and the relative abundance of *Brevundimonas* was significantly increased in those fed the EHCS32 diet (Figure 6E). These findings suggest that different dietary inclusion levels of EHCS significantly modulate the gut microbial community structure in American eels.

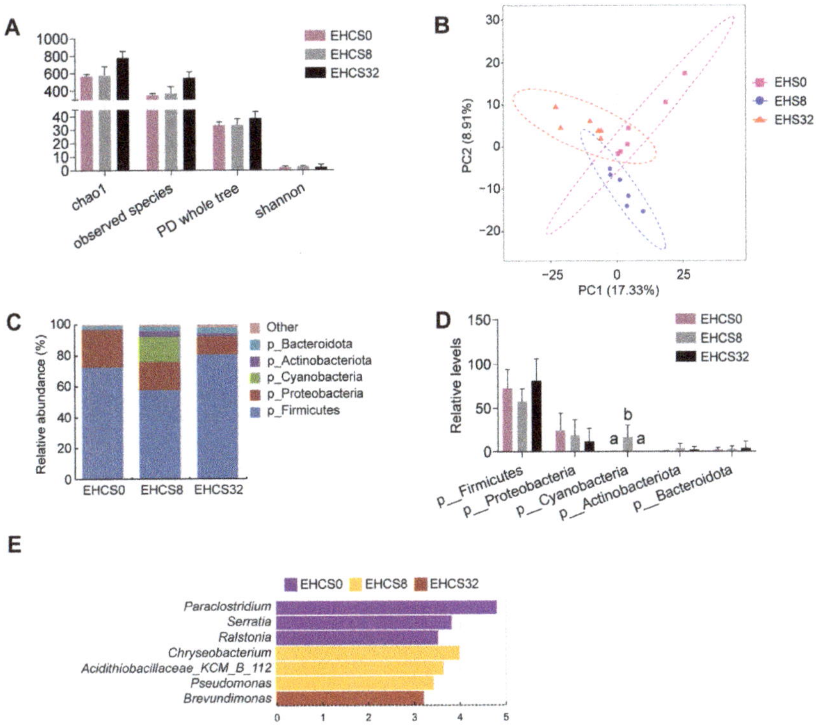

Figure 6. Effects of dietary EHCS inclusion on the intestinal microbiota. (**A**) Alpha diversity of the intestinal flora. Values are shown as mean ± S.D. (**B**) Beta diversity of the intestinal flora. (**C**,**D**) The relative abundance of the intestinal bacterial phyla. Letters (a,b) denote significance at $p < 0.05$. (**E**) The Lefse analysis of the intestinal bacterial genus.

4. Discussion

4.1. Growth Performance

Previous studies on substituting fish meal with enzymatically hydrolyzed soybean protein sources in aquafeeds primarily focused on single protein sources with limited research on the replacement of fish meal with EHCS in the diet of fish. The present study demonstrated that the WGR, SGR, FE, and PER of American eels initially increased and subsequently decreased with incremental levels of EHCS inclusion. Moreover, this study determined that the optimal and maximum levels of dietary EHCS inclusion ranged from 7.78% to 8.33% and from 17.59% to 17.77%, respectively. It was also revealed that appropriate levels of dietary enzymatically hydrolyzed soybean protein as a replacement for fishmeal promoted growth performance, whereas excessive levels inhibited growth in other fish species, including Japanese flounder (*Paralichthys olivaceus*) and starry flounder (*Platichthys stellatus*) [16,17]. An enzymatic hydrolysis of soy protein has been shown to increase the proportions of small peptides and free amino acids, enhancing the nutritional utilization and growth performance of aquatic animals [6,18,19]. However, excessive levels of enzymatically hydrolyzed soybean protein inclusion would lead to an imbalanced amino acid composition and poor palatability of the diet, which adversely affect the growth of fish [16]. Although enzyme treatment reduces the contents of antinutritional factors in enzymatically hydrolyzed soybean protein, increasing its dietary levels still results in high contents of antinutritional factors, ultimately suppressing the growth of fish [19]. Taken together, EHCS is suitable as fishmeal substitute in diets for the growth of American eels and other aquatic animals when used at moderate levels.

4.2. Body Composition

This study demonstrated that increasing levels of dietary EHCS inclusion did not result in significant changes in the whole-body composition of American eels. Research on the effects of EHCS inclusion in fish feed on body composition is limited at present. Previous studies indicated that substituting dietary fish meal with enzymatically hydrolyzed soybean meal did not lead to significant alterations in the whole-body composition of flounder [17]. However, high levels of dietary enzymatically hydrolyzed soy protein replacing fishmeal were shown to significantly increase moisture levels and markedly decrease ash levels in the whole body of totoaba (*Totoaba macdonaldi*) [18]. Therefore, the impact of different dietary sources of enzymatically hydrolyzed soy protein inclusion on the proximate composition of the whole-body of fish exhibits significant variability.

4.3. Serum Biochemical Parameters

Increased serum GOT and GPT activities commonly serve as pivotal signs of liver damage [20]. Our study demonstrated that serum GOT and GPT activities decreased in response to the optimal level of dietary EHCS inclusion, indicating a hepatoprotective effect of optimal dietary EHCS levels. Conversely, excessive levels of dietary EHCS inclusion led to an increase in these two enzymes activities, suggesting a detrimental impact of excessive EHCS on liver function. Song et al. [21] reported that starry flounder fed diets with increasing levels of dietary enzymatically hydrolyzed soy protein replacing fishmeal exhibited analogous changes in serum GOT and GPT activities. Previous studies revealed that an appropriate level of dietary small peptides reduced the serum GOT and GPT activities [22]. Additionally, the enzymatic hydrolysis of soy protein has been shown to increase the proportions of small peptides [17], potentially contributing to the observed reduction in serum GOT and GPT activities of American eels. Our results showed that excessive levels of EHCS inclusion resulted in a deficiency of essential amino acids in the diet, which was shown to result in increased activities of serum GOT and GPT [23]. Furthermore, the antinutritional factors present in EHCS were not completely eliminated. Elevated levels of dietary EHCS inclusion led to an accumulation of these antinutritional factors, which could ultimately result in liver damage [24].

TC and TG are mainly synthesized in the liver and constitute essential components of serum lipids, while LDL-C and HDL-C are critical lipoproteins facilitating cholesterol transport within the body [25,26]. LDL-C is responsible for transporting cholesterol to peripheral tissues, while the primary function of HDL-C is to remove excess cholesterol and low-density lipoproteins from the bloodstream and cells [25]. In the current study, the serum levels of TG initially decreased and then increased as the level of dietary EHCS inclusion increased. The contents of TC and LDL-C were higher, and the HDL-C content was lower, for eels fed diets with excessive inclusion levels of EHCS. These findings suggested that American eels fed a diet with an optimal inclusion level of EHCS exhibited decreased levels of serum lipids and enhanced hepatic lipolysis, whereas an excessive inclusion level of EHCS disturbed hepatic lipid metabolism in American eels. Previous studies revealed that liver injury impaired its ability in lipid metabolism, which ultimately resulted in an increase in serum lipid levels [27,28]. The aforementioned results of serum GOT and GPT activities indicated that an optimal level of dietary EHCS enhanced liver function, while excessive levels of dietary EHCS impaired liver function. These changes in liver function likely contributed to the observed alterations in serum levels of TC, TG, LDL-C, and HDL-C in the present study. However, it was observed that starry flounder fed diets with progressively higher levels of dietary enzymatically hydrolyzed soy protein replacing fishmeal exhibited a reduction in the levels of TG, HDL-C, and LDL-C [17]. Thus, the tolerance of different fish species to dietary enzymatically hydrolyzed soy protein sources in terms of hepatic lipid metabolism varies significantly.

Serum AKP, IgM, C3, and LZM serve as crucial biomarkers for evaluating immune function. AKP contributes to the enhancement of nonspecific immunity [29]. IgM plays a pivotal role in the humoral immune response and is a critical immunoglobulin [30]. C3 not only possesses bactericidal and complement-activating properties but also functions in immune regulation [31]. LZM is involved in the clearance and phagocytosis of pathogenic microorganisms [30]. This investigation revealed that the LZM and AKP activities, along with the concentrations of IgM and C3, initially exhibited an uptrend, which was followed by a subsequent decline as the proportion of dietary EHCS inclusion increased. Therefore, American eels fed diets with an appropriate inclusion level of EHCS exhibited enhanced immunity, whereas an excessive application of EHCS detrimentally affected immune function. Additionally, it was evidenced that Chinese mitten crab (*Eriocheir sinensis*) fed diets wherein fishmeal was increasingly substituted with enzymatically hydrolyzed soybean exhibited similar alterations in immune function [6]. Prior investigations indicated that an optimal level of soybean peptides was able to enhance the nonspecific immunity and cellular viability in immune cells [32]. However, excessive levels of dietary EHCS inclusion resulted in the accumulation of antinutritional factors, which could ultimately lead to immune disorders [33]. Therefore, EHCS in moderate levels serves as a suitable substitute for white fishmeal in diets to support the immune function of American eels and other aquatic animals.

4.4. Intestinal Health

The intestinal tract is crucial for the absorption and digestion of nutrients [14,15]. The structural integrity and permeability, along with the activity of digestive enzymes and antioxidant capacity, serve as critical indicators for assessing intestinal health [14,15]. The thickening of the intestinal muscular layer and the presence of intestinal villi improve the efficiency of intestinal digestion and absorption of nutrients [14,15]. The intestinal lipase, protease, and amylase play a pivotal role in the digestion and absorption of proteins, lipids, and starch in the diets of fish [34]. Additionally, serum DAO activity and D-lac content are commonly utilized to evaluate intestinal mucosal permeability [14,35]. T-AOC serves as an important indicator for evaluating antioxidant capacity [36]. T-SOD, CAT, and GSH-P_X are key antioxidant enzymes responsible for scavenging oxygen free radicals in fish [14,37]. MDA is a lipid peroxidation by-product and a marker for oxidative stress [14,34]. This study demonstrated that American eels fed diets with an appropriate inclusion level of EHCS exhibited enhanced the structural integrity and antioxidant capacity of the intestines. It was revealed that soybean peptides exert protective effects on intestinal mucosa integrity and possessed chemical antioxidant activities, which may underlie the observed anxo-action of an optimal dietary EHCS inclusion level on the structural integrity and permeability, as well as antioxidant capacity, of intestinal tissue [38,39]. Our findings also indicated that excessive dietary EHCS inclusion levels are detrimental to the structural integrity, digestion and absorption of nutrients, and antioxidant capacity of American eels. Dan et al. [40] also reported that turbot juveniles fed diets with excessive replacement levels of fishmeal with α-galactosidase hydrolytic soybean meal showed decreases in MT and VH along with reduced intestinal digestive enzymes activities. Previous studies indicated that the antinutritional factors present in soybean and imbalanced amino acid composition were capable of reducing intestinal antioxidant capacity, leading to an overproduction of reactive oxygen species (ROS) and disruption of the intestinal mucosal structure [33,41–44]. Therefore, optimal levels of dietary EHCS inclusion enhanced the intestinal structure and antioxidant capacity, whereas excessive EHCS inclusion levels compromised it due to the accumulation of antinutritional factors present in EHCS and imbalanced amino acid composition in the diet.

The intestinal flora establishes a stable micro-ecosystem with the host, which performs critical physiological functions in nutrient metabolism, pathogen defense, and immune regulation, thus playing an indispensable role in growth and development [45,46]. In this study, compared to American eels fed the EHCS0 diet, those fed the EHCS8 diet exhibited a lower relative abundance of *Paraclostridium*, *Serratia*, and *Ralstonia*, alongside a higher relative abundance of *Chryseobacterium*, *Acidithiobacillaceae_KCM_B_112*, and *Pseudomonas*, in the intestines. *Paraclostridium* can reduce intestinal mucosal permeability, strengthen the mucosal barrier function, and prevent the colonization and proliferation of pathogenic bacteria [47]. *Serratia* is an opportunistic pathogen that poses a threat to the health of aquatic animals and is one of the primary pathogens responsible for diarrhea [48]. *Ralstonia* consists of various opportunistic pathogens, and its increased relative abundance can induce intestinal inflammation in the host [49]. *Chryseobacterium* and *Pseudomonas* were shown to promote nutrient digestion and absorption [50]. *Acidithiobacillaceae_KCM_B_112* is a member of the family *Acidithiobacillaceae* that is capable of activating elemental sulfur and converting it into bioavailable nutrients [51]. The above results indicate that the optimal dietary EHCS inclusion level is able to decrease the relative abundance of potential pathogenic bacteria while increasing the relative abundance of beneficial bacteria in the intestines of juvenile American eels, improving the intestinal microbiota balance, which may promote nutrient absorption and enhance immunity. Compared to American eels fed the EHCS0 diet, those fed the EHCS32 diet exhibited a lower relative abundance of *Paraclostridium*, *Serratia*, and *Ralstonia* but a higher relative abundance of *Brevundimonas* in the intestines. *Brevundimonas* is a potential pathogen capable of disrupting the intestinal microbiota balance and causing intestinal inflammation [52]. Although the above findings demonstrate that American eels fed diets with excessive EHCS inclusion exhibited an increased abundance of certain probiotics in the intestines compared with those fed the control diet, the aforementioned indicators related to intestinal health suggest that the probiotic effect of the elevated probiotics was insufficient to mitigate the intestinal damage resulting from the excessive dietary EHCS inclusion. Moreover, the elevated abundance of *Brevundimonas* may exacerbate the intestinal damage caused by excessive dietary EHCS inclusion.

In comparison with previous studies, the proportion of fishmeal substituted by enzymatically hydrolyzed soybean in this experiment was relatively modest. Firstly, the fishmeal substituted in this trial was premium-grade white fishmeal. Eels exhibit a stringent requirement for fishmeal quality in their diet, which presents challenges in substituting premium-grade fishmeal in eel diets. Furthermore, our findings indicate that an excessive inclusion of EHCS in eel diets could result in an imbalance in amino acid composition. This underscores the necessity of amino acid supplementation when utilizing EHCS to replace fishmeal in eel diets, ensuring that the amino acid composition optimally supports the growth and health of eels.

5. Conclusions

The current investigation unequivocally demonstrated that dietary EHCS inclusion significantly affects the growth performance and intestinal health of juvenile American eels. In comparison to eels fed diet without EHCS inclusion, those fed diet with EHCS inclusion levels at 8% exhibited superior growth performance and enhanced intestinal health. The diet containing a 16% inclusion level of EHCS had comparable effects on the growth performance and intestinal health of American eels compared to a diet without EHCS inclusion. However, dietary inclusion levels of EHCS reaching 24% were found to have negative impacts on both the growth performance and intestinal health of American eels. Furthermore, the 8% dietary inclusion level of EHCS was found to enhance the abundance of beneficial intestinal microbiota. Overall, it is recommended that the optimal levels of dietary EHCS inclusion should range from 7.78% to 8.33%, while the maximum permissible levels were suggested to be between 17. 59% and 17.77% for American eels. Our findings provide a valuable reference for the inclusion of EHCS in the diets of American eels as well as its potential applicability to other fish species, promoting the advancement of low or non-fishmeal diets.

Author Contributions: Conceptualization, S.Z., Y.X. and C.W.; methodology, S.Z., Y.X. and C.W.; software, Y.X. and C.W.; validation, Y.X., C.W. and P.W.; formal analysis, Y.X. and J.Y.; investigation, Y.X. and C.W.; resources, S.Z.; data curation, Y.X. and C.W.; writing—original draft preparation, Y.X. and J.Y.; writing—review and editing, Y.X. and J.Y.; visualization, Y.X. and X.H.; supervision, X.H.; project administration, S.Z.; funding acquisition, S.Z. All authors have read and agreed to the published version of the manuscript.

Funding: This research was funded by the earmarked fund for China Agriculture Research System (CARS-46), the Regional Development Project for Science and Technology Plan Program of Fujian Province (2022N3002), and the Open Fund of Engineering Research Center of the Modern Industry Technology for Eel, Ministry of Education of China (RE202010).

Institutional Review Board Statement: This study on American eels culture and management adhered to the Management Rule of Laboratory Animals (Chinese Order No. 676 of the State Council, revised 1 March 2017). All necessary measures were taken to minimize animal suffering. The Jimei University Ethics Committee approved the research protocols (identification code: 2021-11 and approved in 11 March 2021).

Informed Consent Statement: Not applicable.

Data Availability Statement: The data that support the findings of this study are available from the corresponding author upon reasonable request.

Conflicts of Interest: The authors declare no conflicts of interest.

References

1. Cho, J.H.; Kim, I.H. Fish meal—Nutritive value. *J. Anim. Physiol. Anim. Nutr.* **2011**, *95*, 685–692. [CrossRef] [PubMed]
2. Naylor, R.L.; Hardy, R.W.; Bureau, D.P.; Chiu, A.; Elliott, M.; Farrell, A.P.; Forster, I.; Gatlin, D.M.; Goldburg, R.J.; Hua, K.; et al. Feeding aquaculture in an era of finite resources. *Proc. Natl. Acad. Sci. USA* **2009**, *106*, 15103–15110. [CrossRef] [PubMed]
3. Maulu, S.; Langi, S.; Hasimuna, O.J.; Missinhoun, D.; Munganga, B.P.; Hampuwo, B.M.; Gabriel, N.N.; Elsabagh, M.; Van Doan, H.; Abdul Kari, Z.; et al. Recent advances in the utilization of insects as an ingredient in aquafeeds: A review. *Anim. Nutr.* **2022**, *11*, 334–349. [CrossRef] [PubMed]
4. Tibaldi, E.; Hakim, Y.; Uni, Z.; Tulli, F.; de Francesco, M.; Luzzana, U.; Harpaz, S. Effects of the partial substitution of dietary fish meal by differently processed soybean meals on growth performance, nutrient digestibility and activity of intestinal brush border enzymes in the European sea bass (*Dicentrarchus labrax*). *Aquaculture* **2006**, *261*, 16. [CrossRef]
5. Zhang, G.G.; Li, X.; Cai, X.B.; Zhang, S.X.; Hua, X.M.; Huang, Z.Y.; Li, N.Y.; Yao, J.T. Effects of enzymatic hydrolyzed soybean meal on growth performance, liver function and metabolism of largemouth bass (*Micropterus salmoides*). *Acta Hydrobiol. Sin.* **2019**, *43*, 1001–1012.
6. Huang, Q.; Miao, Y.; Liu, J.; Wang, H.; Qin, C.; Wang, X.; Li, E.; Qin, J.; Chen, L. Partial replacement of fish meal by enzymatically hydrolyzed soybean does not adversely impact the growth performance, antioxidant capacity, immunity and intestinal health of the juvenile *Eriocheir sinensis*. *Aquacult. Rep.* **2024**, *36*, 102072. [CrossRef]
7. Hart, S.D.; Bharadwaj, A.S.; Brown, P.B. Soybean lectins and trypsin inhibitors, but not oligosaccharides or the interactions of factors, impact weight gain of rainbow trout (*Oncorhynchus mykiss*). *Aquaculture* **2010**, *306*, 310–314. [CrossRef]
8. Nguyen, H.P.; Do, T.V.; Tran, H.D. Dietary replacement of fish meal by defatted and fermented soybean meals with taurine supplementation for pompano fish: Effects on growth performance, nutrient digestibility, and biological parameters in a long-term feeding period. *J. Anim. Sci.* **2020**, *98*, skaa367. [CrossRef]
9. Weng, Z.; Sun, L.; Wang, F.; Sui, X.; Fang, Y.; Tang, X.; Shen, X. Assessment the flavor of soybean meal hydrolyzed with Alcalase enzyme under different hydrolysis conditions by E-nose, E-tongue and HS-SPME-GC-MS. *Food Chem. X* **2021**, *12*, 100141. [CrossRef]
10. Wang, Z.; Cui, Y.; Liu, P.; Zhao, Y.; Wang, L.; Liu, Y.; Xie, J. Small peptides isolated from enzymatic hydrolyzate of fermented soybean meal promote endothelium-independent vasorelaxation and ACE inhibition. *J. Agric. Food Chem.* **2017**, *65*, 10844–10850. [CrossRef]
11. Wang, Y.; Chen, X.; Wu, X.; Cai, G.; Zhai, S. Effects of dietary supplementation of peanut skin proanthocyanidins on growth performance and lipid metabolism of the juvenile American eel (*Anguilla rostrata*). *Animals* **2022**, *12*, 2375. [CrossRef] [PubMed]
12. AOAC. *Official Methods of Analysis of Official Analytical Chemists International*, 17th ed.; Association of Official Analytical Chemists: Arlington, VA, USA, 2002.
13. *GB/T 18246-2019*; Determination of amino acids in feeds, Standardization Administration of China: Beijing, China, 2019.
14. Xu, Y.; Zheng, H.; Guo, J.; Tan, X.; Zhao, T.; Song, Y.F.; Wei, X.; Luo, Z. Effects of different dietary zinc (Zn) sources on growth performance, Zn metabolism, and intestinal health of grass carp. *Antioxidants* **2023**, *12*, 1664. [CrossRef] [PubMed]
15. Lu, W.; Yu, H.; Liang, Y.; Zhai, S. Evaluation of methanotroph (*Methylococcus capsulatus*, bath) bacteria protein as an alternative to fish meal in the diet of juvenile American eel (*Anguilla rostrata*). *Animals* **2023**, *13*, 681. [CrossRef]

16. Mamauag, R.E.P.; Koshio, S.; Ishikawa, M.; Yokoyama, S.; Gao, J.; Nguyen, B.T.; Ragaza, J.A. Soy peptide inclusion levels influence the growth performance, proteolytic enzyme activities, blood biochemical parameters and body composition of Japanese flounder, *Paralichthys olivaceus*. *Aquaculture* **2011**, *321*, 252–258. [CrossRef]
17. Song, Z.D.; Li, H.Y.; Wang, J.Y.; Li, P.Y.; Sun, Y.Z.; Zhang, L.M. Effects of fishmeal replacement with soy protein hydrolysates on growth performance, blood biochemistry, gastrointestinal digestion and muscle composition of juvenile starry flounder (*Platichthys stellatus*). *Aquaculture* **2014**, *426*, 96–104. [CrossRef]
18. Villanueva-Gutiérrez, E.; Rodriguez-Armenta, C.; González-Félix, M.L.; Perez-Velazquez, M. Incorporating hydrolyzed soy protein or black soldier fly (*Hermetia illucens*) larvae meal into feeds for *Totoaba macdonaldi*. *Aquaculture* **2022**, *554*, 738152. [CrossRef]
19. Yu, X.; Luo, K.; Rao, W.; Chen, P.; Lei, K.; Liu, C.; Cui, Z.; Zhang, W.; Mai, K. Effects of replacing dietary fish meal with enzyme-treated soybean meal on growth performance, intestinal microbiota, immunity and mTOR pathway in abalone *Haliotis discus hannai*. *Fish Shellfish Immunol.* **2022**, *30*, 9–21. [CrossRef]
20. Hanley, A.J.; Williams, K.; Festa, A.; Wagenknecht, L.E.; D'Agostino, R.B., Jr.; Haffner, S.M. Liver markers and development of the metabolic syndrome: The insulin resistance atherosclerosis study. *Diabetes* **2005**, *54*, 3140–3147. [CrossRef]
21. Song, Q.; Xiao, Y.; Xiao, Z.; Liu, T.; Li, J.; Li, P.; Han, F. Lysozymes in fish. *J. Agric. Food Chem.* **2021**, *69*, 15039–15051. [CrossRef]
22. Yu, G.C.; Lv, J.; He, H.; Huang, W.; Han, Y. Hepatoprotective effects of corn peptides against carbon tetrachloride-induced liver injury in mice. *J. Food Biochem.* **2012**, *36*, 458–464. [CrossRef]
23. Kuang, S.Y.; Xiao, W.W.; Feng, L.; Liu, Y.; Jiang, J.; Jiang, W.D.; Hu, K.; Li, S.H.; Tang, L.; Zhou, X.Q. Effects of graded levels of dietary methionine hydroxy analogue on immune response and antioxidant status of immune organs in juvenile Jian carp (*Cyprinus carpio* var. Jian). *Fish Shellfish Immunol.* **2012**, *32*, 629–636. [CrossRef] [PubMed]
24. Wójcik, M.; Grabowski, S.; Jarosz, Ł.S.; Szymczak, B.; Longo, V.; Della Croce, C.M.; Hejdysz, M.; Cieślak, A.; Gruszczyński, K.; Marek, A. Liver antioxidant capacity and steatosis in laying hens exposed to various quantities of lupin (*Lupinus angustifolius*) seeds in the diet. *Antioxidants* **2024**, *13*, 251. [CrossRef] [PubMed]
25. Karlson, B.W.; Palmer, M.K.; Nicholls, S.J.; Lundman, P.; Barter, P.J. Doses of rosuvastatin, atorvastatin and simvastatin that induce equal reductions in LDL-C and non-HDL-C: Results from the VOYAGER meta-analysis. *Eur. J. Prev. Cardiol.* **2016**, *23*, 744–777. [CrossRef] [PubMed]
26. Xu, Y.C.; Zheng, H.; Hogstrand, C.; Tan, X.Y.; Zhao, T.; Song, Y.F.; Wei, X.L.; Wu, L.X.; Luo, Z. Novel mechanism for zinc inducing hepatic lipolysis via the HDAC3-mediated deacetylation of β-catenin at lysine 311. *J. Nutr. Biochem.* **2023**, *121*, 109429. [CrossRef]
27. Hosomi, R.; Fukunaga, K.; Arai, H.; Nishiyama, T.; Yoshida, M. Effects of dietary fish protein on serum and liver lipid concentrations in rats and the expression of hepatic genes involved in lipid metabolism. *J. Agric. Food Chem.* **2009**, *57*, 9256–9262. [CrossRef]
28. Luukkonen, P.K.; Zhou, Y.; Nidhina Haridas, P.A.; Dwivedi, O.P.; Hyötyläinen, T.; Ali, A.; Juuti, A.; Leivonen, M.; Tukiainen, T.; Ahonen, L.; et al. Impaired hepatic lipid synthesis from polyunsaturated fatty acids in TM6SF2 E167K variant carriers with NAFLD. *J. Hepatol.* **2017**, *67*, 128–136. [CrossRef]
29. Shi, Z.Y.; Chen, X.W.; Gu, Y.F. Cloning and expression pattern of alkaline phosphatase during the development of *Paralichthys olivaceus*. *Fish Physiol. Biochem.* **2011**, *37*, 411–424. [CrossRef]
30. Yu, Y.; Wang, Q.; Huang, Z.; Ding, L.; Xu, Z. Immunoglobulins, mucosal immunity and vaccination in teleost fish. *Front. Immunol.* **2020**, *11*, 567941. [CrossRef]
31. Najafpour, B.; Cardoso, J.C.R.; Canário, A.V.M.; Power, D.M. Specific evolution and gene family expansion of complement 3 and regulatory factor H in fish. *Front Immunol.* **2020**, *11*, 568631. [CrossRef]
32. Wen, L.; Jiang, Y.; Zhou, X.; Bi, H.; Yang, B. Structure identification of soybean peptides and their immunomodulatory activity. *Food Chem.* **2021**, *359*, 129970. [CrossRef]
33. He, S.; Simpson, B.K.; Sun, H.; Ngadi, M.O.; Ma, Y.; Huang, T. Phaseolus vulgaris lectins: A systematic review of characteristics and health implications. *Crit. Rev. Food Sci. Nutr.* **2018**, *58*, 70–83. [CrossRef] [PubMed]
34. Yin, Q.; Xu, Y.; Ma, D.; Ma, Z.; Zhai, S. The effects of coated inorganic trace minerals on the growth performance, mineral retention, and intestinal health comparing with inorganic and organic trace minerals for juvenile American eels (*Anguilla rostrata*). *Aquacult. Rep.* **2024**, *38*, 102287. [CrossRef]
35. Gilani, S.; Howarth, G.S.; Kitessa, S.M.; Tran, C.D.; Forder, R.E.A.; Hughes, R.J. Intestinal permeability induced by lipopolysaccharide and measured by lactulose, rhamnose and mannitol sugars in chickens. *Animal* **2017**, *11*, 1174–1179. [CrossRef] [PubMed]
36. Li, F.; Zhang, B.; Zhang, Y.; Zhang, X.; Usman, S.; Ding, Z.; Hao, L.; Guo, X. Probiotic effect of ferulic acid esterase-producing Lactobacillus plantarum inoculated alfalfa silage on digestion, antioxidant, and immunity status of lactating dairy goats. *Anim. Nutr.* **2022**, *11*, 38–47. [CrossRef]
37. Juniper, D.T.; Rymer, C.; Briens, M. Bioefficacy of hydroxy-selenomethionine as a selenium supplement in pregnant dairy heifers and on the selenium status of their calves. *J. Dairy Sci.* **2019**, *102*, 7000–7010. [CrossRef]
38. Ren, J.; Yang, B.; Lv, Y.; Guo, S. Protective and reparative effects of peptides from soybean β-conglycinin on mice intestinal mucosa injury. *Int. J. Food Sci. Nutr.* **2014**, *65*, 345–350. [CrossRef]
39. Zhang, Q.; Tong, X.; Sui, X.; Wang, Z.; Qi, B.; Li, Y.; Jiang, L. Antioxidant activity and protective effects of Alcalase-hydrolyzed soybean hydrolysate in human intestinal epithelial Caco-2 cells. *Food Res. Int.* **2018**, *111*, 256–264. [CrossRef]

40. Dan, Z.; Zhang, W.; Zheng, J.; Gong, Y.; Mai, K.; Ai, Q. Effects of fishmeal substitution by α-galactosidase hydrolytic soybean meal (EhSBM) on growth, antioxidant capacity, inflammatory responses and intestinal health of turbot juveniles (*Scophthalmus maximus* L.). *Aquaculture* **2023**, *563*, 738927. [CrossRef]
41. Chelakkot, C.; Ghim, J.; Ryu, S.H. Mechanisms regulating intestinal barrier integrity and its pathological implications. *Exp. Mol. Med.* **2018**, *50*, 1–9. [CrossRef]
42. Jiang, W.D.; Hu, K.; Zhang, J.X.; Liu, Y.; Jiang, J.; Wu, P.; Zhao, J.; Kuang, S.Y.; Tang, L.; Tang, W.N.; et al. Soyabean glycinin depresses intestinal growth and function in juvenile Jian carp (*Cyprinus carpio* var Jian): Protective effects of glutamine. *Br. J. Nutr.* **2015**, *114*, 1569–1583. [CrossRef]
43. Bai, M.; Wang, L.; Liu, H.; Xu, K.; Deng, J.; Huang, R.; Yin, Y. Imbalanced dietary methionine-to-sulfur amino acid ratio can affect amino acid profiles, antioxidant capacity, and intestinal morphology of piglets. *Anim. Nutr.* **2020**, *6*, 447–456. [CrossRef] [PubMed]
44. Zhang, Y.L.; Jiang, W.D.; Duan, X.D.; Feng, L.; Wu, P.; Liu, Y.; Jiang, J.; Kuang, S.Y.; Tang, L.; Zhou, X.Q. Soybean glycinin caused NADPH-oxidase-regulated ROS overproduction and decreased ROS elimination capacity in the mid and distal intestine of juvenile grass carp (*Ctenopharyngodon idella*). *Aquaculture* **2020**, *516*, 734651. [CrossRef]
45. Balcázar, J.L.; Vendrell, D.; de Blas, I.; Ruiz-Zarzuela, I.; Muzquiz, J.L.; Girones, O. Characterization of probiotic properties of lactic acid bacteria isolated from intestinal microbiota of fish. *Aquaculture* **2008**, *278*, 188–191. [CrossRef]
46. Hao, W.Z.; Li, X.J.; Zhang, P.W.; Chen, J.X. A review of antibiotics, depression, and the gut microbiome. *Psychiatry Res.* **2020**, *284*, 112691. [CrossRef] [PubMed]
47. Zhao, H.; Wang, J.; Peng, Y.; Cai, X.; Liu, Y.; Huang, W.; Huang, H.; Nie, Y. Genomic insights from *Paraclostridium bifermentans* HD0315_2: General features and pathogenic potential. *Front. Microbiol.* **2022**, *13*, 928153. [CrossRef]
48. Khaitlina, S.; Bozhokina, E.; Tsaplina, O.; Efremova, T. Bacterial actin-specific endoproteases grimelysin and protealysin as virulence factors contributing to the invasive activities of Serratia. *Int. J. Mol. Sci.* **2020**, *21*, 4025. [CrossRef]
49. Rivera-Zuluaga, K.; Hiles, R.; Barua, P.; Caldwell, D.; Iyer-Pascuzzi, A.S. Getting to the root of *Ralstonia* invasion. *Semin. Cell Dev. Biol.* **2023**, *148–149*, 3–12. [CrossRef]
50. Kang, D.; Shoaie, S.; Jacquiod, S.; Sørensen, S.J.; Ledesma-Amaro, R. Comparative genomics analysis of keratin-degrading *Chryseobacterium* species reveals their keratinolytic potential for secondary metabolite production. *Microorganisms* **2021**, *9*, 1042. [CrossRef]
51. Iturbe-Espinoza, P.; Brown, D.M.; Weedon, J.T.; Braster, M.; Brandt, B.W.; Bonte, M.; van Spanning, R.J. Microbial communities associated with landfarming amendments during bioremediation of crude oil in Niger Delta soils. *Appl. Soil Ecol.* **2023**, *191*, 105058. [CrossRef]
52. Ryan, M.P.; Pembroke, J.T. *Brevundimonas* spp.: Emerging global opportunistic pathogens. *Virulence* **2018**, *9*, 480–493. [CrossRef]

animals

MDPI

Article

Effects of Different Forms of Organic Selenium on Growth Performance, Antioxidant Capacity, and Intestinal Health in Rice Field Eel (*Monopterus albus*)

Denghang Yu †, Yujia Liu †, Jincheng Wan, Jiaxiang Chen, Yangjie Qiu and Chi Zhang *

Hubei Key Laboratory of Animal Nutrition and Feed Science, School of Animal Science and Nutritional Engineering, Wuhan Polytechnic University, Wuhan 430023, China; yudenghang1985@163.com (D.Y.); 18365970228@163.com (Y.L.); m13329798888@163.com (J.W.); chenjiaxiang777@163.com (J.C.); 15286303323@163.com (Y.Q.)

* Correspondence: zhch@whpu.edu.cn

† These authors contributed equally to this work.

Simple Summary

This study systematically evaluated the effects of four organic selenium compounds—diselenoacetic acid, diselenopropionic acid, diselenobutyric acid, and diselenodibutyric acid—on growth performance, antioxidant capacity, and intestinal health in rice field eel. Rice field eels were fed isonitrogenous and isolipidic diets supplemented with one of these organic selenium sources or no selenium (control) for 60 days. The results showed that diselenoacetic acid significantly increased weight gain and specific growth rate, while diselenobutyric acid enhanced hepatic and intestinal antioxidant enzyme activities. Supplementation with diselenopropionic acid notably improved intestinal digestive enzyme activity, preserved villus structure, and promoted the abundance of beneficial gut bacteria, thereby supporting nutrient absorption. These findings demonstrate that targeted organic selenium supplementation can effectively improve growth, antioxidant defense, and gut health in rice field eel. The study provides new insights into functional feed development and offers practical guidelines for optimizing selenium nutrition in intensive aquaculture, with the goal of reducing oxidative stress and mortality in rice field eel production.

Abstract

To evaluate the effects of different organic selenium sources on rice field eel (*Monopterus albus*), four isonitrogenous and isolipidic diets were tested: a control (no supplement) and four diets each supplemented with 0.24 mg/kg selenium as diselenoacetic acid (Se1), diselenopropionic acid (Se2), diselenobutyric acid (Se3), or diselenodibutyric acid (Se4). Rice field eels (initial body weight 18.43 $\pm$ 0.03 g) were fed for 60 days. The Se1 group showed significantly higher WGR and SGR than the control. Whole-body crude protein was significantly increased in the Se3 and Se4 groups, while crude lipid content was significantly decreased in Se1, Se3, and Se4 groups. Selenium concentrations in the liver and kidney were highest in the Se2 group. Serum AST and ALT activities were decreased in Se4. In the Se4 group, intestinal CAT, T-SOD, and GSH-PX activities were enhanced, along with upregulation of Gpx8, CAT, Nrf2, and Keap1. Hepatic CAT and GSH-PX increased in Se2 and Se4, with Gpx8 upregulated in Se1. Intestinal lipase activity was significantly increased in Se2. At the phylum level, *Proteobacteria* predominated. At the genus level, *Cetobacterium* predominated. In summary, diselenoacetic acid enhanced growth, while diselenopropionic acid and diselenodibutyric acid improved antioxidant defense and intestinal health in rice field eel.

Keywords: *Monopterus albus*; organic selenium; growth performance; antioxidant capacity; gut microbiota

1. Introduction

Selenium (Se), an essential trace element in animals, plays a crucial role in maintaining normal physiological functions. In intensive aquaculture systems, selenium is predominantly absorbed by fish via feed additives, where it exists in two main forms: inorganic selenium (e.g., sodium selenite) and organic selenium (e.g., selenomethionine and selenium-enriched yeast). Studies on grass carp (*Cenopharyngodon idellus*) [1] and Wuchang bream (*Megalobrama amblycephala*) [2] have demonstrated that the incorporation of an appropriate amount of sodium selenite significantly enhances the weight gain rate, specific growth rate, and antioxidant enzyme activity in fish. The supplementation of yeast selenium and selenomethionine in the feed of the Atlantic croaker (*Micropogonias undulatus*) [3] and the Siberian sturgeon (*Acipenser baerii*) [4], respectively, can promote fish growth and enhance antioxidant capacity. Numerous studies have demonstrated the superior efficacy of organic selenium, which exhibits higher bioavailability, lower toxicity, reduced environmental impact, and greater assimilation efficiency compared to inorganic forms [5]. At the molecular level, selenium's physiological functions are mediated through two key mechanisms. Regarding growth regulation, selenium is incorporated as selenocysteine into the catalytic center of thyroxine deiodinase (DIO), a critical enzyme regulating fish growth and development [6]. Dietary selenium supplementation has been shown to enhance growth performance in grass carp by modulating the growth hormone–insulin-like growth factor axis and hypothalamic–pituitary–thyroid axis [7]. Supplementation with 4 mg/kg selenium yeast significantly improved weight gain and specific growth rate in rainbow trout [8]. Additionally, digestive enzyme activities are enhanced by selenium supplementation, thereby directly promoting growth in aquatic animals. Studies have shown that selenium supplementation significantly increases digestive enzyme activities, including those of amylase and lipase, in grass carp, which ultimately stimulates growth [9]. Within the antioxidant defense system, selenium is an essential component of major antioxidant enzymes, including glutathione peroxidase (GSH-PX) and thioredoxin reductase (TRXR). Reducing cellular concentrations of hydrogen peroxide (H_2O_2) and organic hydroperoxides helps to mitigate potential free radical damage and prevent lipid peroxidation [10]. Appropriate supplementation with organic selenium has been shown to significantly increase the activities of total superoxide dismutase (T-SOD) and GSH-PX enzymes in *Haliotis discus hannai* [11].

The threshold between optimal and toxic selenium levels in animals is extremely narrow. The National Research Council reports the dietary selenium requirement for fish is approximately 0.1–0.5 mg/kg [12]. Wang et al. [13] documented that selenium requirements among major farmed fish species range from 0.09 mg/kg in channel catfish to 20.26 mg/kg in Chinese sturgeon. Extensive research demonstrates that selenium-deficient diets result in stunted growth, reduced feed efficiency, and impaired antioxidant and immune functions in fish. Conversely, excessive selenium induces toxicity, causing oxidative stress, lipid metabolism disorders, tissue damage, and other adverse effects. Optimal supplementation with organic selenium enhances growth performance and non-specific immunity in various aquaculture species including grass carp [14] and gibel carp (*Carassius auratus gibelio*) [15]. Thus, when administered as a feed additive at appropriate concentrations, organic selenium substantially improves aquatic animal health.

Rice field eel (*Monopterus albus*) is considered a premium freshwater aquaculture species due to its high protein and low lipid content [16]. In 2023, rice field eel production reached 355,200 metric tons, reflecting a 6.28% year-on-year increase compared to 2022. Responding to increasing market demand, the industry has shifted to large-scale, intensive farming methods. However, high-density cultivation environments frequently result in stunted growth, compromised immunity, and increased oxidative stress in rice field eel, thereby heightening susceptibility to disease and elevating mortality rates. Previous studies have shown that organic selenium offers greater advantages compared to inorganic selenium [5]. Currently, our laboratory is focusing on the study of several different forms of selenium, while also aiming to investigate the specific effects of these various selenium forms on the growth of rice field eels. This study investigates the effects of four distinct organic selenium compounds—diselenoacetic acid, diselenopropionic acid, diselenobutyric acid, and diselenodibutyric acid—used as dietary supplements on farmed rice field eel. The research evaluates the effects of these compounds on growth performance, blood biochemical parameters, antioxidant activity, and gut health, aiming to mitigate oxidative stress in aquaculture. By identifying the most effective organic selenium formulation, this study provides both theoretical foundations and practical guidelines for optimizing selenium supplementation in rice field eel diets.

2. Materials and Methods

2.1. Experimental Design and Feed Preparation

Healthy male adult rice field eels (initial body weight of 18.43 ± 0.03 g) were randomly assigned to five groups (one control, four treatments), each with three replicates of 30 eels housed in 1.5 × 2 m net cages. Experimental diets differed only in selenium source: the control (Se0) contained no selenium, while Se1–Se4 had 0.24 mg/kg selenium from diselenoacetic acid, diselenopropionic acid, diselenobutyric acid, and diselenodibutyric acid. The recommended added dose is based on Gong Bing's research [17], which indicates that the optimal amount of selenium atoms in selenomethionine is 0.24 mg/kg. The trial lasted 60 days. Feed composition and nutrient levels appear in Table 1. The basal diet included fish meal (protein), wheat flour (carbohydrate), and a 1:1 mix of fish oil and soybean lecithin (lipid). Other ingredients were sourced from Wuhan CP Aquaculture Co., Ltd. All were ground, sieved (40-mesh), weighed, blended stepwise, then pelleted (2.0 mm), air-dried, labeled, and stored at −5 °C.

Table 1. Composition and nutrient levels of the basal diet (%).

Ingredients	Content (%)
Fish meal	49.0
Soy protein concentrate	17.0
Corn gluten meal	11.9
Soy lecithin	2.0
Wheat meal	15.0
Fish oil	2.0
Choline chloride	0.5
$Ca(H_2PO_4)_2$	1.5
Mineral premix [2]	0.5
Vitamin premix [1]	0.5
Cellulose powder	0.1
Total	100

Table 1. *Cont.*

Ingredients	Content (%)
Nutrient level	
Moisture (%)	9.82
Crude protein (%)	47.47
Crude lipid (%)	17.67
Crude ash (%)	13.46

Note: [1] Vitamin premix (per/kg): VA, 4500 IU; VD3, 1000 IU; VE, 200 mg; VK, 40 mg; VC, 25 mg; VB1, 28 mg; VB2, 80 mg; VB6, 40 mg; VB12, 0.01 mg; VB5, 80 mg; VB7, 0.2 mg; VB9, 4 mg; VB8, 80 mg; and Choline, 500 mg. [2] Mineral premix (per/kg): NaCl, 1000 mg; $CuSO_4{\cdot}5H_2O$, 3.9 mg; $FeSO_4{\cdot}7H_2O$, 180 mg; $ZnSO_4{\cdot}7H_2O$, 70 mg; $MnSO_4{\cdot}H_2O$, 28 mg; $MgSO_4{\cdot}7H_2O$, 50 mg; $CoCl_2$, 0.89 mg; and KI, 0.8 mg.

2.2. Aquaculture Management

The rice field eels were sourced from Hubei Xiantao Rice Field Eel Industry Group Co., Ltd. in Xiantao, China and were acclimated for one week at the Xiantao Rice Field Eel Industry Technology Research Institute. Water hyacinth, provided by the same company, can purify water, provide shade and reduce temperature, alleviate high-temperature stress, offer hiding spaces for rice field eels, and reduce mutual fighting. Culture water came from filtered local wells with controlled parameters: temperature 25–30 °C, pH 6.50–7.50, ammonia $\leq$ 0.15 mg/L, nitrite $\leq$ 0.05 mg/L, and dissolved oxygen > 6 mg/L, meeting Chinese fishery standards. Fish were fed once daily (18:00) following the "four fixed principles." Uneaten feed was removed at 06:00. Water was managed according to conditions. Pest control was applied to water hyacinth as needed.

2.3. Sample Collection

After 24 h of fasting, rice field eels were anesthetized with 30 mg/L MS-222 (Tricaine methanesulfonate, M14788) from AbMole in Shanghai, China, prior to sampling. The eels in each culture tank were counted and weighed to calculate final body weight (FBW), weight gain ratio (WGR), specific growth ratio (SGR), and feed conversion ratio (FCR). In each culture tank, five rice field eels were randomly selected to weigh and measure for length. Following tail amputation for blood sampling, eels were dissected on ice, with visceral and liver weights measured for morphometrics. Liver and intestinal tissues were flash-frozen in liquid nitrogen and stored at −80 °C. Blood was left at 4 °C/24 h; serum was collected and stored at −80 °C for analysis. In each culture tank, three rice field eels were randomly selected for whole-body nutritional analysis.

2.4. Calculation of Growth Performance and Morphological Parameters

$$\text{Specific growth rate (SGR, \%/d)} = (\mathrm{Ln}W_t - \mathrm{Ln}W_0)/t \times 100, \quad (1)$$

$$\text{Feed conversion ratio (FCR)} = F/(W_t - W_0), \quad (2)$$

$$\text{Viscerasomatic index (VSI, \%)} = (VW/W) \times 100, \quad (3)$$

$$\text{Hepatosomatic index (HSI, \%)} = (HW/W) \times 100, \quad (4)$$

$$\text{Condition factor (CF, g/cm}^3\text{)} = W/L^3, \quad (5)$$

$$\text{Weight gain ratio (WGR, \%)} = (W_t - W_0)/W_0 \times 100 \quad (6)$$

In these formulas: W_0 and W_t are the initial and final average weights (g) of the eels; t is the culture duration (days); F is the average feed intake (g); VW, HW, and W represent visceral, liver, and whole-body weights (g), respectively; and L is body length (cm).

2.5. Determination of Nutritional Components

Nutritional components in feed and fish samples were analyzed using standard protocols: moisture (GB/T 6435-2014) [18], crude protein (GB/T 6432-2018) [19], crude fat (GB/T 6433-2006) [20], and crude ash (GB/T 6438-2007) [21].

2.6. Determination of Selenium Content

For selenium determination, feed samples were crushed, sieved (1 mm), and stored at room temperature; viscera samples were dried at 65 °C for 48 h, ground, and kept at −20 °C. All samples were digested using a microwave system (TANK40, Xinyi Microwave Chemistry, Shanghai, China).

2.7. Liver and Intestinal Biochemical Indices

After the feeding trial, rice field eels fasted for 24 h. Three eels per cage were randomly selected, and blood collected from the caudal vein for serum albumin (ALB, A028-2-1), total protein (TP, A045-2-2), total cholesterol (TC, A111-1-1), triglyceride (TG, A110-1-1), glucose (GLU, F006-1-1), high-density lipoprotein (HDL, A112-1-1) or low-density lipoprotein (LDL, A113-1-1), aspartate aminotransferase (AST, C010-2-1), and alanine aminotransferase (ALT, C009-2-1) analyses. Liver and intestine were dissected to assay catalase (CAT, A007-1-1), glutathione peroxidase (GSH-PX, A006-2-1), superoxide dismutase (SOD, A001-3-2), malondialdehyde (MDA, A003-1-2), trypsin (A080-2-2), amylase (AMS, C016-1-2), and lipase (LPS, A054-2-1) activities using kits from Nanjing Jiancheng Bioengineering Institute (Nanjing, China).

2.8. Measurement of Related Antioxidant Gene Expression Levels

The published genome sequence information of rice field eels was obtained from NCBI, leading to the selection of RPL17 as the reference gene. Primers were then designed using Primer 5.0 software and synthesized by Qingke Biotechnology Co., Ltd. (Beijing, China). The gene sequences are shown in Table 2. The RNA was extracted using RNAiso Easy kit (9109) Reverse transcription was conducted to synthesize complementary DNA (cDNA) following the protocol outlined in the PrimeScript RT Reagent Kit with gDNA Eraser (Perfect Real Time) (RR047A). Real-time quantitative PCR (qPCR) was conducted using a Roche LIghtCycler 480 system and TB Green® Premix Ex Taq II kit (RR820A), following the manufacturer's instructions. Cycling conditions were 95 °C for 30 s (pre-denaturation), then 40 cycles of 95 °C for 5 s and 60 °C for 30 s. Relative gene expression was calculated using the $2^{-\Delta\Delta Ct}$ method [22]. All the kits were from TaKaRa Biotechnology Co., Ltd. in Shanghai, China.

Table 2. Composition and nutrient levels of the basal diet (%).

Gene Name	Primer Sequence (5′–3′)	Product Length (bp)
RPL17	F: GTTGTAGCGACGGAAAGGGAC R: GACTAAATCATGCAAGTCGAGGG	160
CAT	F: GTCCAAGTCTAAGGCATCTCC R: CTCCTCTTCGTTCAGCACC	106
GPX8	F: GTCCACTTACGGTGTTACCT R: ATGGGCTCGTCAGTTCTC	180

Table 2. *Cont.*

Gene Name	Primer Sequence (5′–3′)	Product Length (bp)
Keap1	F: AGCCTGGGTGCGATACGA R: CAAGAAATGACTTTGGTGGG	198
Nrf2	F: CTTCAGACAGCGGTGACAGG R: GCCTCATTCAGTTGGTGCTT	260

2.9. Analysis of Intestinal Tissue Structure

Intestinal tissues were fixed in 4% paraformaldehyde for at least 48 h, then sent to Servicebio Company for tissue sectioning. Samples were dehydrated through graded ethanol and xylene. Samples were embedded in paraffin and sectioned at 4 μm thickness using a microtome. Sections were stained with hematoxylin and eosin (H&E) and examined under an optical microscope (Olympus BX53, Tokyo, Japan). Villus width (VW), villus height (VH), and muscle layer thickness (MT) were measured using ImageJ software (2019).

2.10. Analysis of Intestinal Microbiota

After the culture period, rice field eels were fed to satiation. Twenty-four hours later, six eels from each cage in all five groups were randomly selected for dissection. Posterior intestinal contents were collected into sterile 2 mL tubes and stored at −80 °C for subsequent microbiota analysis. Sequencing was performed by Shanghai Majorbio Bio-pharm Technology Co., Ltd. (Shanghai, China). The sequencing experimental procedure involved thawing the samples at −80 °C; cutting 50 mg of the sample into a homogenization tube; adding 725 μL of SLX-Mlus buffer (PD090) for centrifugation; and extracting the total intestinal DNA from the supernatant. The 16S rRNA high-throughput sequencing (Illumina MiSeq PE300 platform, San Diego, CA, USA) was selected, and the obtained sequence data were uploaded to the NCBI SRA database.

2.11. Data Processing and Analysis

The scheme of all animals and experimental timelines appears in Table 3. The data were presented as mean ± standard deviation (S.D., $n = 3$). One-way ANOVA was performed using SPSS 26.0 (SPSS, Chicago, IL, USA), and Duncan's method was used for multiple comparisons. Differences were considered significant at $p < 0.05$.

Table 3. Scheme of all animals and experimental timelines.

Experiment	Time
Aquaculture Experiment	20 July 2024–20 September 2024
Sample Collection	22 September 2024–24 September 2024
Nutritional Components	29 September 2024–30 September 2024
Liver and Intestinal Biochemical Indices	8 October 2024–5 November 2024
Related Antioxidant Gene Expression	10 November 2024–5 December 2024

3. Results

3.1. Growth Performance

As shown in Table 4, there were no significant differences in feed conversion ratios among the groups ($p > 0.05$). Compared with the control and other experimental groups, the Se1 group exhibited significantly higher FW, WGR, and SGR in rice field eels ($p < 0.05$), indicating that dietary supplementation with 0.42 mg/kg diselenoacetic acid could promote the growth of rice field eels.

Table 4. Effects of different forms of organic selenium on growth performance in rice field eel (n = 3).

Parameters	Se0	Se1	Se2	Se3	Se4
IW (g)	18.53 ± 0.18	18.33 ± 0.00	18.40 ± 0.18	18.40 ± 0.12	18.49 ± 0.08
FCR	0.88 ± 0.18	0.99 ± 0.02	1.07 ± 0.04	1.05 ± 0.15	1.04 ± 0.02
FW (g)	61.93 ± 5.63 ^{a}	67.99 ± 5.84 ^{b}	62.74 ± 4.41 ^{a}	61.94 ± 1.55 ^{a}	57.63 ± 6.25 ^{a}
WGR (%)	234.02 ± 27.60 ^{a}	270.84 ± 31.87 ^{b}	240.85 ± 20.65 ^{a}	236.65 ± 9.50 ^{a}	211.80 ± 35.06 ^{a}
SGR (%·d^{-1})	2.01 ± 0.13 ^{a}	2.18 ± 0.15 ^{b}	2.04 ± 0.10 ^{a}	2.02 ± 0.05 ^{a}	1.89 ± 0.19 ^{a}

Note: Values in the same column with different letter superscripts indicate a significant difference ($p < 0.05$), while those with no or the same small letter superscripts mean no significant difference ($p > 0.05$). The same applies below.

3.2. Morphological Parameters

As shown in Table 5, dietary supplementation with different forms of organic selenium did not significantly affect CF, HSI, or VSI of rice field eel ($p > 0.05$). Nevertheless, the Se2 group exhibited higher VSI and HSI than other treatment groups.

Table 5. Effects of different forms of organic selenium on morphological parameters in rice field eel (n = 3).

Parameters	Se0	Se1	Se2	Se3	Se4
CF (g/cm^3)	0.09 ± 0.01	0.10 ± 0.01	0.10 ± 0.01	0.10 ± 0.01	0.09 ± 0.01
HSI (%)	4.64 ± 0.64	4.23 ± 0.17	5.24 ± 0.77	4.38 ± 0.70	4.72 ± 0.22
VSI (%)	7.78 ± 0.65	8.10 ± 0.50	8.99 ± 0.81	8.03 ± 0.33	8.72 ± 0.44

3.3. Whole-Body Composition

As shown in Table 6, the Se3 group exhibited significantly higher crude protein content in rice field eel compared to both the control and other experimental groups, whereas the Se1 and Se2 groups showed significantly lower levels ($p < 0.05$). Relative to the control, crude lipid content was significantly reduced in the Se1, Se3, and Se4 groups ($p < 0.05$), but significantly elevated in the Se2 group ($p < 0.05$). The crude ash content in the Se3 and Se4 groups was significantly greater than that of the control group ($p < 0.05$).

Table 6. Effects of different forms of organic selenium on whole-body composition in rice field eel (n = 3). Note: Values in the same column with different letter superscripts indicate a significant difference ($p < 0.05$), while those with no or the same small letter superscripts mean no significant difference ($p > 0.05$).

Parameters	Se0	Se1	Se2	Se3	Se4
Moisture (%)	68.04 ± 0.23	66.42 ± 0.90	66.13 ± 0.91	69.99 ± 0.20	68.48 ± 0.02
Crude protein (%)	52.46 ± 1.13 ^{b}	50.21 ± 0.88 ^{a}	49.74 ± 0.46 ^{a}	57.79 ± 0.09 ^{c}	51.78±0.96 ^{ab}
Crude lipid (%)	34.36 ± 0.12 ^{d}	31.38 ± 0.50 ^{c}	34.50 ± 0.74 ^{d}	26.59 ± 0.22 ^{b}	24.62 ± 0.08 ^{a}
Ash (%)	7.18 ± 0.69 ^{a}	6.35 ± 0.08 ^{a}	6.25 ± 0.78 ^{a}	8.74 ± 0.56 ^{b}	8.88 ± 0.64 ^{b}

3.4. Se Accumulation

As shown in Figure 1, dietary supplementation with different forms of organic selenium did not significantly affect the whole-body selenium content of rice field eels ($p > 0.05$). Regarding muscle selenium content, the Se2 group exhibited significantly higher levels than both Se1 and Se3 groups ($p < 0.05$), while no significant differences were observed among the other experimental groups ($p > 0.05$). For hepatic selenium content, the Se2, Se3, and Se4 groups all showed significantly higher values than the control group ($p < 0.05$), with the Se2 group being significantly higher than Se1. In renal selenium content, no significant difference was found between the Se4 group and the control ($p > 0.05$), while all other experimental groups demonstrated significantly higher levels than the control ($p < 0.05$). Notably, the Se3 group showed significantly higher selenium content than the other experimental groups ($p < 0.05$).

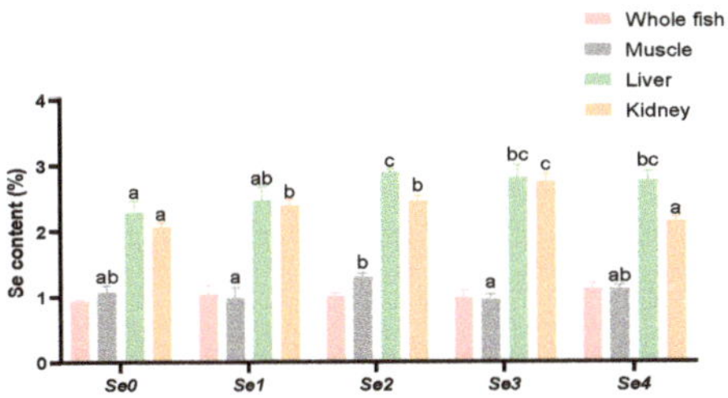

Figure 1. Effects of different forms of organic selenium on Se accumulation in rice field eel tissues. The data were shown as mean ± standard error (n = 3), with different letters indicating significant differences among different groups ($p < 0.05$).

3.5. Serum Biochemical Parameters

As shown in Table 7, no significant differences were observed in TP, ALB, TG, HDL, or LDL levels among the groups ($p > 0.05$). Compared with the control group, serum AST levels were significantly higher in the Se3 group ($p < 0.05$), whereas the Se4 group had significantly lower AST and ALT levels ($p < 0.05$). The Se2 group showed a significant increase in serum TC concentration ($p < 0.05$). Relative to the other experimental groups, the Se1 group had significantly higher serum GLU levels ($p < 0.05$).

Table 7. Effects of different forms of organic selenium on serum biochemical parameters in rice field eel (n = 3). Note: Values in the same column with different letter superscripts indicate a significant difference ($p < 0.05$), while those with no or the same small letter superscripts mean no significant difference ($p > 0.05$).

Parameters	Se0	Se1	Se2	Se3	Se4
TP (g/L)	39.62 ± 3.42	43.22 ± 4.24	46.22 ± 2.69	41.42 ± 2.05	43.65 ± 0.34
ALB (g/L)	13.06 ± 0.47	13.33 ± 1.01	14.20 ± 0.57	12.97 ± 0.36	13.67 ± 0.33
AST (U/L)	63.50 ± 0.50 ^{b}	57.00 ± 1.00 ^{a}	53.33 ± 1.15 ^{a}	93.00 ± 4.36 ^{c}	52.50 ± 0.50 ^{a}
ALT (U/L)	5.00 ± 1.00 ^{bc}	5.33 ± 1.53 ^{bc}	4.67 ± 1.53 ^{bc}	7.00 ± 1.00 ^{c}	2.00 ± 0.00 ^{a}
TC (mmol/L)	3.46 ± 0.35 ^{a}	3.50 ± 0.45 ^{a}	4.39 ± 0.23 ^{b}	3.65 ± 0.28 ^{a}	4.19 ± 0.20 ^{ab}
TG (mmol/L)	1.27 ± 0.28	1.28 ± 0.36	1.55 ± 0.17	1.19 ± 0.27	1.26 ± 0.02
GLU (mmol/L)	4.03 ± 0.49 ^{ab}	5.27 ± 0.50 ^{b}	4.87 ± 0.95 ^{b}	3.60 ± 0.00 ^{a}	3.50 ± 0.30 ^{a}
HDL (mmol/L)	1.64 ± 0.14	1.65 ± 0.25	2.05 ± 0.06	1.69 ± 0.17	1.57 ± 0.69
LDL (mmol/L)	0.54 ± 0.08	0.49 ± 0.09	0.72 ± 0.07	0.51 ± 0.06	0.51 ± 0.24

3.6. Hepatic and Intestinal Antioxidant Parameters

As shown in Figure 2, the experimental groups exhibited significantly higher hepatic CAT activity than the control group ($p < 0.05$), with peak activity observed in the Se2 group. The Se1 and Se4 groups showed markedly elevated hepatic GSH-PX (GPX) activity compared to controls ($p < 0.05$). Notably, hepatic malondialdehyde (MDA) content was significantly increased in the Se1 and Se4 groups relative to both the control and other experimental groups ($p < 0.05$). No significant difference was found in hepatic T-SOD activity between the Se1 and Se4 groups and controls ($p > 0.05$).

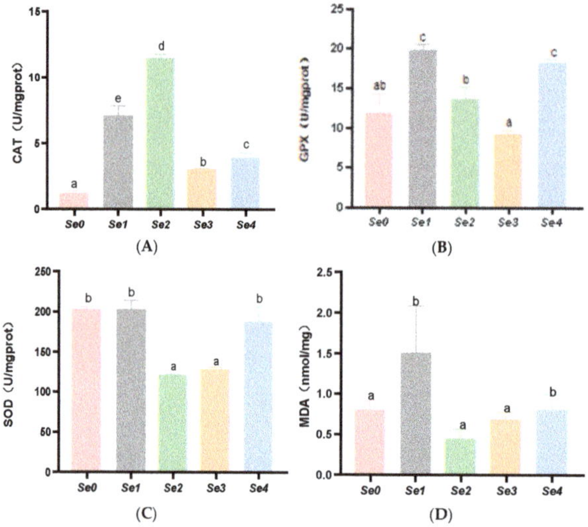

Figure 2. Effects of different forms of organic selenium on antioxidant parameters in the liver in rice field eel. (**A**) CAT; (**B**) GPX; (**C**) SOD; and (**D**) MDA, with different letters indicating significant differences among different groups ($p < 0.05$).

As shown in Figure 3, intestinal CAT and GPX activities were significantly higher in the Se3 and Se4 groups than in the control group ($p < 0.05$). Compared with the control group and other experimental groups, the MDA content in the intestines of the Se2 group was significantly increased ($p < 0.05$). Compared with the other experimental groups, T-SOD levels in the intestines of the Se2 and Se4 groups were significantly elevated ($p < 0.05$), while no significant difference was observed between the control group and the experimental groups ($p > 0.05$).

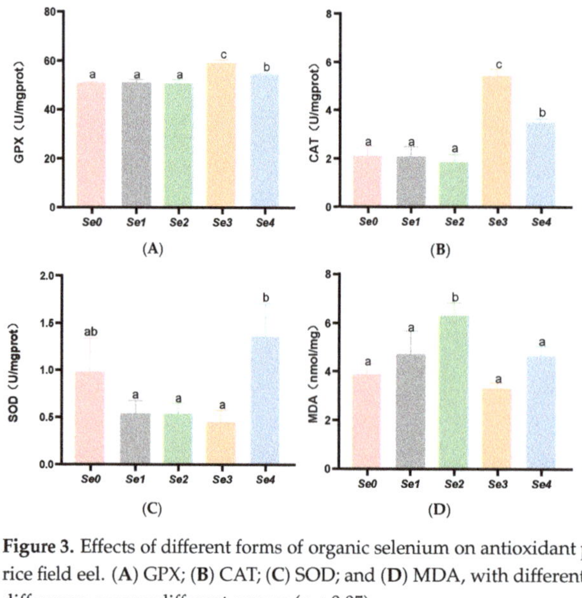

Figure 3. Effects of different forms of organic selenium on antioxidant parameters in the intestine in rice field eel. (**A**) GPX; (**B**) CAT; (**C**) SOD; and (**D**) MDA, with different letters indicating significant differences among different groups ($p < 0.05$).

3.7. Expression Levels of Antioxidant-Related Genes in the Liver and Intestine

As shown in Figure 4, hepatic *Gpx8* expression was significantly elevated in the Se1 group compared to both the control and other experimental groups ($p < 0.05$). Similarly, hepatic *CAT* expression showed significant increases in both Se1 and Se2 groups ($p < 0.05$), while liver *Nrf2* and *Keap1* levels were markedly higher in the Se2 group ($p < 0.05$).

As shown in Figure 5, intestinal *Gpx8*, *CAT*, *Nrf2*, and *Keap1* expression levels were significantly upregulated in the Se3 and Se4 groups compared to controls and other experimental groups ($p < 0.05$).

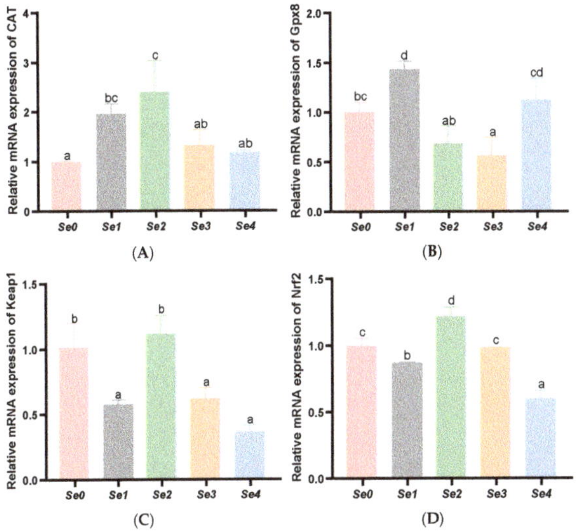

Figure 4. Effects of different forms of organic selenium on the expression of antioxidant genes in the liver in rice field eel. (**A**) *CAT*; (**B**) *Gpx8*; (**C**) *Keap1*; and (**D**) *Nrf2*, with different letters indicating significant differences among different groups ($p < 0.05$).

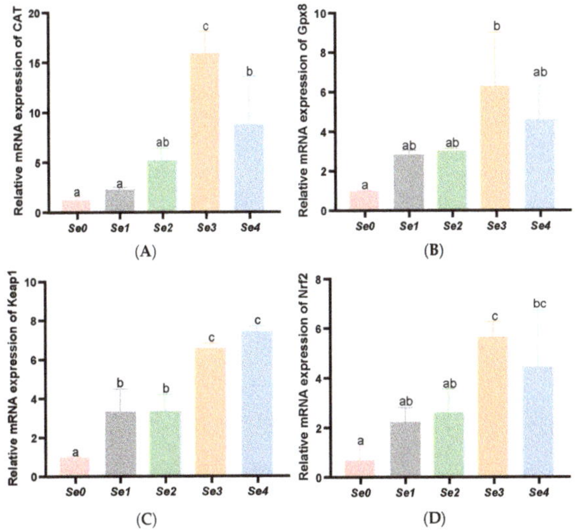

Figure 5. Effects of different forms of organic selenium on in the expression of antioxidant genes in the intestine in rice field eel. (**A**) *CAT*; (**B**) *Gpx8*; (**C**) *Keap1*; and (**D**) *Nrf2*, with different letters indicating significant differences among different groups ($p < 0.05$).

3.8. Intestinal Histomorphology

Figure 6 illustrates the morphological alterations in transverse midgut sections of rice field eels fed different forms of organic selenium. The Se2 group displayed well preserved intestinal villi structure, characterized by slender, finger-shaped projections with optimal morphology. Figure 7 demonstrates the intestinal histomorphological parameters, revealing significantly greater VH in Se2 and Se4 groups compared to controls ($p < 0.05$), whereas Se3 showed no significant difference ($p > 0.05$). VW was significantly increased in Se3 and Se4 groups ($p < 0.05$). MT was significantly enhanced in Se1, Se3, and Se4 groups ($p < 0.05$), with Se2 exhibiting no significant difference from controls ($p > 0.05$).

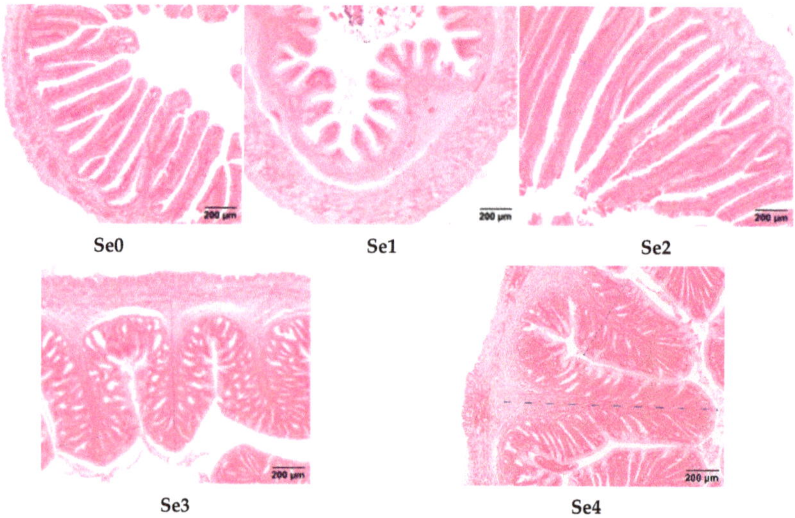

Figure 6. Effects of different forms of organic selenium on the morphological structure in rice field eel intestinal tissue (100×).

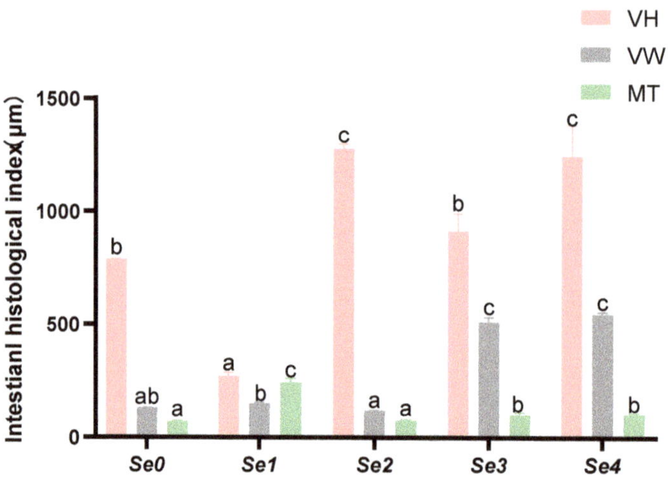

Figure 7. Effects of different forms of organic selenium on the morphological indices of the intestinal tissue in rice field eel, with different letters indicating significant differences among different groups ($p < 0.05$).

3.9. Intestinal Digestive Enzyme Activities

As shown in Figure 8, intestinal lipase activity was significantly higher in the experimental groups compared to the control, with the Se2 group demonstrating the most pronounced enhancement ($p < 0.05$). Trypsin activity showed significant reduction in the Se1, Se2, and Se4 groups ($p < 0.05$). No statistically significant difference was observed in intestinal amylase activity between experimental and control groups ($p > 0.05$).

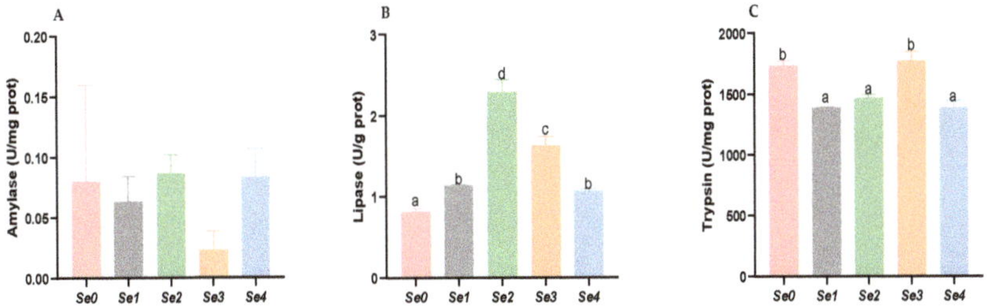

Figure 8. Effects of different forms of organic selenium on the digestive enzyme activity of the intestine in rice field eel. (**A**) Amylase; (**B**) lipase; and (**C**) trypsin, with different letters indicating significant differences among different groups ($p < 0.05$).

3.10. Intestinal Microbial Flora

As shown in Table 8, supplementation with different forms of organic selenium did not significantly affect the ACE, Chao, Shannon, and Simpson index of rice field eel ($p > 0.05$).

As shown in Figure 9, the dominant phyla at the phylum level included *Proteobacteria, Fusobacteria, Firmicutes, Bacteroidetes,* and *Actinobacteria*, with *Proteobacteria, Fusobacteria,* and *Bacteroidetes* being the predominant bacteria. As shown in Figure 10, the dominant populations at the genus level included *Pelomonas, Cetobacterium, Acinetobacter, Pseudomonas,* and *Plesiomonas*.

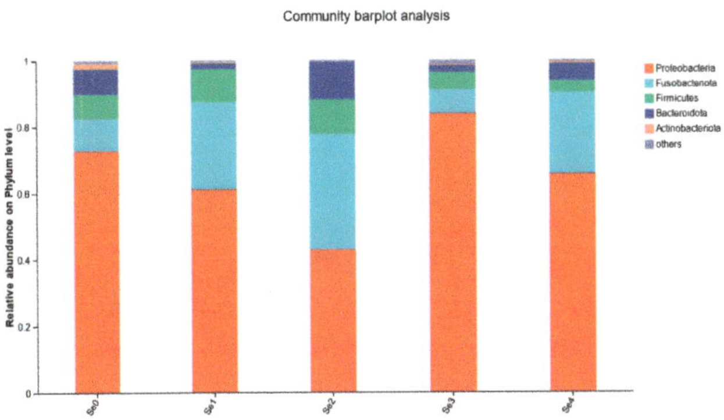

Figure 9. Effects of different forms of organic selenium on the composition of intestinal flora in rice field eel (phylum level).

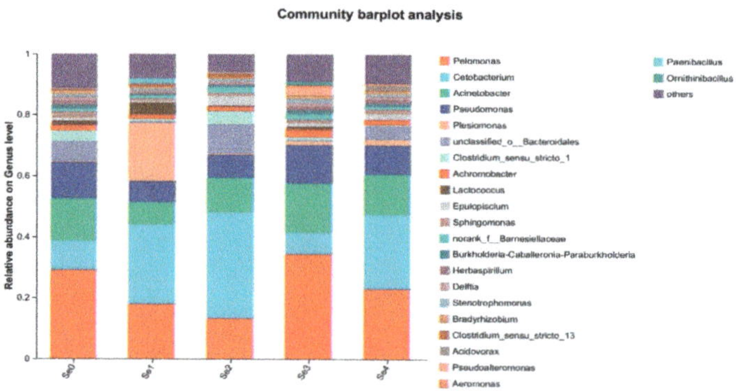

Figure 10. Effects of different forms of organic selenium on the composition of intestinal flora in rice field eel (genus level).

Table 8. Effects of different forms of organic selenium on the alpha diversity of the intestinal microbial flora in rice field eel ($n = 3$).

Parameters	Se0	Se1	Se2	Se3	Se4
ACE	199.26 ± 18.02	227.04 ± 82.67	168.79 ± 43.76	191.27 ± 14.60	178.84 ± 31.90
Chao	198.87 ± 17.73	229.49 ± 79.70	167.19 ± 46.35	191.14 ± 14.79	177.43 ± 32.02
Shannon	2.98 ± 0.35	2.61 ± 0.91	2.67 ± 0.62	2.96 ± 0.35	2.70 ± 0.46
Simpson	0.16 ± 0.10	0.21 ± 0.13	0.18 ± 0.10	0.16 ± 0.07	0.20 ± 0.09

4. Discussion

4.1. Growth Performance

Selenium, as an essential trace element, plays a crucial role in the growth and development of fish. In recent years, selenium has increasingly been incorporated as a nutritional additive in fish feed. In this experiment, the addition of 0.42 mg/kg diselenoacetic acid significantly enhanced the SGR and WGR of rice field eel, while no significant differences were observed between other experimental groups and the control group. This finding is consistent with the effects reported for selenomethionine supplementation in rice field eel. Numerous studies have demonstrated that appropriate dietary selenium supplementation can promote fish growth. For example, supplementing feed with a suitable concentration of selenium significantly enhanced the growth and feed conversion rate of Nile tilapia [23]; Marwa et al. reported that European sea bass fed with 0.5 mg/kg nano-selenium exhibited significantly higher FBW, WG, and SGR compared to the control group [24]. Similarly, Dawood et al. found that supplementing red sea bream diets with 1.0 mg/kg nano-selenium significantly increased FBW, WG, and SGR [25]. In this experiment, diselenoacetic acid promoted growth, while other selenium sources showed no significant effects on the growth of rice field eel. This lack of effect may be attributed to the supplementation levels of these selenium sources being neither deficient nor excessive, resulting in neither growth promotion nor inhibition in the short term. Differences in their biological efficacy were mainly reflected in tissue accumulation rather than growth. However, the specific mechanisms underlying these differences warrant further investigation.

4.2. Whole-Body Composition and Morphological Indices

The evaluation of fish nutritional value primarily depends on the moisture, protein, and lipid content in the muscle. In this experiment, the crude protein content significantly increased with the addition of diselenobutyric acid to the basal diet, indicating that selenium supplementation in feed enhances the protein content in fish meat, thereby improving its nutritional value. Dawood et al. found that dietary supplementation with selenium nanoparticles in red sea bream feed increased the crude protein content of the whole fish at different concentrations, with significant differences compared to the control group [25]. These findings are largely consistent with the results of this study, where the crude protein content in fish significantly decreased with the addition of diselenoacetic acid and diselenopropionic acid, while diselenodibutyric acid showed no significant effect, suggesting that different organic selenium compounds may have varying impacts on the crude protein content in fish at the same concentration. Except for diselenopropionic acid, which had no significant effect, the crude lipid content in fish significantly decreased with other selenium sources. Similarly, Liu et al. reported that dietary selenium supplementation significantly reduced crude lipid content in grass carp, with exosomal miR-22 implicated in the selenium-mediated attenuation of hepatic lipid accumulation [5]. These results align with the findings of this experiment, indicating that appropriate selenium supplementation in feed can influence muscle fat metabolism in fish to some extent. The crude ash content in the diselenobutyric acid and diselenodibutyric acid experimental groups was significantly higher than in the control group, reflecting that selenium can be deposited in fish muscle to some degree.

CF, HSI, and VSI are morphological indices used to assess the energy and nutritional status of fish [26]. After 60 days of the feeding trial in this study, there were no significant differences in CF, VSI, and HSI between the experimental groups and the control group of rice field eel, indicating that the addition of different selenium forms to the feed did not affect the morphological indices in this experiment. This finding aligns with some previous studies [27,28], and it is speculated that variations in feed formulation, cultured species, and rearing environments may contribute to the differing results.

4.3. Se Accumulation

Compared to inorganic selenium, organic selenium is more readily absorbed and converted by the organism, accumulating in tissues. In this study, dietary selenium supplementation did not significantly affect selenium concentrations in the whole fish or muscle tissues of rice field eels, but it did significantly increase selenium levels in the liver and kidneys. This finding may be attributed to the liver being the first organ to encounter selenium post-absorption, as well as its central role in selenoprotein synthesis and selenium metabolism. After ingestion, selenium—primarily present as selenomethionine—is converted into selenocysteine, and through β-lyase activity, generates H_2Se, which participates in selenoprotein synthesis [29]. The remaining selenium is primarily excreted through urine produced by the kidneys and feces, leading to a significantly higher accumulation of selenium in the kidneys compared to other tissues. Studies on groupers have shown that, compared to inorganic selenium, organic selenium is more easily deposited in muscles. At a dietary selenium level of 1.5 mg/kg, the muscle selenium concentration in fish fed organic selenium was more than three times higher than in those fed inorganic selenium [30]. The findings differ from the present study, which may be due to species-specific metabolism; as carnivorous fish, rice field eels may rely more on liver processing and kidney excretion of selenium rather than muscle storage.

4.4. Serum Biochemical Parameters

Serum biochemical indicators are widely used to assess the overall health status of fish in physiological processes [31]. Serum TP and ALB play important roles in immune response mechanisms [32]. In this experiment, supplementation with different forms of organic selenium had no significant effect on the plasma TP and ALB levels in rice field eels. Dawood et al. reported that dietary supplementation with 1 mg/kg nano-selenium significantly increased serum total protein content in Nile tilapia but had no significant effect on serum ALB [23]. Mansour et al. found that dietary supplementation with 2.97–3.98 mg/kg yeast selenium significantly increased serum TP and ALB levels in *Argyrosomus regius* [32]. These results differ from those of the present study, which may be attributed to differences in species, environmental conditions, or the relatively low concentration of organic selenium added.

Serum AST and ALT activities are important indicators of liver health. In this experiment, dietary supplementation with diselenodibutyric acid significantly reduced AST and ALT activities, indicating improved liver function in rice field eels. Soliman et al. reported that nano-selenium particles reduced AST and ALT activities, thereby enhancing the liver health of catfish [33]. These findings are consistent with the present findings. However, dietary supplementation with 0.3–0.9 mg/kg organic selenium had no significant effect on plasma AST and ALT activities in *Carassius auratus* [34], which may be attributed to differences in selenium dosage and its varying effects on the health status of tissues and organs.

The levels of GLU, TC, and TG serum reflect the body's lipid metabolism capacity [32]. In this experiment, TG and GLU showed no significant difference compared to the control group, while the addition of selenopropionic acid increased TC levels. No significant effects were observed in the other treatment groups compared to the control. Deilamy Pour et al. found that nano-selenium supplementation in cobia diets reduced serum GLU, TC, and TG levels [35], which is inconsistent with the results of this study. The discrepancy may be attributed to the carnivorous nature of rice field eels, which rely primarily on protein and fat for energy metabolism and have limited capacity for carbohydrate utilization, leading to differences in GLU regulation mechanisms. Additionally, the high fat content of the feed in this experiment might have masked the effect of selenium on TG.

4.5. Hepatic and Intestinal Antioxidant Capacity

Excessive production of reactive oxygen species can damage the lipid content in cell membranes, induce lipid peroxidation, and impair RNA, which can be assessed by the level of MDA [36]. Selenium, as a functional component of GSH-PX, is incorporated as selenocysteine, serving as an active site within the enzyme. It plays a crucial role in activating the antioxidant defense system, with its antioxidant activity being the most important biological function of selenium. Antioxidant enzymes (CAT, SOD, T-AOC, and GSH-PX) constitute the first line of defense in the antioxidant system of fish [37]. In this study, supplementation with diselenodibutyric acid significantly increased the activities of GSH-PX and CAT in both the intestine and liver, potentially due to enhanced selenium deposition in the liver. Supplementation with different forms of organic selenium enhanced hepatic CAT activity, while diselenoacetic acid and diselenodibutyric acid increased GSH-PX activity. Zhang et al. demonstrated that dietary supplementation with 0.44–0.77 mg/kg selenomethionine in striped sea bass increased liver GSH-PX, T-SOD, and CAT activities in a selenium dose-dependent manner, reaching peak levels [38]. This result is consistent with the findings of the present study. However, in this experiment, supplementation with different forms of organic selenium had no significant effect on MDA levels or T-SOD activity. This may be due to variations in the effective doses of different selenium forms in organisms, and the dosage of organic selenium in this experiment did not reach the threshold required to significantly influence MDA or T-SOD. Differences in the bioavailability of various selenium forms or the organic selenium dosage in this study may not have reached the threshold necessary to significantly affect MDA and T-SOD.

Antioxidant enzyme activity in animals is partly regulated at the gene expression level corresponding to each enzyme. This finding aligns with some previous studies on selenium forms and the upregulated gene expression levels of Gpx8 and CAT, which were largely consistent with the observed enzyme activity trends. Yu et al. reported that hepatic GSH-PX and CAT gene expression in grass carp fed 0.6–0.9 mg/kg nano-selenium was significantly increased [39], which is consistent with the findings of this experiment.

Nuclear factor erythroid 2-related factor 2 (Nrf2) is a key regulator of cellular redox balance and plays an anti-inflammatory role by protecting antioxidants involved in growth factor signaling, nutrient balance, and protein folding [40]. Under oxidative stress, Nrf2 translocates into the nucleus and initiates the transcription of antioxidant genes responsible for ROS clearance [41]. However, the accumulation of Nrf2 in the nucleus can lead to free radical damage, apoptosis, and tumorigenesis. In such cases, Kelch-like ECH-associated protein 1 (Keap1) acts as a suppressor to shut down the Nrf2 response [42]. The results of this experiment showed that dietary supplementation with selenium significantly upregulated mRNA expression levels of both *Nrf2* and *Keap1* in the liver and intestine of rice field eels, with consistent trends observed in both tissues. This aligns with the findings of Itoh et al. [43], who demonstrated that Keap1 directs Nrf2 for lysosomal degradation and observed significant increases in relative expression levels of both *Keap1* and *Nrf2*.

4.6. Intestinal Histomorphology

The intestine is a vital digestive organ in living organisms. Damage to the morphology of intestinal villi and mucosa can compromise the intestinal barrier, leading to immune-related diseases and impaired growth in fish. VH, VW, and MT are important histomorphological indicators of intestinal development and its digestive and absorptive functions [44]. In this experiment, supplementation with diselenopropionic acid and diselenodibutyric acid significantly increased VH, whereas diselenobutyric acid and diselenodibutyric acid significantly increased VW. Khalil et al. reported that dietary supplementation with different levels of selenium yeast in Atlantic croaker juveniles significantly improved intestinal health, as evidenced by increases in villus length, width, and absorptive area [3]. These results are consistent with the findings of this study. Additionally, the optimal dietary selenium concentration for improving intestinal morphology varies among fish species, likely due to differences in feeding habits, physiological characteristics, farming systems, and feed processing methods. In this experiment, supplementation with diselenobutyric acid and diselenodibutyric acid resulted in fewer intestinal villi and caused villus damage in rice field eels. These compounds may activate pathways such as NF-κB and the NLRP3 inflammasome, leading to the release of inflammatory factors such as IL-1β and TNF-α, which trigger chronic intestinal mucosal inflammation and ultimately result in villus atrophy. Further studies are needed to confirm these mechanisms.

4.7. Intestinal Digestive Enzyme Activities

Digestive enzyme activity is a key determinant of a fish's ability to utilize dietary nutrients and directly influences growth [45]. This study demonstrated that the addition of different organic selenium significantly enhanced the intestinal lipase activity in rice field eels, while it markedly reduced trypsin activity in the Se1, Se2, and Se4 groups and had minimal effect on amylase activity. Dietary supplementation with 0.5 mg/kg and 1.0 mg/kg selenium-enriched yeast in *Apostichopus japonicus* significantly increased the activities of protease, lipase, and amylase in the digestive tract [46]. Similar findings were reported by

Li et al., who showed that 0.8 mg/kg selenium-enriched yeast increased digestive tract protease and amylase activity in *Apostichopus japonicus* [47]. The results of this experiment are largely consistent with the above findings, except that selenium supplementation did not affect intestinal amylase activities in rice field eels. This discrepancy may be due to differences in selenium dosage, feeding habits, physiological characteristics, aquaculture practices, or feed processing methods.

4.8. Intestinal Microbial Flora

The gut microbiota is a complex microbial ecosystem composed of different bacterial species that interact with each other and play key roles in maintaining host growth and development, resisting pathogenic invasion, and regulating physiological functions [44].

The results of this study have indicated that different forms of organic selenium did not significantly affect the ACE, Chao, Shannon, and Simpson index, which differs from the findings in juvenile turbot (*Scophthalmus maximus* L.) [48]. This discrepancy may be attributed to differences in species, feed composition, and aquaculture environments. Numerous studies have shown that at the phylum level, the dominant gut microbiota in fish primarily consists of *Proteobacteria*, *Fusobacteria*, *Firmicutes*, *Bacteroidetes*, and *Actinobacteria* [49,50]. Among these, *Proteobacteria* is the most predominant phylum, and some of its members are closely associated with host digestive functions. *Proteobacteria* can produce polyhydroxybutyrate as an energy source for intestinal epithelial cells, thereby supporting gut health and promoting fish growth [51]. In this study, both the experimental and control groups were dominated by *Proteobacteria*, *Fusobacteria*, *Firmicutes*, *Bacteroidetes*, and *Actinobacteria*, with *Proteobacteria* being the most prevalent. Other studies have reported that dietary sodium selenite supplementation in juvenile turbot resulted in *Firmicutes*, *Proteobacteria*, *Bacteroidetes*, and *Actinobacteria* as the dominant gut microbiota, with *Firmicutes* showing the highest abundance among all groups [48]. The present findings are consistent with previous studies, indicating that selenium supplementation can maintain the stability of beneficial gut microbiota in fish. However, in this study, supplementation with diselenobutyric acid and diselenodibutyric acid decreased the relative abundance of *Firmicutes* in the gut. This reduction may be associated with decreased ecological requirements for *Firmicutes* resulting from elevated digestive enzyme activities, leading to a lower relative abundance [52]. At the genus level, the abundance of *Cetobacterium* was significantly higher in groups supplemented with diselenoacetic acid and diselenopropionic acid compared to the control and other experimental groups. *Cetobacterium* is commonly found in the intestines of freshwater fish and facilitates the breakdown of peptides and carbohydrates, producing vitamin B12 [53]. Overall, these results suggest that selenium-supplemented feed modulates the intestinal microbiota of rice field eels by decreasing the abundance of potentially harmful bacteria and increasing beneficial bacteria. This modulation may reduce disease incidence and enhance growth performance in rice field eels.

5. Conclusions

In summary, dietary supplementation with diselenoacetic acid demonstrates a significant growth-promoting effect in rice field eel. Supplementation with diselenodibutyric acid improves the antioxidant capacity of the liver and intestine, thereby enhancing stress resistance. Dietary supplementation with diselenopropionic acid increases intestinal digestive enzyme activity, preserves villus structure integrity, and increases the relative abundance of beneficial bacteria, thereby enhancing nutrient digestion and absorption in rice field eel. This study presents novel compound feeds and provides important insights for promoting the growth of rice field eel, reducing oxidative stress, decreasing mortality rates, and advancing rice field eel aquaculture.

Author Contributions: Conceptualization, D.Y.; data curation, J.W., J.C. and Y.Q.; writing—original draft preparation, Y.L.; writing—review and editing, C.Z. All authors have read and agreed to the published version of the manuscript.

Funding: This work was supported by the National Natural Science Foundation of China (grant no. 31902389), the project of Hubei Key Laboratory of Animal Nutrition and Feed Science, Wuhan Polytechnic University (202405); The Research and Innovation Initiatives of WHPU (2023).

Institutional Review Board Statement: Wuhan University of Light Industry Animal Welfare Ethics Committee approved the animal study protocol with approval number WPU202407004 (date of approval; 4 June 2024).

Informed Consent Statement: Informed consent was obtained from all subjects involved in the study.

Data Availability Statement: The data that support the findings of this study are available from the corresponding author upon reasonable request.

Conflicts of Interest: The authors declare no conflicts of interest.

References

1. Su, C.F.; Luo, L.; Li, Q.; Wen, H.; Wang, F.B. Effects of dietary selenium on the anti-oxidation capability and histological structure of grass carpce (*Cenopharyngodon idellus*). *J. Southwest China Norm. Univ. (Nat. Sci. Ed.)* **2008**, *33*, 69–75. (In Chinese) [CrossRef]
2. Hao, J.Y. Effects of Dietary Selenium on Growth, Glycolipid Metabolism and Antioxidant Capacity and the Relaxation Effects on Excess Selenium of Bamboo Charcoal in Blunt Snout Bream (*Megalobrama amblycephala*). Master's Thesis, Nanjing Agricultural University, Nanjing, China, 2020. (In Chinese) [CrossRef]
3. Khalil, H.S.; Mansour, A.T.; Goda, A.M.A.; Omar, E.A. Effect of selenium yeast supplementation on growth performance, feed utilization, lipid profile, liver and intestine histological changes, and economic benefit in meagre (*Argyrosomus regius*), fingerlings. *Aquaculture* **2019**, *501*, 135–143. [CrossRef]
4. Foroshani, N.S.; Islami, H.R.; Mousavi, S.A.; Khara, H.; Siahkalroodi, Y.S. Supplementation of juvenile Siberian sturgeon (*Acipenser baerii*) diet with different levels of selenium: Growth performance, digestive enzyme activities, and intestinal histomorphology. *Aquac. Rep.* **2024**, *38*, 102325. [CrossRef]
5. Liu, L.W.; Liang, X.F.; Li, J.; Yuan, X.H.; Alam, M.S.; Fang, J. Effects of dietary selenium on growth performance and oxidative stress in juvenile grass carp (*Ctenopharyngodon idellus*). *Aquac. Nutr.* **2018**, *24*, 1296–1303. [CrossRef]
6. Aman, A.J.; Kim, M.; Saunders, L.M.; Parichy, D.M. Thyroid hormone regulates abrupt skin morphogenesis during zebrafish postembryonic development. *Dev. Biol.* **2021**, *477*, 205–218. [CrossRef]
7. Ma, P.; Hu, Z.Y.; Li, L.; Li, D.P.; Tang, R. Dietary selenium promotes the growth performance through growth hormone–insulin-like growth factor and hypothalamic–pituitary–thyroid axes in grass carp (*Ctenopharyngodon idella*). *Fish Physiol. Biochem.* **2021**, *47*, 1313–1327. [CrossRef]
8. Wang, L.; Wang, L.; Zhang, D.F.; Li, S.; Yin, J.J.; Xu, Z.; Zhang, X.Z. Effect of dietary selenium on postprandial protein deposition in the muscle of juvenile rainbow trout (*Oncorhynchus mykiss*). *Br. J. Nutr.* **2021**, *125*, 721–731. [CrossRef]
9. Adineh, H.; Naderi, M.; Nazer, A.; Yousefi, M.; Ahmadifar, E. Interactive effects of stocking density and dietary supplementation with Nano selenium and garlic extract on growth, feed utilization, digestive enzymes, stress responses, and antioxidant capacity of grass carp (*Ctenopharyngodon idella*). *J. Word Aquac. Soc.* **2021**, *52*, 789–804. [CrossRef]
10. Betancor, M.B.; Caballero, M.J.; Terova, G.; Saleh, R.; Atalah, E.; Benítez-Santana, T.; Bell, J.G.; Izquierdo, M. Selenium inclusion decreases oxidative stress indicators and muscle injuries in sea bass larvae fed high DHA microdiets. *Br. J. Nutr.* **2012**, *108*, 2115–2128. [CrossRef]
11. Kong, Y.; Li, S.; Liu, M.; Li, M.; Yang, X.; Yao, C.; Zhao, N. Effect of dietary organic selenium on survival, growth, antioxidation, immunity and gene expressions of selenoproteins in abalone *Haliotis discus hannai*. *Aquac. Res.* **2019**, *50*, 847–855. [CrossRef]
12. Jobling, M. National Research Council (NRC): Nutrient requirements of fish and shrimp. *Aquac. Int.* **2012**, *20*, 601–602. [CrossRef]
13. Wang, L.; Sagada, G.; Wang, R.L.; Li, P.W.; Xu, B.Y.; Zhang, C.; Qiao, J.L.; Yan, Y. Different forms of selenium supplementation in fish feed: The bioavailability, nutritional functions, and potential toxicity. *Aquaculture* **2022**, *549*, 737819. [CrossRef]

14. Zheng, L.; Feng, L.; Jiang, W.D.; Kuang, S.Y.; Liu, Y.; Tang, L.; Wu, P.; Zeng, Y.Y.; Zhou, X.Q. Selenium deficiency impaired immune function of the immune organs in young grass carp (*Ctenopharyngodon idella*). *Fish Shellfish Immunol.* **2018**, *77*, 53–70. [CrossRef]
15. Zhou, X.; Wang, Y.B.; Gu, Q.; Li, Y.F. Effects of different dietary selenium sources (selenium nanoparticle and selenomethionine) on growth performance, muscle composition and glutathione peroxidase enzyme activity of crucian carp (*Carassius auratus gibelio*). *Aquaculture* **2009**, *291*, 78–81. [CrossRef]
16. Yang, D.Q.; Chen, F.R.; Guo, L. Effects of dietary choline supplementation on growth, tissue fat content, and digestive enzyme activity in rice-field eels. *J. Fish.* **2006**, *5*, 676–682.
17. Gong, B.; Ju, D.D.; Xu, H.M.; Xiao, M.S. Effects of Selenomethionine on Growth Performance, Selenium Content in Muscle Tissue and Serum Biochemical Indicators of rice field eels. *Feed Res.* **2023**, *46*, 61–64. [CrossRef]
18. *GB/T 6435-2014*; Meng, F.S.; Zhang, S.; Su, S.L.; Liu, J.M.; Liang, M.; Li, H.; Guo, J.; Gong, L.L. National Standard of the People's Republic of China: Animal feeding stuff—Determination of crude moisture. China Standards Press: Shanghai, China, 2014.
19. *GB/T 6432-2018*; Xiao, Z.M.; Fan, X.; Ma, D.X.; Li, L.B.; Wang, S.; Jia, Z.; Liu, X.X.; Wang, Z.G.; Liu, J. National Standard of the People's Republic of China: Determination of crude protein in feeds—Kjeldahl method. China Standards Press: Shanghai, China, 2018.
20. *GB/T 6433-2006*; Zhang, S.; Li, L.P.; Fan, Z.Y.; Yang, S.M.; Su, X.O. National Standard of the People's Republic of China: Animal feeding stuff—Determination of crude fat. China Standards Press: Shanghai, China, 2016.
21. *GB/T 6438-2007*; Wu, R.L.; Yang, L.; He, Y.F.; Qian, L.; Zhong, M.; Yang, A.H.; Zhang, C.F. National Standard of the People's Republic of China: *Animal feeding stuff*—Determination of crude ash. China Standards Press: Shanghai, China, 2007.
22. Eggers, D.M. Some recent methods for estimating food-consumption by fish-reply. *J. Fish. Res. Board Can.* **1979**, *36*, 1018–1019.
23. Lee, S.; Nambi, R.W.; Won, S.; Bai, S.C.; Katya, K. Dietary selenium requirement and toxicity levels in juvenile Nile tilapia, *Oreochromis niloticus*. *Aquaculture* **2016**, *464*, 153–158. [CrossRef]
24. Abd El-Kader, M.F.; Fath El-Bab, A.F.; Abd-Elghany, M.F.; Abdel-warith, A.W.A.; Dawood, M.A.O.; Younis, E.M. Selenium nanoparticles act potentially on the growth performance, hemato-biochemical indices, antioxidative, and immune-related genes of European seabass (*Dicentrarchus labrax*). *Biol. Trace Elem. Res.* **2021**, *199*, 3126–3134. [CrossRef]
25. Dawood, M.A.; Koshio, S.; Zaineldin, A.I.; Abdel-daim, M.M.; Elsabagh, M.; Van Doan, H.; Ahmed, H.A. An evaluation of dietary selenium nanoparticles for red sea bream (*Pagrus major*) aquaculture: Growth, tissue bioaccumulation, and antioxidative responses. *Environ. Sci. Pollut. Res.* **2019**, *26*, 30876–30884. [CrossRef]
26. Dawood, M.A.O.; Koshio, S.; Ishikawa, M.; Yokoyama, S.; El Basuini, M.F.; Hossain, M.S.; Nhu, T.H.; Moss, A.S.; Dossou, S.; Wei, H. Dietary supplementation of β-glucan improves growth performance, the innate immune response and stress resistance of red sea bream (*Pagrus major*). *Aquac. Nutr.* **2017**, *23*, 148–159. [CrossRef]
27. Izquierdo, M.S.; Ghrab, W.; Roo, J.; Hamre, K.; Hernández-Cruz, C.M.; Bernardini, G.; Terova, G.; Saleh, R. Organic, inorganic and nanoparticles of Se, Zn and Mn in early weaning diets for gilthead seabream (*Sparus aurata*). *Aquac. Res.* **2017**, *48*, 2852–2867. [CrossRef]
28. Zhu, Y.; Chen, Y.; Liu, Y.; Yang, H.; Liang, G.; Tian, L. Effect of dietary selenium level on growth performance, body composition and hepatic glutathione peroxidase activities of largemouth bass (*Micropterus salmoide*). *Aquac. Res.* **2012**, *43*, 1660–1668. [CrossRef]
29. Brandt-Kjelsen, A.; Govasmark, E.; Haug, A.; Salub, B. Turnover of Se in adequately fed chickens using Se-75 as a tracer. *J. Anim. Physiol. Anim. Nutr.* **2014**, *98*, 547–558. [CrossRef]
30. Lin, Y.H. Effects of dietary organic and inorganic selenium on the growth, selenium concentration and meat quality of juvenile grouper *Epinephelus malabaricus*. *Aquaculture* **2014**, *430*, 114–119. [CrossRef]
31. Davis, K.B. Temperature affects physiological stress responses to acute confinement in sunshine bass (*Morone chrysops* × *Morone saxatilis*). *Comp. Biochem. Physiol. Part A Mol. Integr. Physiol.* **2004**, *139*, 433–440. [CrossRef]
32. Mansour, A.T.E.; Goda, A.A.; Omar, E.A.; Khalil, H.S.; Esteban, M.A. Dietary supplementation of organic selenium improves growth, survival, antioxidant and immune status of meagre, *Argyrosomus regius*, juveniles. *Fish Shellfish Immunol.* **2017**, *68*, 516–524. [CrossRef] [PubMed]
33. El-Sharawy, M.E.; Hamouda, M.; Soliman, A.A.; Amer, A.A.; El-Zayat, A.M.; Sewilam, H.; Younis, E.M.; Abdel-Warith, A.W.A.; Dawood, M.A.O. Selenium nanoparticles are required for the optimum growth behavior, antioxidative capacity, and liver wellbeing of Stripped catfish (*Pangasianodon hypophthalmus*). *Saudi J. Biol. Sci.* **2021**, *28*, 7241–7247. [CrossRef]
34. Jahanbakhshi, A.; Pourmozaffar, S.; Adeshina, I.; Erfanifar, E.; Mahmoudi, R.; Ajdari, A. Selenium nanoparticle and selenomethionine as feed additives: Effects on growth performance, hepatic enzymes' activity, mucosal immune parameters, liver histology, and appetite-related gene transcript in goldfish (*Carassius auratus*). *Fish Physiol. Biochem.* **2021**, *47*, 639–652. [CrossRef]

35. Pour, H.D.; Mousavi, S.M.; Zakeri, M.; Kochanian, P. Synergistic Effects of Selenium and Magnesium Nanoparticles on Growth, Digestive Enzymes, Some Serum Biochemical Parameters and Immunity of Asian Sea Bass (*Lates calcarifer*). *Biol. Trace Elem. Res.* **2021**, *199*, 3102–3111. [CrossRef]
36. Martínez-Álvarez, R.M.; Morales, A.E.; Sanz, A. Antioxidant defenses in fish: Biotic and abiotic factors. *Rev. Fish Biol. Fish.* **2005**, *15*, 75–88. [CrossRef]
37. Pacini, N.; Abete, M.C.; Dörr, A.J.M.; Elia, A.C.; Natali, M.; Prearo, M. Detoxifying response in juvenile tench fed by selenium diet. *Environ. Toxicol. Pharmacol.* **2012**, *33*, 46–52. [CrossRef] [PubMed]
38. Zhang, Q.W.; Chen, C.; Shao, Y.X.; ZHAO, T.T. Effects of Selenomethionine on Growth Performance, Tissue Selenium Content, Antioxidant Capacity, and Serum Biochemical Parameters in Striped Bass. *Chin. J. Anim. Nutr.* **2018**, *30*, 4746–4756. Available online: https://link.cnki.net/urlid/11.5461.S.20180831.1100.028 (accessed on 31 August 2018). (In Chinese)
39. Yu, H.; Zhang, C.; Zhang, X. Dietary nano-selenium enhances antioxidant capacity and hypoxia tolerance of grass carp (*Ctenopharyngodon idella*) fed with high-fat diet. *Aquac. Nutr.* **2020**, *26*, 545–557. [CrossRef]
40. Tsai, C.F.; Wu, J.Y.; Hsu, Y.W. Protective effects of rosmarinic acid against selenite-induced cataract and oxidative damage in rats. *Int. J. Med. Sci.* **2019**, *16*, 729. [CrossRef] [PubMed]
41. Jia, R.; Gu, Z.; He, Q.; Ga, L.P.; Du, J.L.; Jeney, G.; Xu, P.; Yin, J.G. Anti-oxidative, anti-inflammatory and hepatoprotective effects of Radix Bupleuri extract against oxidative damage in tilapia (*Oreochromis niloticus*) via Nrf2 and TLRs signaling pathway. *Fish Shellfish Immunol.* **2019**, *93*, 395–405. [CrossRef]
42. Copple, I.M.; Goldring, C.E.; Kitteringham, N.R.; Park, K.B. The Nrf2–Keap1 defence pathway: Role in protection against drug-induced toxicity. *Toxicology* **2008**, *246*, 24–33. [CrossRef]
43. Itoh, K.; Chiba, T.; Takahashi, S. An Nrf2/small Maf heterodimer mediates the induction of phase II detoxifying enzyme genes through antioxidant response elements. *Biochem. Biophys. Res. Commun.* **1997**, *236*, 313–322. [CrossRef]
44. Sivieri, K.; Bassan, J.; Peixoto, G.; Monti, R. Gut microbiota and antimicrobial peptides. *Curr. Opin. Food Sci.* **2017**, *13*, 56–62. [CrossRef]
45. Trenzado, C.E.; Carmona, R.; Merino, R.; Domezain, A.; Furne, M.; García-gallego, M.; Sanz, A. Effect of dietary lipid content and stocking density on digestive enzymes profile and intestinal histology of rainbow trout (*Oncorhynchus mykiss*). *Aquaculture* **2018**, *497*, 10–16. [CrossRef]
46. Ning, Y.; Wu, X.; Zhou, X.; Zhou, R.; Yang, Z.; Chang, Y.; Ding, J.; Huang, Z. An evaluation on the selenium yeast supplementation in the practical diets of early juvenile sea cucumber (*Apostichopus japonicus*): Growth performance, digestive enzyme activities, immune and antioxidant capacity, and body composition. *Aquac. Nutr.* **2021**, *27*, 2142–2153. [CrossRef]
47. Li, X.M.; Xu, T.; Sun, G.H.; Yang, J.M.; Li, G.D.; Li, H.Z. Effects of Exogenous Selenium on the Activity of Key Physiological Enzymes and Selenium Levels in Sea Cucumber (*Apostichopus japonicus*). *Prog. Fish. Sci.* **2017**, *38*, 154–163.
48. Hao, T.T.; Yu, C.L.; Wang, C.Q.; Wang, X.Y.; Li, L. Effects of Selenium Supplementation in High Plant Protein Feed on Liver and Intestinal Health of Juvenile Turbot (*Scophthalmus maximus* L.). *J. Guangdong Ocean Univ.* **2024**, *44*, 1–11. Available online: https://link.cnki.net/urlid/44.1635.n.20241031.0839.002 (accessed on 31 October 2024). (In Chinese)
49. Chen, Q.; Li, G.Y.; Luo, K.; Jie, K.; Mo, Z.L. Icrobial Diversity in Water During the Breeding Period of Litopenaeus vannamei Broodstock. *Oceanol. Limnol. Sin.* **2017**, *48*, 130–138.
50. Zhao, X.G.; Luo, H.; Liu, Q.G.; Zhao, L.J.; Cai, L.R.; Dai, L.L.; Zhang, Z. Effects of Rice-Fish Farming with Odontobutis obscura on Microbial Community Structure and Diversity in Paddy Field Water and Sediment. *Freshw. Fish.* **2017**, *47*, 8–14.
51. Ma, Y.; Su, Z.; Chen, F. Terrestrial compound protein replacing dietary fishmeal improved digestive enzymeactivity, immune response, intestinal microflora composition, and protein metabolism of golden pompano (*Trachinotus ovatus*). *Aquac. Nutr.* **2023**, *2023*, 2716724. [CrossRef]
52. Perry, W.B.; Lindsay, E.; Payne, C.J.; Brodie, C.; Kazlauskaite, R. The role of the gut microbiome in sustainable teleost aquaculture. *Proc. Biol. Sci.* **2020**, *287*, 20200184. [CrossRef]
53. Tsuchiya, C.; Sakata, T.; Sugita, H. Novel ecological niche of *Cetobacterium* somerae, an anaerobic bacterium in the intestinal tracts of freshwater fish. *Lett. Appl. Microbiol.* **2008**, *46*, 43–48. [CrossRef]

animals

MDPI

Article

Genomic Characterization of Four Novel Probiotic Strains with Enzymatic Activity and Their Effects on Carp (*Cyprinus carpio*)

Evgeniya Valeryevna Prazdnova *, Maria Sergeevna Mazanko, Victoria Nikolaevna Shevchenko, Radomir Viktorovich Skripnichenko, Maksim Pavlovich Kulikov, Lilia Sergeevna Golovko, Vadim Alexeevich Grigoriev, Tatiana Alexandrovna Maltseva, Daria Borisovna Kulikova and Dmitry Vladimirovich Rudoy

Research Laboratory «Agrobiotechnology Center», Don State Technical University, Gagarina Sq. 1, Rostov-on-Don 344002, Russia; vshevchenko@donstu.ru (V.N.S.); aqua-group@yandex.ru (V.A.G.); tamalceva@donstu.ru (T.A.M.); rudoy.d@gs.donstu.ru (D.V.R.)

* Correspondence: prazdnova@sfedu.ru

Simple Summary

This study evaluates four novel *Bacillus* strains as probiotics to enhance plant-based feed utilization in carp aquaculture. Two preparations based on two pairs of synergistic strains significantly improved the weight gain of carp when added to feed on early stages of fish life, and genome analysis confirmed that these bacteria have the necessary traits to survive in the fish gut, such as resistance to gastrointestinal enzymes and acidic medium; to help digest food, while they have enzymatic activity that promotes the splitting of complex plant components; and fight harmful microbes by producing antimicrobial substances. No significant gene expression changes occurred in juveniles, though one of two preparations transiently modulated fry expression profiles. The results indicate that the mechanism of action of these probiotics is primarily mediated by their enzymatic activity. Beyond characterizing novel aquaculture probiotics, this work identifies genetic markers for the selection of potential probiotic strains in the future.

Abstract

The shift toward plant-based feeds in aquaculture necessitates a search for probiotics to improve the digestibility of such feeds and fish growth. This study characterizes four *Bacillus* strains for their probiotic potential in carp (*Cyprinus carpio*). Strains isolated from river sediments underwent in vitro screening for proteolytic/amylolytic activity, antibiotic sensitivity, and biofilm formation. Whole-genome sequencing was performed and functional annotation identified probiotic-related genes. All strains were identified as *Bacillus velezensis* and harbored genes providing thermal, osmotic, and oxidative stress resistance, alongside enzymes and bioactive metabolite synthesis pathways. Two non-antagonistic strain pairs were solid-phase cultivated on soybeans and incorporated into carp feed (0.1% *w/w*) for a 76-day trial, and growth metrics and gene expression were analyzed in juvenile fish. Group 1 (strains MT14 + MT42) exhibited a 40.75% higher biomass increase than the control group while Group 2 (strains MT141 + MT142) exhibited a 56.62% higher biomass growth. No significant gene expression changes occurred in juveniles, though MT141 + MT142 transiently modulated fry expression profiles. Therefore, these *Bacillus velezensis* strains synergistically enhance growth in carp that are fed plant-rich diets, highlighting their viability as aquaculture probiotics.

Keywords: probiotic; *Bacillus*; aquaculture; lytic enzymes; non-ribosomal peptides; microbiota; feed conversion ratio

1. Introduction

Aquaculture is a rapidly developing area playing a highly important role in the provision of food security, especially in developing countries. It provides people with an accessible protein source and ensures the economic growth of coastal and rural regions [1,2]. The scarcity of fish in natural water bodies caused by overfishing and pollution further emphasizes the need to develop intensive and sustainable fish farming [3]. Demand for seafood is expected to double by 2050, and aquaculture has a crucial role to play in meeting this demand [4].

Intensive fish farming comes with certain challenges, such as high stress levels in farmed fish [3]. The economic side of commercial aquaculture is also an important issue. A significant part of the costs in intensive fish farming (40–70%) is spent on feed [5]. In carp farming, feed costs have increased by 60–80% in recent years [6].

Another challenge is growth stimulation. Rapid individual weight gain in fish is known to improve winter survival under conditions of boreal and continental climates when water temperatures drop to critical values and hydrobionts' life preservation depends solely on reserves accumulated during the growing season. In European carp farming, there are two wintering periods before the fish reach commercial size [7].

For a long time, the main source of protein in feed has been fishmeal, but its limitations force us to look for alternatives. Therefore, the use of vegetable ingredients (soy, rapeseed, and various grains) is increasing: [8–10]. Fish are unable to produce cellulolytic enzymes, and the presence of cellulose in feed can be a problem affecting the digestibility of food. Adding probiotic bacteria to food that produces cellulolytic exoenzymes can improve the digestibility and assimilation of feed with plant components in the composition [11–13]. The prevention of pathological conditions by functional feeds is a preferable strategy for the aquaculture industry than treating existing diseases [14]. Emphasis is now being placed on developing more environmentally friendly and sustainable methods of disease management, such as the use of probiotic microorganisms, included in feed formulations [15].

Probiotics are live microorganisms that, when administered in sufficient quantities, have beneficial effects on the host organism's health [16]. In aquaculture, their application shows promising results, improving various aspects of fish health and ecosystem stability. Probiotic microorganisms stimulate the immune system, increasing the resistance of macroorganisms to infectious diseases. Additionally, they improve digestion, facilitating more efficient nutrient absorption from feed and resulting in enhanced growth rates and feed conversion ratios [17,18].

Bacteria of the genus *Bacillus* rank among the most commonly used probiotics in aquaculture [19]. One of the most important features of *Bacillus* is its ability to form endospores. These spores exhibit remarkable resilience to adverse environmental conditions such as heat, drying, and exposure to disinfectants. Their durability ensures long-term bacterial viability, which is an important factor for aquaculture applications [15].

Bacillus species are widespread in nature and can be found in soil [20], water [21], and fish intestines [22]. Strains used in aquaculture are isolated from various sources: soil, water from fish farms, fish intestines, and commercial preparations [20,21,23–26]. Notably, many *Bacillus* species inhabiting the intestines of fish are autochthonous microorganisms, i.e., they are natural inhabitants of this environment and form part of the normal microflora [3,27]. *Bacillus* strains are suitable probiotics for aquaculture because they are frequently found in both freshwater and seawater and the gastrointestinal tract of animals [28]. Some *Bacillus* probiotics were isolated from the gastrointestinal tract of healthy fish [29–34]. This indicates that they are well adapted and tend to colonize rather than just transit through the gastrointestinal tract [35], underscoring the safety and efficacy of *Bacillus*-based probiotics in aquaculture [3].

Probiotics containing enzymes such as protease, amylase, cellulase, and xylanase can significantly improve feed efficiency and growth performance in various aquaculture species such as white shrimp (*Litopenaeus vannamei*), zebrafish (*Danio rerio*), common pangasius (*Pangasius bocourti*), European trachinotus (*Trachinotus ovatus*), and Nile tilapia (*Oreochromis niloticus*) [30,36–40]. Various studies indicate that *Bacillus* can produce digestive enzymes such as amylase, protease, lipase, lipase, cellulase, and xylanase [35,41,42]. Studies demonstrate that enzymes secreted by bacteria of the genus *Bacillus* enhance nutrient assimilation in marine animals from cost-effective plant-based resources [43]. The introduction of probiotic strains, in addition to realizing antagonistic and immunomodulatory effects, also improves nutrient uptake efficiency, including in the larval stages of ontogenesis. Probiotic strains can improve activity of digestive enzymes such as lipase, protease, and amylase, leading to better growth and survival of fish [44,45].

Bacilli are recognized as natural inhabitants of the carp gut microbiota [46]. Therefore, bacilli-based probiotics are more likely to successfully colonize and exert beneficial effects in fish than species uncharacteristic of carp. Furthermore, bacilli are able to form spores that survive drying, which greatly facilitates the production, storage, and industrial application of probiotic formulations [47]. For these reasons, bacilli were selected as the focus of our search for new probiotics.

Thus, probiotics with an enzymatic activity contribute to more efficient feed digestibility, increasing the growth rate in commercial aquaculture facilities [12]. The objective of this work is to isolate new probiotic strains with enzymatic activity for aquaculture, characterize their properties through microbiological assays and full-genome sequencing, and evaluate their effects on carp.

2. Materials and Methods

2.1. Strain Isolation and Growth

Bottom sediments were collected from the Don River, at the experimental research base of the Southern Scientific Center of the Russian Academy of Sciences, Kagalnik, Rostov Region, Russia (47.09619 north latitude, 39.30526 east longitude). Samples were taken at the depth 0–5 cm, pasteurized (95 °C, 5 min) and immediately inoculated on test media.

The bacteria were incubated for 24 h at 30 °C. Proteolytic activity was tested on solid nutrient LB (Luria–Bertani) medium supplemented with 200 mL/liter milk (Groupe Danone, Krasnodar, Russia). Colonies surrounded by clear zones (casein hydrolysis) were selected. Amylolytic activity was tested on LB agar with the addition of 10 g/L starch (SP-Don, Rostov-on-Don, Russia). Colonies with clear zones (starch degradation) after iodine staining were selected. Cellulolytic activity was tested on solid LB supplemented with 10 g/L carboxymethylcellulose (LenReactiv, St. Petersburg, Russia). After Congo red (LenReactiv, St. Petersburg, Russia) staining, colonies with unstained zones (CMC hydrolysis) were selected. Inoculation was carried out in 20 replications on each medium.

During isolation of probiotic *Bacillus* from natural sources and intestinal contents, colony morphology on solid media was the primary selection criterion. *Bacillus* usually exhibit considerable morphological diversity, with colonies on the same nutrient medium varying in size, edge shape, colony height, color, density, surface texture, presence of wrinkles, their number, thickness, height, and location [48–50]. To maximize genetic diversity, morphologically identical strains from a single sample were excluded from further analysis.

Colonies that showed a high degree of activity (≥8 mm) were streaked onto LB agar and examined for strain purity using a depleting stroke. Purified strains were re-tested on respective media (4 colonies/plate), and clear zone diameters were measured. Reference strains *B. amyloliquefaciens* B1895 (GenBank number JMEG00000000) and *B. subtilis* KATMIRA, which have repeatedly demonstrated their probiotic activity in our previous studies, were used as controls [51–53].

2.2. Assessment of Strain Safety

To evaluate the safety of using the strains as probiotics, we investigated their antibiotic resistance as well as hemolytic activity.

Standard antibiotic disks were used for antibiotic resistance study: azithromycin 15 μg; amoxicillin/clavulanate 20/10 μg; ampicillin 10 μg; tetracycline 30 μg; gentamicin 10 μg; oxacillin 10 μg; clarithromycin 15 μg; cefazolin 30 μg; and ceftriaxone 30 μg (NICF, Tver, Russia). The tested strains were lawn-inoculated on Muller–Hinton agar (LenReactiv, St. Petersburg, Russia), then antibiotic disks were placed three per plate. Incubation was done for one day at 37 °C. After a day, inhibition zones were measured. Only strains with inhibition zones > 8 mm for any clinically important antibiotic were retained. Each test was carried out in three replications.

To assess hemolytic activity, selected strains were streaked onto blood agar (7% bull blood) and incubated for one day at 37 °C. Levels of β-hemolysis (complete lysis) or α-hemolysis (partial lysis) were recorded by color and width of halo. Non-hemolytic strains were selected for further work.

2.3. Strain Compatibility Testing

On LB agar plates, two strains of potential probiotic bacilli were streaked in parallel lines on undried dishes 5 mm apart. All pairwise combinations of strains were tested. Plates were incubated for 48 h at 37 °C and then interaction zones were analyzed. Each test was carried out in two replications.

2.4. Probiotic Preparation Using Solid-Phase Fermentation

Preparations were produced based on *B. velezensis* MT14, *B. velezensis* MT42, *B. velezensis* MT141, and *B. velezensis* MT142 strains, each strain prepared separately. Petri dishes containing LB medium were inoculated with the corresponding probiotic strain and incubated for 48 h at 37 °C. The resulting plates served as the starting material for the preparation of the biopreparation. Soybeans were taken on the basis that 1.2 kg of beans is required to obtain 1 kg of the final probiotic product. Soybeans were washed and soaked for 12 h at 20–22 °C. Then the hydrated soybeans were sterilized in an autoclave at 121 °C for 40 min in a food-grade aluminum container.

After cooling to 60 °C (controlled by a mercury thermometer disinfected with 70% ethanol), the starters were added to the containers and mixed thoroughly with a sterile food-grade stainless steel spoon, then incubated for 48 h at 40 °C. During this time, the biofilm of bacilli grew on a soybean.

After incubation, the substrate was mixed using a sterile spoon and ground on a pre-sterilized hand grinder. The resulting mass was placed on food-grade plastic pallets sterilized with 70% ethyl alcohol and 3% hydrogen peroxide. The preparation was carefully leveled with a layer of about 1 cm and dried on thermal mats at 40–45 °C for four days. Then the mass was turned over and dried for another three to five days until drying was complete. The preparation was ground on a ML-50B mill (Biobase, Jinan, China), the bowl of which was pre-treated with 70% ethyl alcohol and 3% hydrogen peroxide.

The bacterial count in the preparation was determined by grinding it with water (1:10) and preparing a series of decimal dilutions, which were inoculated superficially onto solid nutrient medium LB. The resulting powders were mixed in pairs to achieve the same amount of each strain in the preparation. The final probiotic spores count in the preparations was $3.7 \cdot 10^8$ CFU/g for MT42 + MT14, and $6.4 \cdot 10^8$ CFU/g for MT141 + MT142.

2.5. Experimental Diet

The diet of individuals included specialized starter mixed feeds for young carp with a pellet size of 2 mm. Production of experimental feeds consisted of the following stages: weighing of components in accordance with the recipe (laboratory scales ADAM HCB 1002, Dongguan, China, were used for weighing), mixing of dry components of feeds (including the probiotic supplement and moistening to a mass fraction of moisture 12% (mixing was carried out in a horizontal SG-1.5 mixer, (manufacturer AgroPostavka LLC, Nizhny Novgorod, Russia), water was added to the mixer through nozzles mounted on the mixer body), pelleting (carried out on a ZLSP-150 compound feed granulator with a 2 mm matrix installed, manufactured by AgroPostavka LLC, Nizhny Novgorod, Russia), cooling and sieving of pellets (carried out in a cooling column and sieving of KO 5.5 granules, manufactured by AgroPostavka LLC, Nizhny Novgorod, Russia), and oiling of pellets under vacuum (carried out in a vacuum granule oiler, manufactured by IE Shelkunov, Krasnodar Krai, Russia). Feed for the experiment was made from the following components: fish meal (27%) (Livadia-N LLC, Belgorod Region, Russia); wheat flour (17.24%) (Yug Russi LLC, Rostov-on-Don, Russia); chicken meat and bone meal (12.5%) (Livadia-N LLC, Belgorod Region, Russia); soybean meal (10%) (ROSAGROCORM LLC, Novosibirsk, Russia); pork meat meal (8%) (Livadia-N LLC, Belgorod Region, Russia); blood meal (6%) (Trading House "Feed Flour Factory" LLC, Republic of Tatarstan, Russia); corn meal (5%) (ROSAGROCORM LLC, Novosibirsk, Russia); corn gluten (3%) (ROSAGROCORM LLC, Novosibirsk, Russia); carp fish premix (4%) (manufacturer VetAgroSnab LLC, Rostov-on-Don, Russia); linseed oil (3%) (Len OK LLC, Nizhny Novgorod Region, Russia); fish oil (3%) (manufacturer NVC Agrovetzashita LLC, Moscow, Russia); table salt (NaCl) (1%) (manufacturer Russol LLC, Orenburg, Russia); calcium lignosulfonate (pellet fixer) (0.1%) (manufacturer Polyplast JSC, Russia); probiotic preparation (0.1%) (produced in the Russia, Rostov-on-Don, DSTU); and astaxanthin (0.06%) (manufacturer VetAgroSnab LLC, Rostov-on-Don, Russia). Two probiotic preparations were used: the first consisted of a mixture of probiotic strains *Bacillus velezensis* MT14 and *B. velezensis* MT42 (the same concentration), the second consisted of a mixture of probiotic strains *B. velezensis* MT141 and *B. velezensis* MT142 (the same concentration). The nutritional composition of the feed was developed taking into account the following target indicators: crude protein content—approximately 47% (±1.5%), crude fat content—approximately 13% (±1.5%), crude ash content—approximately 5% (±0.6%), and crude fiber content—approximately 2% (±0.5%). The water resistance of the granule was 35 min (±0.5 min).

The daily feeding rate was 3% of the fish biomass in the pool. Feeding was carried out every 4 h daily. Probiotic strains *Bacillus velezensis* MT14 and *Bacillus velezensis* MT42 were included in the diet of experimental group No. 1. *Bacillus velezensis* MT141 and *Bacillus velezensis* MT142 were included as part of the diet of experimental group No. 2. The content of probiotic preparation in the total composition of feed components was 0.1%. The diet of the control group included feed without any probiotic added. To confirm the persistence of bacilli in the feed, the feed was rubbed with water (1:10) and a series of decimal dilutions were prepared and inoculated superficially onto LB solid nutrient medium. The dishes were incubated for 24 h at 37 °C and then counted.

2.6. Experimental Design and Fish Conditions

The study was approved by the Local Independent Ethical Committee of the Don State Technical University (Rostov-on-Don, Russia). During the experiments, the recommendations for humane and ethical treatment of laboratory animals were followed, including in accordance with Directive 2010/63/EU of the European Parliament and of the Council of 22 September 2010 "On the Protection of Animals Used for Scientific Purposes".

Juvenile carp *Cyprinus carpio* Linnaeus, 1758 in the amount of 450 specimens obtained in July 2024 from producers kept in the pond fish farm (Rostov Region, Russia) were used for the study. At the start of experiment the fish were clinically healthy, with no external signs of diseases. Health status (excess mucous secretion, normal coloration, erosion of scales or fins, skin, exophthalmos and presence of cysts, spots, or areas on the body and gills) and behavioral signs (swimming and feeding reflexes) were controlled by physical examination [54]. Fish were randomly distributed into three experimental groups: control (150 specimens), experiment No. 1 (150 specimens) and experiment No. 2 (150 specimens). The duration of the experiment was 76 days.

The fish were kept in round tanks as part of a recirculating aquaculture system (RAS), which included mechanical and biological filtration of recycled water, ultraviolet radiation treatment for disinfection, and an oxygenator for aerating purified water. Each tank was made of polypropylene and had a diameter of 150 cm and height of the water column of 70 cm. At the beginning of the experiment, 150 juvenile carp were placed in each tank (121 specimens/m^3). Mortalities were recorded daily. The biomass of fish at the beginning of the experiment was 0.57 kg in the control group, 0.55 kg in experimental group No. 1, and 0.44 kg in experimental group No. 2.

During the experiment, the water temperature in the experimental pools averaged 21.95 ± 0.19/(21.58–22.37)°C [55]. The concentration of oxygen (O_2) dissolved in water during the experiment stayed within the optimal range for juvenile carp and averaged 8.98 ± 0.21(8.53–9.46) mg/L. The pH of the water was on average 7.62 ± 0.45/(7.49–8.01), which are normal values for carp content [56].

To minimize handling, control measurements of the size and mass values of fish were carried out twice during the experiment period: at the beginning and at the end. At the end of the experiment, feeding was stopped for 24 h. A random sample of 60 specimens was taken from each group, and they were anesthetized using clove powder (150 mg/L) [57] to determine biometric parameters. The length of the fish was measured to the end of the scale cover (L, cm) using a Gigant 0–300 mm GSMI-300 digital vernier caliper (LLC Gigant-Pribor, Izhevsk, Russia) with a measurement accuracy of 0.01 mm. The mass of fish (m, g) was determined using laboratory scales of accuracy class II BEL LG-2202i with a discreteness of 0.01 g (ChangZhou XingYun Electronic Equipment Co., Changzhou, China).

To evaluate efficiency of feeding, calculation of the following parameters was made [58–62]

- Fulton's fatness coefficient (QF, conventional units) according to the following formula:

$$Q_F = \frac{m * 100}{l^3}$$

- individual biomass growth (WGi, g) using the following formula:

$$WGi = m_1 - m_0$$

- total biomass growth (WGt, kg) according to the following formula:

$$WGt = M_1 - M_0$$

- feed conversion rate (FCR, kg/kg) basing on the following formula:

$$FCR = \frac{M_f}{WGt}$$

- fish survival rate (S, %) estimated with the following formula:

$$S = 100 * \frac{N_1}{N_0}$$

- specific rate of weight growth (SGRW, %/day) determined according to the following formula:

$$SGRW = 100 * \frac{In(m_1) - In(m_0)}{t}$$

- specific growth rate of fish in length (SGRL, %/day) using the following formula:

$$SGRL = 100 * \frac{In(l_1) - In(l_0)}{t}$$

where t—duration of the experiment, days; l_1—length at the end of the experiment, cm; l_0—length at the beginning of the experiment, cm; m_1—weight at the end of the experiment, g; m_0—weight at the beginning of the experiment, g; N_1—number of fish at the end of the experiment, specimens; N_0—the number of fish at the beginning of the experiment, specimens; M_f—feed weight, kg; M_1—feed weight, kg; M_0—fish biomass at the beginning of the experiment, kg; m—weight, g; l—length, cm.

The significance of the differences in the obtained values was determined using Student's parametric criterion.

2.7. Determination of the Probiotic Bacteria Amount in Fish Intestinal Contents

The intestines were cooled and transported to the laboratory within 2 h. The posterior intestine was separated using sterile instruments, and intestinal contents were extracted into a microtube. Contents from the intestines of five individuals were pooled, homogenized, and averaged.

Microbiological analysis was performed to estimate both vegetative forms of bacilli and spores. To count total bacilli, undiluted samples were serially diluted (1:10 in sterile saline), plated on LB agar, and incubated 48 h at 37 °C. For spore count, homogenates were pasteurized (95 °C, 5 min) to kill vegetative cells, followed by dilution and plating. Colonies were counted after 48 h. Spore counts were derived from post-pasteurization plates, while total bacilli (vegetative and spores) were enumerated from pre-pasteurization plates. Colony morphology (wrinkled, dry, opaque) and Gram staining (rod-shaped, spore-forming) were used to confirm Bacillus identity. Vegetative cell counts were calculated as total CFU—spore CFU.

2.8. Genomic DNA Extraction, Sequencing, Assembly and Annotation

DNA was isolated according to the protocol described by Gautam [63]. Sequencing was performed on the MinION device using the Rapid Barcoding Sequencing Kit V14 reagent kit in accordance with the instructions. Quality control of the readings was carried out using the NanoPlot v1.42.0 program, and filtering and rejection of low-quality readings using the Filtlong v0.2.1 (—min_length 200; —keep_percent 90), discarding the worst 10% of bases by quality and retaining reads $\geq$ 200 bp.

Genomes were assembled using Flye 2.9.5 with "—nano-hq" flag. The quality of the genome assembly was assessed using the CheckM v1.2.3 software (assessment based on the presence/absence of single-copy genes) and CheckM2 v1.0.2 (assessment program based on machine learning) and by visual assessment of the assemblage to the ring chromosome using Bandage v0.9.0. The circular genome map was built using the Proksee service [64]. The annotation was conducted using Bakta v1.8.2 (DB: v5.0—Light) [65].

The analysis of antibiotic resistance genes was carried out using CARD RGI (The Comprehensive Antibiotic Resistance Database (CARD) Resistance Gene Identifier (RGI)) [66]. The functional analysis of the metabolic pathways was performed using Blast KOALA [67].

The analysis of genes for the synthesis of secondary metabolites was carried out using AntiSMASH with "—relaxed" detection strictness parameter [68].

2.9. Phylogenetic Analysis

The taxonomic identification was performed on the basis of ANI (Average Nucleotide Identity), according to the GTDB taxonomy (Genome Taxonomy Database) using the classify_wf GTDB-Tk protocol v4.2.0 [69] based on The Genome Taxonomy Database (GTDB) (release 220) [70]. The phylogenetic tree was built using the de_novo_wf protocol based on concatenated alignment with 120 phylogenetically informative markers. The tree was visualized using the Interactive Tree of Life (iTOL) online [71].

2.10. Analysis of the Differential Expression in Carp Tissues

Fragments of carp muscle tissue were used to analyze the differential expression of genes. The analysis was performed in four biological replicates (cDNA from the muscles of different fish) and three technical (cDNA from the muscles of one fish). Fish were divided by age into fry (14 days from the start of active nutrition) and adults (one month from the start of active nutrition). Based on probiotic treatment, two groups were formed: Group 1 (strains MT14 + MT42) and Group 2 (strains MT141 + MT142). During transportation and storage, tissue fragments were fixed using the IntactRNA reagent (Evrogen, Moscow, Russia), and subsequent RNA isolation was performed using the ExtractRNA kit (Evrogen, Moscow, Russia) in accordance with the manufacturer's recommendations. The reverse transcription was performed using the MMLV RT kit (Evrogen, Moscow, Russia) in accordance with the manufacturer's recommendation. The resulting cDNA was analyzed by real-time PCR using a 5X qPCRmix-HS kit (Evrogen, Moscow, Russia) on the RotorGene Q device (Qiagen, Hilden, Germany). Several stress- and immunity-related genes were investigated, namely *cxc* (CXC chemokine gene), *lyz* (lysozyme gene), *hsp70* (heat shock protein 70 gene), *il-1β* (interleukin 1 beta gene), *il-10* (interleukin 10 gene), *igf-1* (insulin-like growth factor 1 gene), *gst* (glutathione S-transferase gene), *β2m* (beta 2 microglobulin gene), and *mt1* (metallothionein 1 gene); *act* (actin 1 beta gene) was used as the reference gene. The primers used in the study are shown in Table 1.

Reaction conditions were optimized in accordance with the selected primers and kit guidelines. Differential expression of target genes was evaluated using $2^{-\Delta\Delta Ct}$) method.

2.11. Statistical Analysis

Data normality was verified using the Shapiro–Wilk test, statistical significance was determined by two-tailed Student's *t*-test (for normally distributed data) and the Wilcoxon test (if the data distribution was non- normal). No multiple testing corrections were applied in gene-expression analyses. Visualization of the expression data was performed using the ggplot2 package. All calculations were performed in Rstudio (R 4.2.1).

Table 1. Primers for studying the expression of stress and immunity genes in carp.

Target Gene	Forward Primer Sequence (F, 5′-3′)	Reverse Primer Sequence (R, 5′-3′)	Product Length	Melting Point of Primers, °C	GC Content, %
act	GTCTACCACTTCGCCCTCATC	CAGTGTACAGAGACACCCTGG	247	F: 60.20 R: 59.73	F:57.14 R:57.14
cxc	CTGGGATTCCTGACCATTGGT	GTTGGCTCTCTGTTTCAATGCA	88	F: 55.87 R: 57.46	F: 52 R: 45
lyz	GTGTCTGATGTGGCTGTGCT	TTCCCCAGGTATCCCATGAT	359	F: 60.38 R: 57.46	F: 55 R: 50
hsp70	TGAGAACATCAACGAGCCCA	TTGTCAAAGTCCTCCCCACC	195	F: 59.59 R: 59.62	F: 50 R: 55
il-1β	ACTGGAGCTGTCTTCGCATC	CTCCAAGATGAAGCCGAGCA	136	F: 58.48 R: 59.25	F: 55 R: 55
il-10	GCTGTCACGTCATGAACGAGAT	CCCGCTTGAGATCCTGAAATAT	132	F: 58.89 R: 57.01	F: 50 R: 45
igf-1	CCGTCTCCTGTTCGCTAAATCT	CTTTGGTGTCCTGGGACTGT	554	F: 60.86 R: 59.02	F: 50 R: 55
gst	TACAATACTTTCACGCTTTCCC	GGCTCAACACCTCCTTCAC	149	F: 54.41 R: 54.25	F: 41 R: 58
β2m	CCAAATACCCAGCAGACGGA	CAGTTGCTAGGCAGACGTTTA	784	F: 56.75 R: 57.2	F: 55 R: 48
mt1	ATGGATCCTTGCGATTGCGCCA	CGAACAGGTTCACATAGGTGA	232	F: 62.94 R: 55.04	F: 55 R: 48

3. Results

3.1. In Vitro Screening and Selection of Probiotic Strains

To identify strains that possess at least one of the potential probiotic properties of interest, suspension from bottom sediments was inoculated immediately onto dishes with test media to determine enzymatic activity. A total of 51 strains, secreting exoproteases, and 52 strains, secreting exoamylases, were isolated. They were then cross-tested for both types of activity.

Of the 103 strains selected at the first stage, only 14 strains exhibited high exolytic activity: four strains for one type activity, six strains for one type activity and four strains for all three type activities (Table 2). Some of the strains had only high proteolytic or only amylolytic activity, some combined both types of activity. For data conciseness, Table 2 presents only strains with a clearance zone around the colony greater than 8 mm on at least one test medium, alongside reference probiotic strains *B. amyloliquefaciens* B1895 and *B. subtilis* KATMIRA. They were included since in our prior studies they demonstrated probiotic effects in poultry [51,52] and fish [53].

All strains listed in Table 2 were tested for antibiotic resistance (against azithromycin; amoxicillin/clavulanate; ampicillin; tetracycline; gentamicin; oxacillin; clarithromycin; cefazolin; ceftriaxone) and hemolysis. All strains were found to be safe for further use in aquaculture.

To assess the possibility of their joint use in one multi-strain preparation, pairwise co-culture assays were conducted on nutrient agar plates. Strains were inoculated 5 mm apart from each other to allow the colonies to fully develop. Plates were incubated at optimal growth conditions for 48 h without pre-drying to facilitate unrestricted colony development. The resulting interactions could be divided into three types (Table 3):

1. Synergistic Merging: Colonies exhibited unrestricted growth into each other's territories, forming confluent biomass without visible separation zone (Figure 1a, marked with "+" in Table 3);
2. Neutral Coexistence: Colonies halted growth approximately 0.5–1 mm apart, while continuing to grow in opposite direction (Figure 1b, marked with "=" in Table 3)
3. Antagonism: Growth suppression at the interaction interface, characterized by reduced biomass density or partial clearance zones (Figure 1c, marked with "−" in Table 3).

Table 2. Values of the width of the clearance zone around strains grown on the milk medium (proteolytic activity) and starch medium (amylolytic activity).

Strain	Proteolytic Activity (Width of the Clearing Zone in the Milk Medium, mm)	Amylolytic Activity (Width of the Clearing Zone in the Starch Medium, mm)	Cellulolytic Activity (Width of the Clearing Zone in the Carboxymethilcellulose Medium, mm)
MT14	8 ± 0.3	8 ± 0.6	16 ± 0.3
MT27	8 ± 0.6	8 ± 1.0	-
MT42	8 ± 0.3	11 ± 1.0	17 ± 1.0
MT48	6 ± 1.0	11 ± 0.3	16 ± 0.3
MT49	6 ± 0.6	10 ± 1.0	-
MT56	5 ± 1.0	10 ± 0.3	-
MT64	9 ± 0.3	4 ± 1.0	12 ± 0.6
MT73	5 ± 0.6	10 ± 0.3	-
MT74	8 ± 1.0	8 ± 1.0	16 ± 1.0
MT77	3 ± 0.6	10 ± 0.3	6 ± 0.3
MT84	9 ± 0.3	2 ± 0.6	-
MT102	10 ± 0.6	4 ± 1.0	18 ± 1.0
MT141	11 ± 0.3	6 ± 0.3	14 ± 0.3
MT142	10 ± 1.0	5 ± 1.0	15 ± 0.6
B1895a	5 ± 0.3	7 ± 0.6	-
Katmira	5 ± 1.0	4 ± 0.3	-

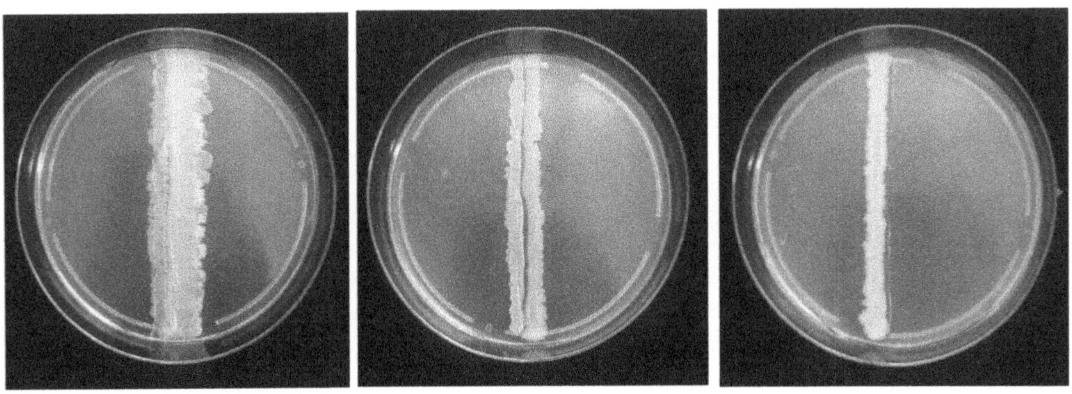

(a) (b) (c)

Figure 1. Interaction of *Bacillus* strains: (**a**). synergism (right: MT42, left: MT14); (**b**). neutrality (right: MT42, left: MT27), (**c**). antagonism (right: MT102, left: MT27).

Table 3. The interaction among strains.

	MT14	MT27	MT42	MT48	MT49	MT56	MT64	MT73	MT74	MT77	MT84	MT102	MT141	MT142
MT14	%	=	+	+	=	=	+	=	–	–	–	–	+	+
MT27	=	%	=	=	=	=	=	=	–	–	=	–	=	=
MT42	+	=	%	=	=	=	=	=	=	=	=	=	+	=
MT48	+	=	=	%	=	=	=	=	=	=	–	=	+	=
MT49	=	=	=	=	%	=	=	=	=	=	=	–	=	=
MT56	=	=	=	=	=	%	=	=	=	=	=	=	=	+
MT64	+	=	=	=	=	=	%	+	=	–	–	–	=	=
MT73	=	=	=	=	=	=	+	%	–	=	–	=	+	=
MT74	–	–	=	=	=	=	=	–	%	–	–	=	–	=
MT77	–	–	=	=	=	=	–	=	–	%	–	–	–	–
MT84	–	=	=	–	=	=	–	–	–	–	%	–	–	=
MT102	–	–	=	=	–	=	–	=	=	–	–	%	–	+
MT141	+	=	+	+	=	=	=	+	–	–	–	–	%	+
MT142	+	=	=	=	=	+	=	=	=	–	=	+	+	%

Signs "+", "–", and "=" marks interaction types described above. The % sign marks self-comparison.

Of 182 pairwise combinations, 22 showed synergy, 112 neutrality, and 48 antagonism. When developing the multistrain formulation, we avoided antagonistic pairs and prioritized synergistic pairs over neutrally interacting ones.

Among the strains with enzymatic activity, strains with both high exoamylase and exoprotease activity were selected. Among these six strains (MT14, MT27, MT42 and MT74, MT141, MT142), pairs MT14 + MT42 and MT141 + MT142 were selected, since they were able to form a joint biofilm.

3.2. *Analysis of the Results of the Whole-Genome Sequencing*

As a result of the ONT sequencing and bioinformatic reading processing, full-genomic nucleotide sequences of 4 strains—MT14, MT42, MT141, and MT42—were obtained, with 26× coverage for MT14, 81× coverage for MT42, 54× coverage for MT141, 83× coverage for MT142. Assembly statistics obtained in the CheckM2 and Bakta programs is presented in Table 4.

Table 4. Assembly statistics and genome features of *B. velezensis* strains.

Strain	MT14	MT42	MT141	MT142
Length b.p.	4,080,572	4,115,985	4,432,692	4,434,607
Contigs count	6	3	3	3
GC %	46.2	46.1	45.4	45.4
N50 b.p.	1,035,922	4,080,892	4,303,216	4,305,438
Coding density %	88.8	89.4	88.8	88.9
tRNAs	87	92	97	93
tmRNAs	1	1	1	1
rRNAs	26	30	31	31
ncRNAs	31	31	34	34
ncRNA regions	58	58	58	58
CRISPR arrays	0	0	0	0

Table 4. *Cont.*

Strain	MT14	MT42	MT141	MT142
CDSs	4571	4433	4918	4813
pseudogenes	343	245	280	215
hypotheticals proteins	515	388	605	522
signal peptides	0	0	0	0
sORFs	3	4	5	5
gaps	0	0	0	0
oriCs	2	2	2	2
Assembly completeness %	97.35	99.97	99.94	99.91
Assembly contamination %	2.19	2.1	1.19	2.32

The results of taxonomic identification using GTDB-tk based on ANI assigned all four strains to *Bacillus velezensis*. The taxonomic affiliation of the strains is shown in Figure 2.

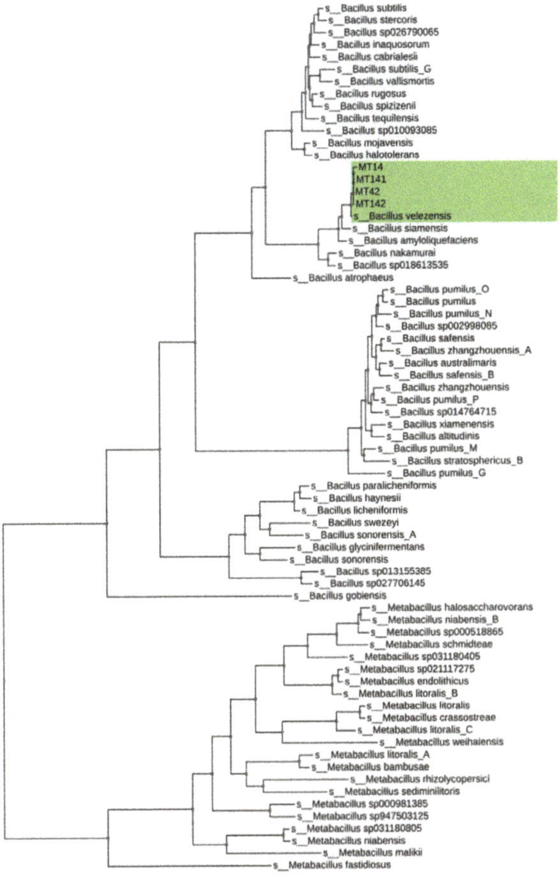

Figure 2. The phylogenetic tree for the studied strains and reference genomes from the GTDB database, based on the alignment of 120 taxonomically significant markers. The genus *Metabacillus* was selected as an outgroup for comparison. The strains used in this study, together with reference *B. velezensis* strain, are highlighted in green.

The genetic apparatus of these bacteria is represented by the ring chromosome, the visualization of which is shown in Figures 3–6.

Figure 3. Circular graphical map of *Bacillus velezensis* MT14. From outer to inner rings: ORFs on the forward strand, CDS on the forward strand, CDS on the reverse strand, ORFs on the reverse strand, GC content, contig(s). Ribosomal RNA genes are marked in red between all other CDS (violet), and ORFs are green.

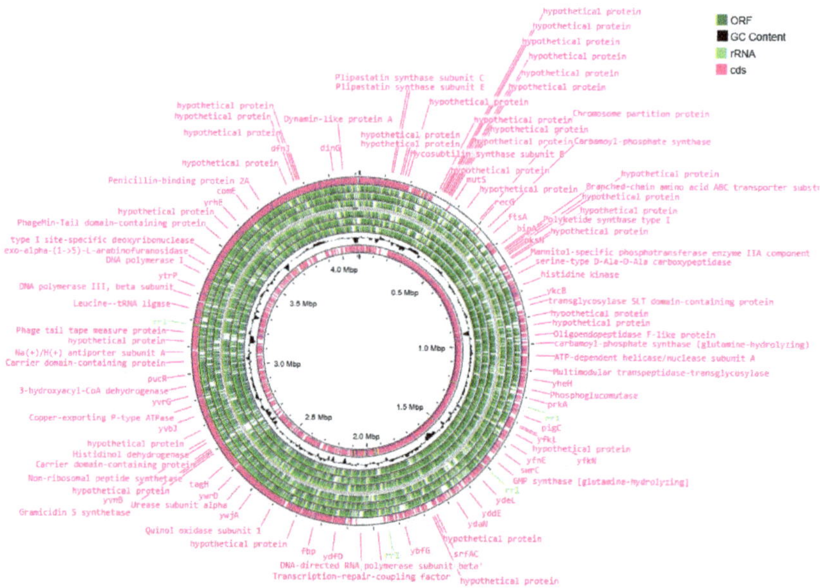

Figure 4. Circular graphical map of *Bacillus velezensis* MT42. From outer to inner rings: ORFs on the forward strand, CDS on the forward strand, CDS on the reverse strand, ORFs on the reverse strand, GC content, contig(s). Ribosomal RNA genes are marked in cyan between all other CDS (purple), and ORFs are green.

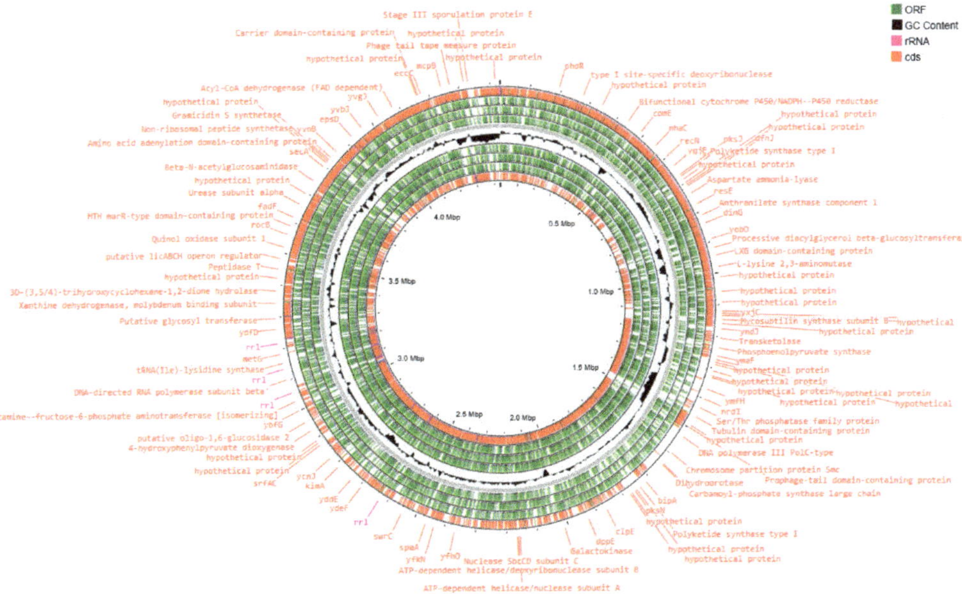

Figure 5. Circular graphical map of *Bacillus velezensis* MT141. From outer to inner rings: ORFs on the forward strand, CDS on the forward strand, CDS on the reverse strand, ORFs on the reverse strand, GC content, contig(s). Ribosomal RNA genes are marked in purple between all other CDS (orange), and ORFs are green.

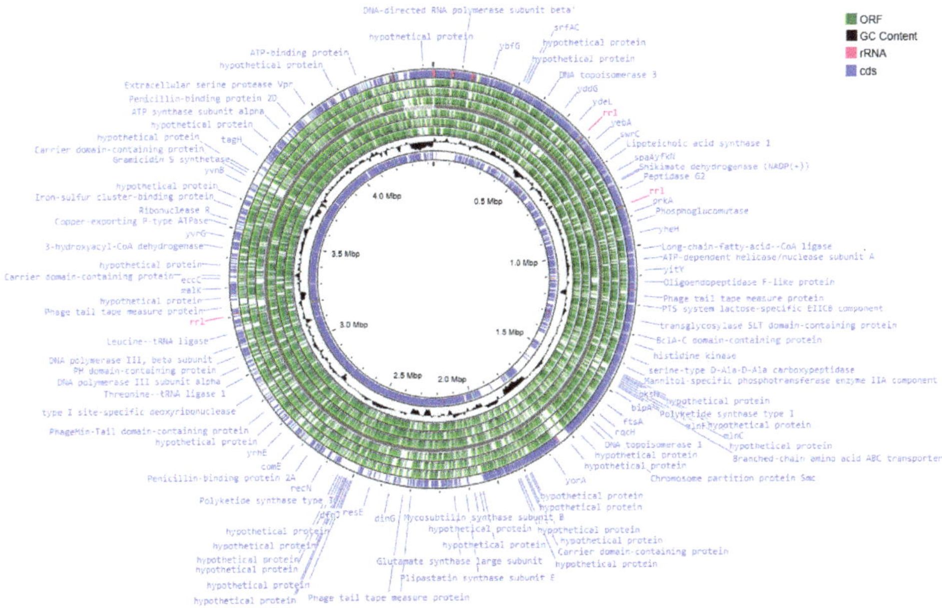

Figure 6. Circular graphical map of *Bacillus velezensis* MT142. From outer to inner rings: ORFs on the forward strand, CDS on the forward strand, CDS on the reverse strand, ORFs on the reverse strand, GC content, contig(s). Ribosomal RNA genes are marked in purple between all other CDS (blue), and ORFs are green.

The genome assembly quality was assessed using CheckM, which revealed highly complete and minimally contaminated drafts for all four strains. MT42, MT141, and MT142 exhibited near-complete assemblies with >99.9% completeness and low contamination (1.19–2.32%), while MT14 showed slightly reduced completeness (97.35%) but remained within acceptable limits for high-quality drafts. All assemblies displayed strong structural integrity, with full rRNA operons, complete tRNA sets further supporting their reliability for downstream genomic analyses. The high contiguity (N50 > 4 Mbp for MT42, MT141, and MT142 and N50 > 1Mb) and low contamination confirm that these assemblies are well-suited for comparative genomics and functional studies.

3.3. Functional Analysis of Genomes

The functional annotation of strain genomes was performed using KEGG Annotation—BlastKOALA. Of the 4571 CDSs of the MT 14 strain, 2266 were annotated (49.5%); of the 4433 CDSs of the MT 42 strain, 2292 were annotated (51.7%); of the 4918 CDSs of the MT 141 strain, 2331 were annotated (47.3%); of the 4813 CDSs of the MT 142 strain, 2332 were annotated (48.4%) with KO identifiers of the KEGG database. Based on the obtained data, the functional groups of genes of the studied strains were reconstructed using KEGG Mapper Reconstruct (Figure 7).

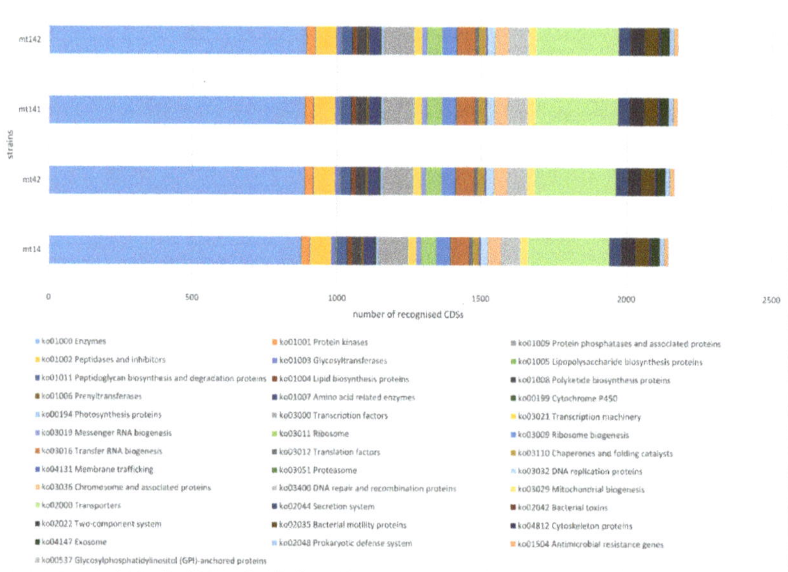

Figure 7. Groups of genes of strains MT14–142 annotated using KEGG Mapper Reconstruct.

The dominant functional categories were Enzymes (ko01000) and Transporters (ko02000), which had the highest numbers across all samples (879–896 and 279–285 CDSs, respectively). Other well-represented pathways included Transcription factors (ko03000), DNA repair and recombination proteins (ko03400), and Ribosome-related functions (ko03011/ko03009). There are no significant differences in KEGG functional categories between the strains. The vast majority of categories show identical or nearly identical CDS counts; variations are minimal and unlikely to be biologically meaningful.

3.3.1. Genes of Probiotic Properties

The genes that can determine probiotic traits were analyzed relying primarily on three functional categories: (1) providing resistance under stressful conditions.; (2) having antimicrobial properties; (3) improving adhesion and/or biofilm formation.

The first group included genes providing antioxidant effects (peroxidases—*ahpC-F*, *gpx*, *ohr*, *tpx*; catalases—*katE*, *ydbD*; superoxide dismutases—*SOD1-2*; thioredoxins—*trxA-B*; bacilliredoxins specific for Gram-positive bacteria—*brxA_B-C*; and bacillithiols—*bdr*, *bshA-C*), maintain proteome stability (temperature shock proteins and chaperones—*cspA*, *hslO*, *HSP20*, *HSP90A*, *htpX*, *dnaJ*, *dnaK*, *groEL*, *groES*, *GRPE*; intracellular proteases—*clpC-X*), provide adaptation to gastrointestinal conditions (membrane pumps-regulators of pH homeostasis—*mnhA-E*, *nhaCl*; bile acid transporters—*TC.BASS*; osmoprotectors—*opuA-D*), as well as the corresponding regulatory genes (*cssR*, *ctsR*, *hrcA*, *opcR*, *tuf*) (Figure 8).

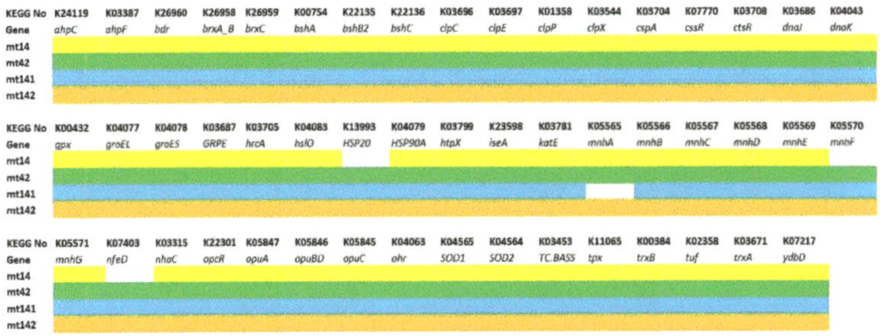

Figure 8. Stress response genes annotated in MT14–142 strains using KEGG Annotation—BlastKOALA.

The MT14 strain lacks nfeD, which is responsible for membrane-associated serine proteases, the heat shock protein HSP20, and mnhF (multicomponent Na+:H+ antiporter subunit F). The MT42 strain lacks mnhA (multicomponent Na+:H+ antiporter subunit A). These deficiencies may slightly reduce the resistance of strains MT14 and MT42 to acid and alkaline stress.

The analysis shows that all four strains have key genes for resistance to thermal, osmotic, and oxidative stress, as well as to acid and bile, which is important for survival in the gastrointestinal tract.

Genes responsible for the synthesis of antimicrobial metabolites (polyketides—*acpK*, *bacA-G*, *dfnD-J*, *mlnB-H*, *pksC-S*; non-ribosomal peptides and lipopeptides—*besA*, *bpsA-B*, *ituB-C*, *nisE-R*, *ppsA-E*, *srfAA-ATE*) and bacteriocins (subtilisin—*aprE*; bacillolysin—*NprE*) are presented in Figure 9.

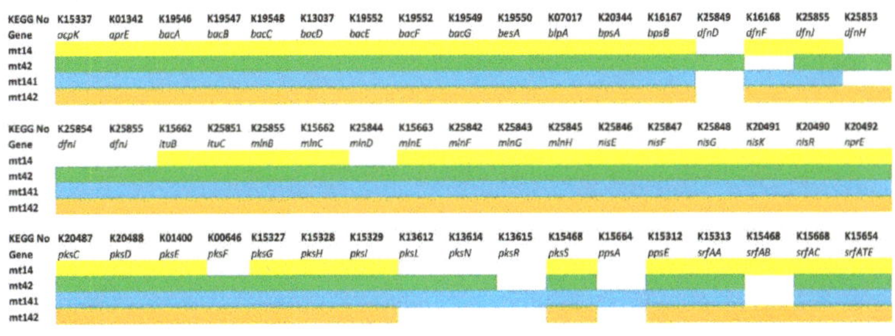

Figure 9. Antimicrobial genes annotated in MT14–142 strains using KEGG Annotation—BlastKOALA.

All strains have been found to have genetic markers suggesting their ability to synthesize bacillolysin, subtilisin. The synthesis of plipastatin in MT14, MT42, and MT142 is disrupted due to the absence of the key gene ppsA. All strains lack at least one dnf gene (difficidin), and all except MT141 lack at least one pks gene (bacillaene). However, this could be a false-negative result associated with the high variability of PKS genes, which include the aforementioned genes.

The third group included genes responsible for adhesion (*bmpA*, *nfeD*, *srtA*, *tapA*), extracellular matrix formation (polyglutamic acid synthesis—*capA-C*; exopolysaccharide—*epsA-O*), ensuring cell mobility (flagella assembly and function—*fdbD*, *flgB-M*, *flhA-G*, *fliA-T*, *motA-B*; chemotaxis—*cheA-Y*; *mcp*) and biofilm population density control (*cidA*, *lrgA-B*), and regulatory genes (*remB*, *sinR*, *sirA-R*) (Figure 10).

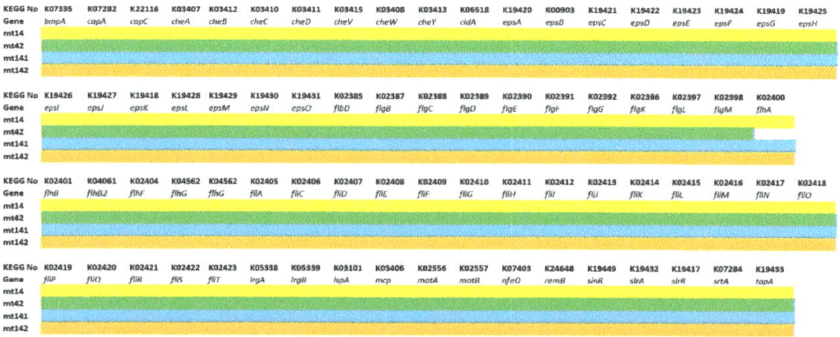

Figure 10. Adhesion, motility, and biofilm formation genes annotated in MT14–142 strains using KEGG Annotation—BlastKOALA.

The strains revealed an almost identical set of genes responsible for the formation and functioning of flagella, chemotaxis, extracellular matrix synthesis, and adhesion. The MT42 strain lacks the flagellar biosynthesis protein FlhA. However, it is difficult to say how much this will affect the motility of the strain.The presence of these genes suggests that the strains have the ability to effectively form biofilms and colonize the gastrointestinal tract of the host organism.

Thereby, the strains have genes for resistance to stress and adaptation to the conditions of the gastrointestinal tract. Their genomes contain adhesion and chemotaxis genes, which are important for colonization of the intestine. The presence of genes for the biosynthesis of secondary metabolites indicates the ability to synthesize antimicrobial compounds and, thus, show antagonism against pathogenic microorganisms.

3.3.2. Genes of Lytic Enzymes

The genes of hydrolytic enzymes (cellulases, proteases, chitinases) were also analyzed (Table 5).

The highest number of endoglucanase copies (two) was observed in strain MT141, while the other strains had one copy each. Among proteases, the highest number of copies was found in strain MT42, with four copies of various hemolysins. Strain MT142 possessed two copies of a CwlT-like protein with a lysozyme domain, while the other strains had one copy. Other enzymes with a high copy number included 6-phospho-beta-glucosidase (four copies in MT14 and MT142) and maltose-6′-phosphate glucosidase (three copies in MT141). The data demonstrate differences in the genetic potential of the strains related to the synthesis of lytic enzymes.

Table 5. Lytic enzymes of the studied strains. The number of "+" is the number of detected copies of genes.

Enzyme	MT14	MT42	MT141	MT142
Glycosyl hydrolases of the alpha-amylase family	+++	+	+++	+
Alpha-amylase	++	++	+	+
Glucosidase/amylase (phosphorylase)	+	+	+	
Beta-glucanase	+	+	+	+
6-phosphorus-beta-glucosidase	++++	++	++	++++
6-beta-D-glucan-glucanohydrolase	+	++	++	+
Alpha-D-1,4-glucosidase	+	+		
Oligo-1,6-glucosidase				+
Aryl-phosphorus-beta-D-glucosidase BglC	+	+	++	+
Aryl-phosphorus-beta-D-glucosidase BglH			+	
Maltodextrin-glucosidase	+	+		+
Maltose-6′-phosphate glucosidase	+	+	+++	+
Chitobiose-specific 6-phosphorus-beta-glucosidase ChbF	+	++	+	
Neopullalanase	+	+	+	
Levansaccharase	+	+	++	++
Alpha, alpha-phosphotregalase			+	+
DNA-recombinase SpoIVCA/DNA-invertase PinE			+	+
DNA-invertase hin		+		+
Cellulose		+		+
Endoglucanase	+	+	++	+
Protein of the lysozyme family	+	+	+	
CwlT-like protein containing lysozyme domain	+	+	+	++
Hemolysin C	+		+	
Hemolysin III family protein	+	+	+	+
Hemolysins and related proteins containing CBS domains		+++		+

All strains had endoglucanase genes in their genomes, and some had other cellulase genes (MT42, MT142). These enzymes are crucial for probiotics that facilitate the digestibility of plant-based feeds in aquaculture [72,73].

Core metabolic enzymes identified in all or almost all strains, such as alpha-amylases, beta-glucanases and glucosidases, glucan hydrolases, which are involved in the assimilation of complex polysaccharides.

Cellulases, found only in MT42 and MT142, together with endoglucanases provide these probiotics ability to degrade cellulose. Cellulases break down cellulose while endoglucanases break the internal β-glycoside bonds in cellulose to facilitate further cleavage by other enzymes.

Lytic enzymes include lysozyme family proteins and CwlT-like proteins with a lysozyme domain. Their antimicrobial activity is due to the hydrolysis of peptidoglycans. CwlT-like proteins cleave the bonds between N-acetylmuramic acid (NAM) and N-acetylglucosamine (NAG) in the peptidoglycan of gram-positive bacteria.

Levansaccharase, which is present in all the studied strains, produces fructo-oligosaccharides, which are prebiotics for lactobacilli [74]. Thus, the enzymatic profile of these four strains provides effective breakdown of complex carbohydrates, cellulose, and cellular structures to improve digestion, protect the body from pathogens, and adapt to aquaculture conditions.

3.4. Antibiotic Resistance Gene Screening

The genomes of the strains were additionally analyzed for the presence of antibiotic resistance genes to assess their biosafety for potential use. Although several clusters of resistance genes have been found, as mentioned above, the data from in vivo tests demonstrate the sensitivity of these bacteria to the main classes of antibiotics. This suggests that the detected resistance genes may be non-functional or epigenetically repressed.

Further analysis using KEGG mapper revealed that most of these clusters of resistance genes are incomplete (Table 6).

Table 6. Resistance genes clusters.

Drug Resistance Cluster	Strain			
	14	42	141	142
M00627 beta-Lactam resistance, Bla system (1)	(2 blocks missing 1/3)	(2 blocks missing 1/3)	(2 blocks missing 1/3)	(2 blocks missing 1/3)
M00704 Tetracycline resistance, efflux pump Tet38 (1)	(1 block missing 1/2)	(1 block missing 1/2)	(1 block missing 1/2)	(1 block missing 1/2)
M00726 Cationic antimicrobial peptide (CAMP) resistance, lysyl-phosphatidylglycerol (L-PG) synthase MprF (1)	(2 blocks missing 1/3)	(2 blocks missing 1/3)	(2 blocks missing 1/3)	(2 blocks missing 1/3)
M00700 Efflux pump AbcA (2)	(complete 2/2)	(complete 2/2)	(complete 2/2)	(complete 2/2)
M00702 Multidrug resistance, efflux pump NorB (1)	(1 block missing 1/2)	(1 block missing 1/2)	(1 block missing 1/2)	(1 block missing 1/2)
M00714 Multidrug resistance, efflux pump QacA (1)	(1 block missing 1/2)	(1 block missing 1/2)	(1 block missing 1/2)	(1 block missing 1/2)

The analysis of four MT strains revealed several clusters of antimicrobial resistance (AMR) genes, including both complete and incomplete orthologous groups from the KEGG database. *AbcA* (M00700), encoding an ABC-type efflux pump, was fully conserved (2/2 blocks) in all strains. Among other determinants of multidrug resistance, efflux pumps *norB* (M00702) and *qacA* (M00714) were identified, each missing half of their functional domains (1 of 2 blocks). Partial deletions (missing 2 of 3 blocks) were observed in all strains in the *bla* (M00627) system, responsible for beta-lactam antibiotic resistance. The *tet38* (M00704) gene, encoding a tetracycline efflux pump, contained only one of the two required blocks. The *mprF* (M00726) cluster, responsible for resistance to cationic antibiotics, was also incomplete (2 of 3 blocks were missing).

These findings imply that although the presence of partial resistance clusters indicates the need for strain control in biotechnological production to prevent the acquisition of missing genes through horizontal transfer, in general, fully functional clusters of resistance genes are limited to non-specific efflux pumps, which are characteristic of many bacteria and do not pose significant safety risks [75].

Secondary Metabolite Production: Non-Ribosomal Peptides and Polyketides

To identify genetic determinants of probiotic activity, we analyzed the genomes for biosynthetic gene clusters (BGCs) associated with bioactive compounds. This included genes encoding non-ribosomal peptides (NRPs) with antimicrobial, antifungal, or immunomodulatory properties, as well as polyketides and other secondary metabolites.

The identified secondary metabolites are summarized in Table 7. The numbers indicate the percentage similarity of the detected substance to reference in AntiSMASH database.

Table 7. Genes for the synthesis of secondary metabolites found in the analyzed genomes.

Substance/Strain	*B. velezensis* MT14	*B. velezensis* MT42	*B. velezensis* MT141	*B. velezensis* MT142	Possible Activity
Bacilysin	100	100	100	100	Antimicrobial [76]
Surfactin	78	78	78	78	Antimicrobial, immunostimulation [77,78]
Microlactin (H)	88	100	90	90	Antiviral [79]
Bacillaene	100	100	100	71	Antibacterial [79]
Mycosubtilin	100	-	100	-	Antifungal [80]
Difficidin	8	100	100	93	Antibacterial [81]
Bacillomycin	-	100	-	100	Antifungal [82]

These metabolites can provide some of the effects of probiotics in aquaculture: to suppress harmful bacteria and fungi, enhance the immune responses of fish, increasing resistance to infections.

It can be noted that 100% of the analyzed strains carry the genes for the synthesis of bacilysin and bacillaene. However, the presence of these genes does not yet ensure their synthesis, since, as shown by us [83] and other authors [84,85], the expression of NPP genes may strongly depend on environmental conditions.

While most strains share 100% similarity for bacillaene (except MT142 at 71%), MT14 stands out with an unusually low 8% similarity for difficidin, contrasting sharply with the >90% seen in other strains. MT42 uniquely carries a 100% identical macrolactin cluster, whereas the remaining strains range about 90%. Additionally, mycosubtilin is fully conserved in MT14 and MT141 but absent in MT42 and MT142, while bacillomycin shows the opposite pattern—missing in MT14 and MT141 but present at 100% in MT42 and MT142. These disparities suggest strain-specific adaptations that could influence their bioactive potential, particularly in antimicrobial, antiviral, and antifungal functions.

In general, the production of these bioactive compounds makes these strains of *B. velezensis* and *B. subtilis* promising probiotics for improving the health and productivity of aquaculture facilities. However, there is a possibility that their antimicrobial or immunostimulating effect is not detectable in vitro but is present in vivo. The study of this will be the subject of our further research.

3.5. Fish Growth Parameters

Two probiotic formulations were developed based on 1) *B. velezensis* MT14 and MT42 strains, and 2) *B. velezensis* MT141 and MT142, which were added to the feed during the manufacturing process.

The final probiotic cells count in the feed was $3.7 \pm 0.4 \cdot 10^6$ CFU/g for MT42 + MT14, and $6.4 \pm 0.3 \cdot 10^6$ CFU/g for MT141 + MT142.

During in vivo studies, assessment of growth, size, feed efficiency, and survival in carp subjected to different probiotic treatments was conducted, allowing for a detailed comparison of the effects of each probiotic formulation on fish performance.

As it can be seen from the results provided in Tables 8 and 9, supplementation of carp diets with probiotic preparations led to notable improvements in several key fish-breeding performance indicators compared to the control group at the end of the experiment.

Table 8. Size and weight characteristics of juvenile carp *Cyprinus carpio* Linnaeus, 1758 when adding multistrain probiotic feed additives (0.1%) based on four new strains of *Bacillus velezensis* to the diet.

Parameter	Control Group		Experimental Group No. 1 MT14 + MT42		Experimental Group No. 2 MT141 + MT142	
	Day 1	Day 76	Day 1	Day 76	Day 1	Day 76
Individual weight (g)						
min	1.65	6.51	1.78	10.54	1.37	7.82
max	6.69	32.87	6.42	44.71	5.69	45.63
Average value ± SD	3.79 ± 0.96	17.97 ± 3.73	3.67 ± 0.86	23.82 ± 6.23	2.91 ± 0.72	21.61 ± 7.11
Individual length (cm)						
min	4.26	8.91	4.19	9.51	3.64	7.48
max	6.51	12.46	6.32	13.49	5.78	13.10
Average value ± SD	5.05 ± 0.44	10.41 ± 0.73	5.02 ± 0.41	11.32 ± 0.73	4.51 ± 0.36	10.61 ± 0.99
Fulton's fatness coefficient, standard units	2.94	1.59	2.91	1.64	3.17	1.81

Table 9. Fish-breeding indices of juvenile carp *Cyprinus carpio* Linnaeus, 1758 when added to the diet of multi-strain probiotic feed additives (0.1%) based on four new strains of *Bacillus velezensis*.

Criteria	Control Group	Experimental Group No. 1 MT14 + MT42	Experimental Group No. 2 MT141 + MT142
Fish biomass, kg (during the end of the experiment)	2.55	3.48	3.07
Total increase in biomass, kg	1.98	2.93	2.63
Biomass growth, %	448.85	631.74	703.00
Individual biomass growth, g	14.18	20.15	18.7
Individual biomass growth, %	474.14	649.05	742.61
Specific mass growth rate, %/day	2.04	2.47	2.63
Specific growth rate of fish by length, %/day	0.95	1.08	1.12
Feed conversion rate, kg/kg	1.39	1.32	1.30
Survival rate, %	94.67	97.33	94.67

During the experiment, optimal water quality parameters were observed. The concentration of nitrite nitrogen was below the detection limit (<0.005 mg/L). The concentration of ammonium nitrogen did not exceed 0.04 mg/L in all treatments. The concentration of nitrate nitrogen did not exceed 0.013 mg/L in all tanks. The level of dissolved oxygen in water did not decrease below 8 mg/L due to constant aeration and water quality control.

The average individual weight at the end of the experiment was significantly higher in both experimental groups compared to the control ($p < 0.05$). Group 2 (MT141 + MT142) exhibited a 56.62% higher biomass increase than in the control group, while Group 1 (MT14 + MT42) exhibited a 40.75% higher biomass growth (Tables 8 and 9). However, no statistically significant difference was observed between experimental groups 1 (MT14 + MT42) and 2 (MT141 + MT142).

Differences in size distribution were observed (Figure 11). In experimental Group 1 (Figure 11a), individuals with a body weight of 20 to 25 g represented the largest proportion (25.00%). Individuals with a body weight in the range of 15–20 g and 25–30 g (20.31% each) were less common. At the end of the experiment, no individuals with a body size ≤ 10 g were recorded in the first group. In experimental Group 2 (Figure 11b), 5.06% of the individuals had a body weight of less than 10 g. The size of 15–20 g was more common than others, accounting for 27.85% of the population. In the control group (Figure 11c), the proportion of individuals with a body weight $\leq$ 10 g was 2.96%, and the size group of 15–20 g was 43.7%.

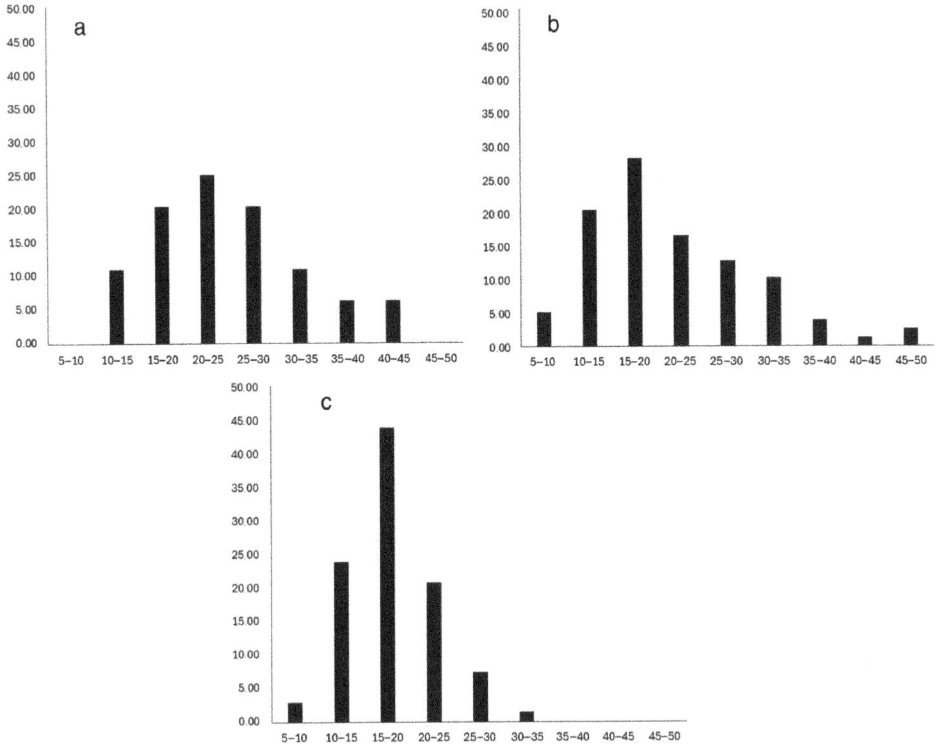

Figure 11. Distribution of juvenile carp *Cyprinus carpio* Linnaeus, 1758 into nine size groups: axis X–size group (g), axis Y–share of individuals in the size group (%): (**a**)—experimental group No. 1 (MT14 + MT42), (**b**)—experimental group No. 2 (MT141 + MT142), (**c**)—control group.

According to a number of other performance metrics (individual biomass increase, %; specific growth rate, feed conversion rate), these strains proved to be promising for use in aquaculture. Both probiotic groups demonstrated enhanced growth performance, feed efficiency, and survival compared to the control, with Group 2 showing the highest values in several parameters.

3.6. Microbiological Tracking of Probiotic Colonization

To track probiotic colonization, we compared colonies from intestinal content cultures to reference plates of pure probiotic strains. Probiotic-derived colonies were identified by their morphological congruence with reference cultures and their absence in control groups receiving alternative probiotics. Their characteristic morphological features were the shape of the surface and edges, the location of wrinkles, and the production of pigment in the medium.

Quantitative analysis revealed significant colonization by the administered probiotic strains in treated carp (Table 10). Minor amounts of other bacilli with a different morphology were also found in the intestinal contents, but their number did not exceed 10^4 CFU/g.

Table 10. The content of bacteria morphologically similar to the corresponding probiotics in the intestinal contents of carp fingerlings of different ages, CFU/g.

Probiotic Used	Total Amount	Spore Form
Control	$3.2 \pm 0.2 \cdot 10^4$	$5.3 \pm 0.4 \cdot 10^3$
Experiment 1 (MT14 + MT42)	$2.7 \pm 0.4 \cdot 10^6$	$1.5 \pm 0.2 \cdot 10^4$
Experiment 2 (MT141 + MT142)	$2.3 \pm 0.3 \cdot 10^6$	$1.1 \pm 0.4 \cdot 10^4$

A sufficiently high number of bacilli of various morphologies was observed in the control (a total of eight different morphotypes). They are likely representing native microbiota of catfish acquired through environmental exposure (water and fee). In contrast, probiotic-supplemented carp exhibited a near-exclusive dominance of two morphotypes, consistent with the administered *B. velezensis* strains (MT14 + MT42 or MT141 + MT142). Minor inclusions of other morphotypes were rare, confirming effective displacement of native bacilli by the probiotics.

Despite no net increase in total bacilli compared to feed concentrations, spore-to-vegetative cell conversion rates reached 94% (Group 1: MT14 + MT42) and 95% (Group 2: MT141 + MT142). This indicates that most spores germinated into metabolically active cells within the carp intestine, enabling sustained probiotic function.

3.7. Analysis of the Differential Expression of Fish Genes

Results of the differential expression analysis of genes is shown in Figure 12 (carp model) and Figure 13 (full-grown fish).

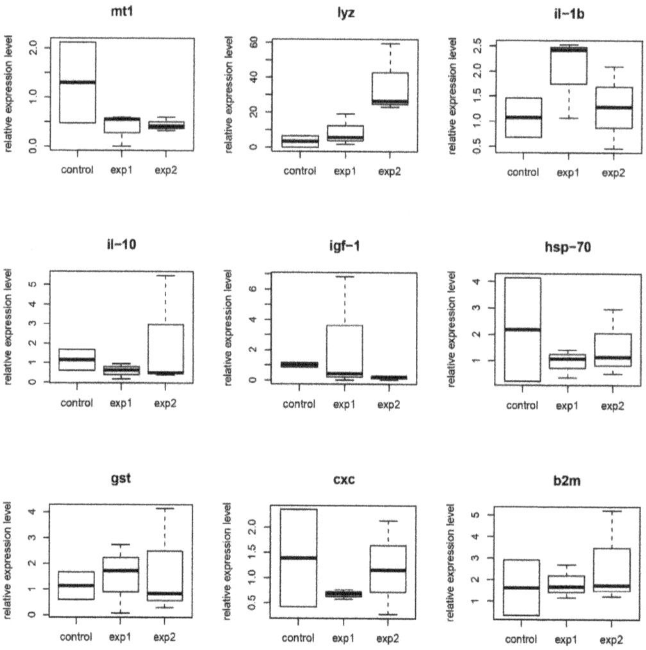

Figure 12. Relative gene expression rates in the muscle tissue of the full-grown *C. carpio*. No statistically significant changes were observed in group 1 relative to the control group ($p > 0.05$).

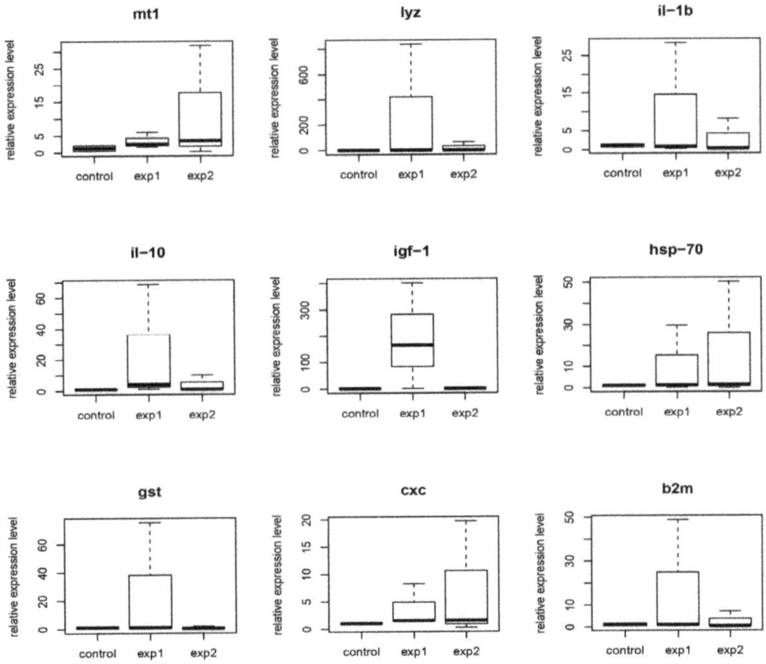

Figure 13. Relative gene expression rates in the muscle tissue of the mature *C. carpio*. No statistically significant changes in expression level were observed for the studied genes relative to the control group ($p > 0.05$).

There are no statistically significant results ($p > 0.05$) of changes in the expression of most genes in the muscles of carp fry in group 1, while group 2 demonstrated an increase in *lyz* gene expression and a decrease in *igf-1* expression ($p < 0.05$).

There are no statistically significant changes in expression in the muscles of full-grown carp.

The data suggest that probiotics have a greater effect on carp fry than on adults. A probiotic based on MT141 and MT142 strains had a minor effect on gene expression only in carp fry. This contrasts with our earlier findings in African catfish *Clarias gariepinus*, where it was shown that probiotics can cause significant changes in the expression of some of the listed genes [86]. The disparity implies that the mechanisms underlying these probiotics' action are different, potentially involving enzymatic activity, nutrient absorption, or competitive exclusion of pathogens rather than direct transcriptional modulation.

4. Discussion

Using the pipeline described above, we have succeeded to identify and characterize novel *Bacillus velezensis* strains with enzymatic activity to improve the digestibility of feeds in carp aquaculture. The most notable result was an increase in average individual weight in carp receiving the MT14 + MT42 probiotic supplement compared to controls, demonstrating the potential of these strains to enhance growth and feed efficiency.

4.1. Bacilli as Probiotics for Aquaculture

Probiotics play an important role in animal husbandry and poultry farming, where their use has long been shown to increase productivity and reduce the incidence of livestock [87–89]. However, probiotics in aquaculture are less common and often have less effect than in warm-blooded animals [12]. This can largely be attributed to the fact that many probiotics used in aquaculture are derived from animal husbandry, without sufficient consideration of the physiological and ecological differences between aquatic and terrestrial hosts. But fish differ significantly from warm-blooded animals in terms of physiological parameters and feeding preferences, as well as vulnerabilities and requirements [90–92]

Carp specifically has its own maintenance features. The carp (*Cyprinus carpio*), like other members of the Cyprinidae family, is a stomachless fish (agromastric species). Its digestive system includes an expanded cardiac section of the intestine (morphologically similar to the stomach, but not secreting pepsin and hydrochloric acid), which is typical of phylogenetically ancient fish [93]. Notably, a long intestine compensates for limited stomach digestion, aiding nutrient uptake [94].

Recognizing these differences, we selected bottom sediments as the source of potential probiotic strains. We consider the "fish–water–bottom sediments" system, wherein microorganisms excreted from the fish's body (primarily through feces) are incorporated into the bottom sediments. Subsequently, these microbes can re-enter fish intestines via water movement, gill ventilation, ingestion of water, and through their association with food particles, thereby facilitating their reintroduction into the fish gut. So, the microbiota of bottom sediments and the intestines of aquatic organisms largely overlap. Studies show that 27.8% of shrimp intestinal microbiota overlaps with the microbial composition of bottom sediments. In fish, this percentage can reach up to 71.7%. It has been demonstrated that sediment microbiota is the primary source of fish microbiota, directly shaping the microbiome, especially for benthic fish. Strains derived from bottom sediments successfully colonize the host's intestine [95,96]. Therefore, we hypothesize that bacteria isolated from bottom sediments will be better adapted to life in the intestines of fish than bacteria isolated from other sources.

Genomic analysis revealed that all four strains belong to the species *Bacillus velezensis*. This species is characterized by a wide range of antimicrobial activity against various pathogens [97,98], the ability to produce many lytic enzymes such as protease, cellulase, amylase, and glucanase [48], the ability to improve blood parameters related to inflammation and immunity as well as improve the stability of the intestinal microbiota and the integrity of the gut barrier [98].

The mechanisms underlying *Bacillus* probiotics benefits are multifaceted. Firstly, this probiotic has been shown to enhance the activity of digestive enzymes in aquatic animals. Secondly, it demonstrated stimulation of immunity and antagonism to pathogens. Finally, the production of its own enzymes allows it to improve the digestion of dietary components by the host.

Our findings suggest that our strains exhibit at least some of those effects. Digestive enzyme enhancement was clearly demonstrated in our in vitro screening, where strains MT14, MT42, MT141, and MT142 produced substantial clearance zones on test media—proteolytic activity with clearance zones of 5–11 mm and cellulolytic activity with clearance zones of 14–17 mm—suggesting these strains can effectively break down complex plant polysaccharides in fish diets.

Endogenous enzyme production capacity is also supported by our genomic analysis, which identified numerous lytic enzyme genes in all four strains. The genomes contained multiple copies of glycosyl hydrolases, proteases, and cellulases. The presence of these genes explains the observed growth enhancement (31.9–42.1% increase in individual biomass) despite modest improvements in feed conversion ratio, which will be discussed in more detail below.

Immune stimulation mechanisms, while not directly evidenced in our muscle tissue gene expression analysis, may still be occurring in other immunologically relevant tissues. Although we observed no statistically significant changes in expression of immune-related genes (*cxc*, *il-1β*, *il-10*, *hsp70*) in muscle samples, it's important to note that immune modulation by *Bacillus* probiotics can also manifest in tissues with higher immunological activity such as spleen, head kidney, or intestinal mucosa [99].

Although we used muscle tissue for gene expression analysis, which aligns with established practices in fish nutritional research for assessing growth-related effects [100,101], future studies should examine gene expression in these tissues. The slight increase in survival rate in our experimental groups (97.33% vs. 94.67% in control) may reflect subtle immune benefits not captured in our muscle tissue analysis.

4.2. Enzymatic Activity and Its Role in Improving Feed Digestibility

The introduction of probiotics with the potential to produce various hydrolytic enzymes, including cellulase, protease, and amylase, can improve the digestibility of feed components and enhance nutrient absorption [30,36–38,40]. Bacteria of the genus *Bacillus* are able to secrete various hydrolytic enzymes, such as β-1,3-glucanases, proteases, and cellulases, which helps to improve the digestive processes in fish [19,102]. Various studies confirm the ability of various *Bacillus* strains to produce key enzymes for the proper digestion of feed [35,103] and high proteolytic activity of *B. amyloliquefaciens* and *B. subtilis*, which secrete proteases, amylases, and lipases [103]. The work [35] expands this list to include such enzymes as α-glucosidase, naphthol-AS-BI-phosphohydrolase, esterase lipase, acid and alkaline phosphatases, caseinase and lecithinase [35]. A study [104] demonstrated the positive effect of *B. safensis* NPUST1 on the Nile tilapia *Oreochromis niloticus* (Linnaeus, 1758), increasing feed efficiency and improving the growth rate of individuals [104]. Another study experimentally confirmed the ability of *B. cereus* YB1 to increase the activity of protease, amylase, and lipase in juvenile black sea bass *Sebastes schlegelii* Hilgendorf, 1880, which led to a noticeable increase in growth rates [98].

Our goal was to obtain a multi-strain antimicrobial preparation with lytic properties. We focused on the production of exogenous proteolytic, amylolytic, and cellulolytic enzymes [105]. Media used in screening (containing milk, starch, and carboxymethyl cellulose) clearly show the lytic activity via a light zone around the colony.

Among all isolated strains, 14 had a high degree of exolytic activity (Table 1), of which nine had all studied activities simultaneously, and therefore were selected for enhancing feed digestion in fish.

We used multi-strain probiotic drugs to broaden enzymatic diversity and functional synergy. However, bacilli can inhibit each other through the production of antimicrobial compounds and various toxin-antitoxin systems [106]. Therefore, it was necessary to perform compatibility testing. Strain-specific interactions resulted in some pairs showing mutual growth inhibition, while others formed stable biofilms when co-cultured (Table 2). Two compatible pairs of strains were determined and used in Groups 1 and 2, respectively.

Functional enzyme genes (endoglucanases, cellulases, alpha-amylases, beta-gluconases and glucosidases, glucan hydrolases, lysozyme family proteins and CwlT-like proteins with a lysozyme domain) were found in the genome of our strains, which coincides with in vitro test data. This confirms the effectiveness of chosen screening methodology.

Proteolytic activity—the clearance zones observed on milk medium (8–11 mm) directly correspond to the presence of multiple protease genes identified in the genome sequences of all four strains.

Amylolytic activity—the starch degradation zones (5–11 mm) correlate with the presence of alpha-amylase family genes, alpha-amylases, and glucosidase/amylase genes in the genomes.

Cellulolytic activity—the substantial clearance zones on carboxymethylcellulose medium (14–17 mm) align with the presence of endoglucanase genes in all strains and cellulase genes in MT42 and MT142.

The genomic analysis revealed strain-specific patterns in enzyme gene distribution that perfectly matched the observed phenotypic differences in enzymatic activity. Strains with higher proteolytic activity in vitro (MT141, MT142) showed more copies of glycosyl hydrolases of the alpha-amylase family. Strains with stronger amylolytic activity (MT42) contained more copies of beta-glucanase and 6-beta-D-glucan-glucanohydrolase genes.

The exceptional cellulolytic activity observed in MT42 and MT142 was confirmed by the presence of dedicated cellulase genes in these strains only, while all strains possessed endoglucanase genes.

Levansaccharase, which is also present in all the studied strains, produces fructo-oligosaccharides, which are prebiotics for lactobacilli [74]. Studies show that bacilli, indeed, improve the abundance and activity of lactobacilli [107,108]. This was attributed to the synthesis of heme by bacilli, which is necessary for the heme-dependent catalysis of lactobacilli [107]. However, it seems that there might be another mechanism—providing lactobacilli with prebiotic sugars that are difficult for others to access, i.e., energy. In our prior works [109], we have shown that the probiotic activity of lactobacilli strongly depends on the availability of available sugars in the medium. It is possible that the provision of fructo-oligosaccharides to lactobacilli also contributes to a more active production of lactic acid and metabolites by lactobacilli, improving their probiotic activity.

Notably, our analysis revealed no unexpected absences or exceptions in the correlation between phenotypic screening and genomic confirmation, which confirms the validity of our screening approach.

4.3. Effect on Fish Growth and Feed Conversion

Accelerated somatic growth is extremely significant for fish during early ontogenesis, as it increases the individual competitiveness and survival rates after wintering [110]. The use of probiotics stabilizes intestinal microbiota composition, which improves the activity of digestive enzymes and, as a result, feed digestibility [111].

In aquaculture species, *B. velezensis* was shown to induce improved growth. Hybrid mudcat showed better growth rates when fed diets containing *B. velezensis* YFI-E109 [112]. Pacific white shrimp showed an increase in final body length, weight gain rate, and specific growth rate when fed diets supplemented with *B. velezensis* BV007 [113]. Similar growth stimulation effects were observed in carp [114] and grouper [115].

It aligns with observed effects of our *B. velezensis* strains. Our two groups of probiotics induced a 31.9–42.1% increase in individual biomass growth and a 5.04–6.47% reduction in feed conversion ratio (FCR). Notably, we observed reduction in feed conversion rate in experimental groups. In terms of industrial production, this corresponds to a saving of 6.47 kg of feed per 100 kg of live weight gain. For a fish farm with an annual production volume of 100 tons of carp, the use of a probiotic preparation can reduce feed consumption by 12.94 tons. With an average market price of carp feed compound of USD1.2/kg, the potential annual savings will be about USD15,528. The results obtained demonstrate the practical value of the study for improving the economic efficiency of commercial carp farming.

In comparison, hybrid yellow catfish (*Pelteobagrus* spp.) fed with *B. velezensis* YFI-E109 showed an 18% weight gain increase and 15.8% FCR reduction [112]. In a study [115] *B. velezensis* induced a 15.0% increase in weight gain rate compared to control.

The observed growth improvement seems to be directly related to the activity of the probiotics used: additional lytic enzymes produced by bacilli improved the overall digestibility and the breakdown of complex feed components, allowing fish to extract more energy and nutrients per unit of feed than the control group. As stated above, our genomic and functional analyses of new strains confirmed the presence of genes for key digestive enzymes and stress resistance, supporting the hypothesis that these strains act mainly by improving digestion rather than directly modulating host gene expression.

It should be noted, though, that the observed growth improvements substantially exceed typical *Bacillus velezensis* effects reported in literature, which validates our screening approach. However, the FCR reductions fall within the lower range of reported improvements, while some studies achieve 15–20% FCR reductions.

The observed pattern seems to reflect the distinct underlying mechanisms that operate independently in aquaculture probiotic supplementation. Growth promotion and feed conversion efficiency could represent distinct probiotic mechanisms with different optimization requirements. Our new strains likely enhanced growth performance through enzymatic activity and nutrient absorption rather than feed efficiency optimization.

A meta-analysis [116] that examined the effects of probiotics on specific growth rate (SGR) and feed conversion ratio (FCR) in commercial fish species revealed that the minimum effective concentration of probiotics for improving SGR was 10^7 log CFU/g, while 10^8 log CFU/g was needed to enhance FCR. A supplementation at 0.1% (approximately 10^6–10^7 CFU/g) may have been optimal for growth but suboptimal for maximizing feed conversion efficiency.

Since in our experimental setup no force-feeding methods were employed and all consumption was based on the fish's natural feeding behavior, one possible interpretation for our results is that the *Bacillus velezensis* MT strains enhanced voluntary feed intake through appetite. It is known that probiotics can influence the gut–brain axis by modulating the expression of key appetite-regulating neuropeptides [117], and also can reduce stress levels, which is associated with improved appetite and feeding behavior [118]. These assumptions open up some new promising directions for future research. In addition, a promising area of further research is to evaluate the effectiveness of the probiotic strains under study when they are included in compound feeds with a higher content of plant components.

4.4. *Colonization and Stability of Preparations In Vivo*

One of the important issues in probiotics efficacy is the survival of probiotic bacteria within the host body. Carps lack a developed stomach with acidic conditions, eliminating concerns about probiotic viability in a low-pH environment. However, we investigated whether sediment-derived *Bacillus* spores could germinate and proliferate in the intestinal lumen.

As can be seen from Table 10, the number of probiotic strains in the intestinal contents was comparable to the amount obtained with feed (10^6 CFU/g). That is, the amount of probiotic in the intestinal contents did not increase, but the vast majority of the population was in a vegetative state.

In the control group, the carp intestine contained only 10^4 CFU/g of bacilli, which is significantly less than the amount introduced with the feed and observed in the intestines of the experimental groups. It is likely that the suppression of the growth of the probiotic bacilli population, even considering their vegetative state, is related to their excess in the intestine. It should be noted that some studies have even reported a decrease in the number of probiotic bacilli in the intestine relative to their content in the feed [105,119]. We can assume that this suppression is due both to the action of the rest of the intestinal microbiota of the carp and to the host organism's response, but these assumptions require further verification in future studies.

4.5. Secondary Metabolites and Antimicrobial Activity

The ability to synthesize non-ribosomal peptides is a relatively new and rarely discussed feature of probiotics. However, accumulating evidence suggests that NRP may significantly contribute to the effects of probiotic *Bacillus* strains due to their antimicrobial, immunomodulatory, and intestinal health-enhancing properties.

The genomic analysis of four new strains revealed several clusters of genes for non-ribosomal synthesis of bacilysin, difficidin, macrolactin, bacillaene, mycosubtilin, surfactin, and bacillomycin. Bacilysin, surfactin, macrolactin H, and bacillaene were found in all four strains.

This class of substances is primarily known for antimicrobial and antifungal effects, but some representatives have other properties, for example, antiviral, anti-cancer, immunomodulatory, antioxidant, and quorum quenching against pathogens [120–122]. The main effects of the relevant substances are shown in Table 7.

Notably, only bacilysin and bacillaene had 100% similarity to reference peptides in the AntiSMASH database, while for other substances there were deviations from 100% identity. This suggests that these strains are capable of producing modified variants of non-ribosomal peptides, such as different isoforms of surfactin, which can provide them with a broader spectrum of activity. Such variability in NRP structure allows producing strains to better adapt to different antagonists. It has been shown that a *Bacillus* strain producing only one isoform of surfactin is less effective than a strain producing several isoforms [123].

Bacillysin is a peptide with antimicrobial activity: it inhibits glucosamine synthetase, disrupting peptidoglycan synthesis and causing cell lysis in pathogens. It suppresses genes involved in pathogen virulence and cell wall biosynthesis [124] and is effective against many gram-negative pathogens [125].

Bacillaene is a molecule of internal signaling in the bacterial cell. In low concentrations (0.1–1.0 micrograms/mL), it accelerates the formation of biofilm in *Bacillus*, and in higher concentrations it suppresses competing microbes due to structural instability, forming decomposition products that destroy pathogens [126].

The combined action of these two compounds can create a selective advantage for our introduced strains, explaining the microbiological data showing that spore-to-vegetative cell conversion rates reached 94–95% while displacing native bacilli. Their selective antimicrobial activity, rather than broad-spectrum inhibition, explains why the overall microbial diversity remained largely unchanged despite the significant shift in *Bacillus* populations.

Difficidin is a broad-spectrum antibiotic effective against gram-negative bacteria [124,127]. Macrolactin is antibacterial and anti-inflammatory substance; it suppresses the activity of bacterial ATPase and reduces oxidative stress in host cells. It is also known to contribute to the modulation of the microbiota [128]. Mycosubtilin and bacillomycin are primarily antifungal agents [82,129]. In *Bacillus*, bacillomycin also enhances biofilm formation and increases the binding of siderophores [130] and acts against several pathogens [131]. Surfactin is a multifunctional compound antagonistic to bacteria, fungi, and viruses, that has an anti-inflammatory effect and recognized as a biomarker of probiotic activity, as high surfactin production correlates with the effectiveness of probiotics in colonization of the intestine [132].

Based on our data, we believe that bacillisin and bacillaene may also be markers of probiotic activity in *Bacillus* strains alongside surfactin.

4.6. Gene Expression Analysis and Mechanisms of Probiotic Action

Bacterial metabolites can also directly or indirectly affect gene expression. For example, we previously demonstrated that probiotic bacilli are able to modulate gene expression in the clariid catfish [133]. There is also evidence that *B. velezensis* probiotics affect gene expression in fish and activate genes associated with immunity: IgM (increase in 2.5 times in the spleen), TNF-α (increase in 3 times in the kidneys), MHCI and IgD in lymphoid organs, as well as modulate the expression of antioxidant genes: superoxide dismutase (SOD) and catalase (CAT) [104,114].

All of these effects correlate with increased survival and other positive phenotypic effects in hosts [76].

Guided by the hypothesis that our probiotics may have a similar effect, we included measurement of gene expression in the design of this study. However, no significant transcriptional changes were observed. Apparently, these strains can be attributed to "direct-acting" probiotics rather than to "regulatory" ones, and their effect is ensured by optimizing digestion due to enzyme activity and antimicrobial action, including through the activity of non-ribosomally synthesized peptides. This conclusion is further supported by the observed reduction in feed conversion ratio.

The effects of probiotics are inherently strain-specific [104]. Therefore, it is especially important to monitor the effects and mechanisms of action of individual strains.

This may be due to the fact that for aquaculture species, bacilli are a transient (allochthonous) microbiota that does not remain in the fish intestine for more than a few days after the start of feeding [134–136]. Therefore, unlike autochthonous microbiota, they have not developed a unified strategy for symbiotic interaction with fish hosts, and when they enter the intestinal microbial community, they demonstrate flexibility by choosing one of the many strategies.

4.7. Limitations and Future Directions

Several important limitations should be acknowledged in our experimental design and methodology.

While the study provides compelling evidence for the growth-promoting effects of these probiotics, the gene expression analysis may have been limited to a selected set of muscle genes and may not capture broader immunological or metabolic changes. We conclude that these four *Bacillus velezensis* strains are "direct-acting" rather than "regulatory" probiotics based on the absence of significant transcriptional changes in muscle tissue, but this interpretation must be considered within the context of our experimental design limitations.

The choice of muscle tissue for gene expression analysis, while practically convenient and commonly used in aquaculture research [137], may not represent all probiotic-mediated regulatory effects. Recent studies have demonstrated that some probiotics influence gene expression in the gastrointestinal tract of fish [138], but that modulation of genes involved in immune response, metabolic pathways, and stress regulation sometimes are not detectable in peripheral tissues [139].

Additionally, examining gene expression in other immunologically relevant tissues such as spleen, head kidney, and liver would provide a more complete understanding of whether these strains truly lack regulatory properties or whether our current methodology simply failed to detect tissue-specific and temporally dynamic transcriptional responses.

Furthermore, the temporal dynamics of probiotic effects may vary significantly between tissues, with gut-associated changes potentially occurring earlier or later than those in muscle tissue. The timing of our sampling could have missed critical windows of gene expression modulation, as studies have shown that probiotic-induced transcriptional changes can be highly time-dependent, with some effects manifesting within days of administration while others require weeks to become apparent [118].

Another limitation that should be taken into account is that our experimental design utilized a single tank per treatment group. This design, while presenting statistical considerations regarding pseudoreplication, follows established practices in aquaculture research, where resource constraints and practical considerations necessitate this approach. For example, in work [140] to maintain appropriate production densities of fish, experiments were conducted in one tank per diet until day 86. A similar design has been successfully employed in aquaculture studies, including research on feeding rates, probiotic supplementation in various fish species, and growth performance evaluations [141,142].

Studies have demonstrated that while tank effects can influence results, this experimental approach remains valid when environmental conditions are carefully controlled and monitored, as was maintained throughout our 76-day trial [143]. Many successful aquaculture nutrition studies have utilized similar single-tank-per-treatment designs while acknowledging the statistical implications. Recent surveys of aquaculture growth studies indicate that approximately 80% of published research applies treatments in limited replication scenarios, with researchers recognizing that practical constraints often necessitate balancing statistical power with available resources [144]. This approach helps us to focus on demonstrating biological efficacy, particularly in preliminary screening studies.

The other possible limitation—the absence of quantitative metabolite analysis in our study—reflects a common limitation in probiotic research, where genomic potential identification often precedes comprehensive metabolomic validation [145]. Current research in aquaculture probiotics frequently relies on genomic screening to identify bioactive compound potential, with metabolite quantification representing a subsequent research phase that requires specialized analytical equipment and methodologies [146].

The correlation between genetic potential and actual in vivo production levels can vary considerably in different environmental conditions, host factors, and bacterial growth phases [147]. This identifies another possible future research direction—studying the expression of these genes in vivo. Our genomic identification of key metabolite biosynthetic pathways provides a solid foundation for future metabolomic studies, following the established research progression from genetic screening to functional validation that characterizes the field of aquaculture probiotic development.

The gap between genomic predictions and metabolite quantification is acknowledged as a current frontier in probiotic research, with studies increasingly implementing both approaches to provide better understanding of probiotic mechanisms.

Therefore, we believe that future research on these strains should prioritize transcriptomic analysis of gut mucosa and other tissue samples collected at multiple time points to capture the full spectrum of potential regulatory mechanisms. We would also like to suggest focusing on elucidating additional mechanisms of probiotics action, including regulatory interactions with the gut microbiota and host metabolism. Metabolomic profiling of lipopeptides and other secondary metabolites in intestinal contents should be implemented using advanced analytical platforms such as mass spectrometry to quantify the actual production levels of bioactive compounds identified through genomic analysis.

Multi-location validation trials across different aquaculture facilities would address the generalizability of our findings and account for environmental variability that could influence probiotic efficacy. Additional research directions should include longitudinal studies of gut microbiota dynamics using 16S rRNA sequencing combined with host transcriptomics to understand the temporal establishment of probiotic strains and their impact on indigenous microbial communities.

Finally, the development of standardized protocols for probiotic delivery, dosage optimization, and effect monitoring would facilitate the translation of these research findings into practical aquaculture applications.

4.8. Practical Implications for Sustainable Aquaculture

Despite the acknowledged limitations in our experimental design and methodology, we believe that this study has successfully accomplished its primary objectives, establishing a comprehensive pipeline for the initial screening of novel probiotic strains and demonstrating the correlation between genomic predictions and in vivo performance. Most importantly, our in vivo trials demonstrated that the genomic potential identified through bioinformatics analysis translated into measurable biological effects. This work also provides validated genetic markers that can guide the selection of similar strains for aquaculture.

The demonstrated efficacy of these *Bacillus* strains in enhancing growth on plant-based feeds supports their application as sustainable alternatives to antibiotics in aquaculture. The integration of genomic screening in probiotic development, as exemplified in this study, offers a powerful approach for identifying strains with optimal traits for specific aquaculture challenges.

Potentially, these probiotic strains could also offer substantial environmental advantages by improving water quality through enhanced breakdown of organic matter and reduced nutrient loading in aquaculture effluents.

The enzymatic capabilities of the characterized strains, particularly their cellulolytic and proteolytic activities, directly address aquaculture's transition toward plant-based feed systems by improving digestibility of complex plant materials and reducing dependence on fishmeal-based feeds. This enzymatic enhancement makes plant-based diets more economically viable while supporting sustainable feed formulations using locally available materials.

The integration of probiotic supplementation provides significant cost reductions through improved feed conversion ratios, reduced veterinary treatment expenses, and enhanced survival rates. The stable spore-forming nature of *Bacillus* species ensures longer shelf life and easier storage, improving commercial viability. Additionally, these probiotics support global efforts to reduce antibiotic use in aquaculture, enhancing regulatory compliance and market access for aquaculture products, particularly in markets with strict antibiotic residue standards.

The practical implementation is facilitated by scalable solid-phase fermentation using readily available substrates like soybeans, making the technology accessible and cost-effective.

5. Conclusions

In this study, four new strains of *Bacillus velezensis* were characterized as potential probiotics for aquaculture. These strains were isolated from river sediments and selected based on their high enzymatic activity and antibiotic sensitivity profile. Probiotic formulations were developed based on compatibility testing and included two pairs of probiotics that were able to form joint biofilms. In vivo experiments on carp have shown that probiotics supplementation increases fish weight gain and reduces feed conversion ratio, with MT14 and MT42 strains combination yielding the highest average weight achieved.

Functional genomic analysis of the revealed key genes responsible for probiotic activity, including those conferring resistance to various types of stress and synthesis of bioactive metabolites. The results also indicate that the mechanism of action of these probiotics is primarily mediated by their enzymatic activity.

Beyond characterizing novel aquaculture probiotics, this work identifies genetic markers for the selection of similar strains in the future. These genetic markers and pipeline can guide the rational design of multi-strain probiotics tailored to specific aquaculture systems, ultimately reducing reliance on fishmeal and antibiotics.

Author Contributions: Conceptualization, E.V.P. and M.S.M.; Methodology, M.S.M., M.P.K., V.N.S. and V.A.G.; Investigation, M.S.M., M.P.K., R.V.S., V.A.G. and D.B.K.; Validation, V.N.S. and T.A.M.; Formal analysis, M.P.K., V.N.S. and R.V.S.; Data curation, D.B.K.; Writing—original draft, L.S.G., E.V.P., M.S.M. and V.N.S.; Writing—review & editing, E.V.P., M.S.M., V.N.S. and D.B.K.; Project administration, Funding Acquisition, D.V.R. All authors have read and agreed to the published version of the manuscript.

Funding: The research was funded by the Russian Science Foundation, grant No. 23-76-30006 "Molecular aquaculture strategy in the design of novel synbiotic preparations for improvement of health and quality in fishery".

Institutional Review Board Statement: The study was approved by the Local Independent Ethics Committee of Don State Technical University (Rostov-on-Don, Russian Federation). During the experiments, recommendations for the humane and ethical treatment of laboratory animals were followed, including in accordance with Directive 2010/63/EU of the European Parliament and of the Council of 22 September 2010 on the protection of animals used for scientific purposes.

Informed Consent Statement: Not applicable.

Data Availability Statement: The original genomic data presented in the study are openly available in the NCBI database under accession numbers JBOZKB000000000 (MT14), JBOZKA000000000 (MT42), JBPIYB000000000 (MT141), and JBPIYC000000000 (MT142), or through BioProject PRJNA1269758. Other experimental data are available upon request.

Acknowledgments: Authors express gratitude to E.N. Ponomareva (Southern Scientific Center of the Russian Academy of Sciences) for the resources and materials provided.

Conflicts of Interest: The authors declare no conflict of interest.

References

1. Food And Agriculture Organization (FAO). *The State of World Fisheries and Aquaculture 2018 (SOFIA)*; FAO: Rome, Italy, 2018; ISBN 978-92-5-130562-1.
2. Food And Agriculture Organization (FAO). *The State of World Fisheries and Aquaculture 2020*; FAO: Rome, Italy, 2020; ISBN 978-92-5-132692-3.
3. El-Saadony, M.T.; Alagawany, M.; Patra, A.K.; Kar, I.; Tiwari, R.; Dawood, M.A.O.; Dhama, K.; Abdel-Latif, H.M.R. The Functionality of Probiotics in Aquaculture: An Overview. *Fish Shellfish Immunol.* **2021**, *117*, 36–52. [CrossRef] [PubMed]
4. Naylor, R.L.; Kishore, A.; Sumaila, U.R.; Issifu, I.; Hunter, B.P.; Belton, B.; Bush, S.R.; Cao, L.; Gelcich, S.; Gephart, J.A.; et al. Blue Food Demand across Geographic and Temporal Scales. *Nat. Commun.* **2021**, *12*, 5413. [CrossRef]
5. Kroeckel, S.; Harjes, A.-G.E.; Roth, I.; Katz, H.; Wuertz, S.; Susenbeth, A.; Schulz, C. When a Turbot Catches a Fly: Evaluation of a Pre-Pupae Meal of the Black Soldier Fly (*Hermetia illucens*) as Fish Meal Substitute—Growth Performance and Chitin Degradation in Juvenile Turbot (*Psetta maxima*). *Aquaculture* **2012**, *364*, 345–352. [CrossRef]
6. Zhou, C.; Xu, D.; Lin, K.; Sun, C.; Yang, X. Intelligent Feeding Control Methods in Aquaculture with an Emphasis on Fish: A Review. *Rev. Aquac.* **2018**, *10*, 975–993. [CrossRef]

7. Eljasik, P.; Panicz, R.; Sobczak, M.; Sadowski, J. Key Performance Indicators of Common Carp (*Cyprinus carpio* L.) Wintering in a Pond and RAS under Different Feeding Schemes. *Sustainability* **2022**, *14*, 3724. [CrossRef]
8. Nasr, M.A.F.; Reda, R.M.; Ismail, T.A.; Moustafa, A. Growth, Hemato-Biochemical Parameters, Body Composition, and Myostatin Gene Expression of *Clarias gariepinus* Fed by Replacing Fishmeal with Plant Protein. *Animals* **2021**, *11*, 889. [CrossRef] [PubMed]
9. Zhao, W.; Liu, Z.-L.; Niu, J. Growth Performance, Intestinal Histomorphology, Body Composition, Hematological and Antioxidant Parameters of *Oncorhynchus mykiss* Were Not Detrimentally Affected by Replacement of Fish Meal with Concentrated Dephenolization Cottonseed Protein. *Aquac. Rep.* **2021**, *19*, 100557. [CrossRef]
10. Tusche, K.; Nagel, F.; Arning, S.; Wuertz, S.; Susenbeth, A.; Schulz, C. Effect of Different Dietary Levels of Potato Protein Concentrate Supplemented with Feed Attractants on Growth Performance of Rainbow Trout (*Oncorhynchus mykiss*). *Anim. Feed. Sci. Technol.* **2013**, *183*, 202–209. [CrossRef]
11. Manhar, A.K.; Bashir, Y.; Saikia, D.; Nath, D.; Gupta, K.; Konwar, B.K.; Mandal, M. Cellulolytic potential of probiotic *Bacillus subtilis* AMS6 isolated from traditional fermented soybean (Churpi): An in-vitro study with regards to application as an animal feed additive. *Microbiol. Res.* **2016**, *186*, 62–70. [CrossRef]
12. Wuertz, S.; Schroeder, A.; Wanka, K.M. Probiotics in fish nutrition—Long-standing household remedy or native nutraceuticals? *Water* **2021**, *13*, 1348. [CrossRef]
13. Maity, J.; Kundu, J.; Pramanik, A.; Patra, B.C. Effect of cellulolytic gut bacteria as a feed supplement on the growth performance and nutrient digestibility of Asian seabass (*Lates calcarifer*). *Int. J. Aquat. Sci.* **2011**, *2*, 3–15.
14. Ibrahem, M.D. Evolution of Probiotics in Aquatic World: Potential Effects, the Current Status in Egypt and Recent Prospectives. *J. Adv. Res.* **2015**, *6*, 765–791. [CrossRef]
15. Kuebutornye, F.K.A.; Abarike, E.D.; Lu, Y. A Review on the Application of Bacillus as Probiotics in Aquaculture. *Fish Shellfish Immunol.* **2019**, *87*, 820–828. [CrossRef] [PubMed]
16. Hill, C.; Guarner, F.; Reid, G.; Gibson, G.R.; Merenstein, D.J.; Pot, B.; Morelli, L.; Canani, R.B.; Flint, H.J.; Salminen, S.; et al. The International Scientific Association for Probiotics and Prebiotics Consensus Statement on the Scope and Appropriate Use of the Term Probiotic. *Nat. Rev. Gastroenterol. Hepatol.* **2014**, *11*, 506–514. [CrossRef]
17. Monzón-Atienza, L.; Bravo, J.; Serradell, A.; Montero, D.; Gómez-Mercader, A.; Acosta, F. Current Status of Probiotics in European Sea Bass Aquaculture as One Important Mediterranean and Atlantic Commercial Species: A Review. *Animals* **2023**, *13*, 2369. [CrossRef]
18. Monzón-Atienza, L.; Bravo, J.; Torrecillas, S.; Gómez-Mercader, A.; Montero, D.; Ramos-Vivas, J.; Galindo-Villegas, J.; Acosta, F. An In-Depth Study on the Inhibition of Quorum Sensing by *Bacillus velezensis* D-18: Its Significant Impact on Vibrio Biofilm Formation in Aquaculture. *Microorganisms* **2024**, *12*, 890. [CrossRef]
19. Kuebutornye, F.K.A.; Abarike, E.D.; Lu, Y.; Hlordzi, V.; Sakyi, M.E.; Afriyie, G.; Wang, Z.; Li, Y.; Xie, C.X. Mechanisms and the Role of Probiotic *Bacillus* in Mitigating Fish Pathogens in Aquaculture. *Fish Physiol. Biochem.* **2020**, *46*, 819–841. [CrossRef]
20. Zhou, X.; Wang, Y.; Li, W. Effect of Probiotic on Larvae Shrimp (*Penaeus vannamei*) Based on Water Quality, Survival Rate and Digestive Enzyme Activities. *Aquaculture* **2009**, *287*, 349–353. [CrossRef]
21. Liu, H.; Wang, S.; Cai, Y.; Guo, X.; Cao, Z.; Zhang, Y.; Liu, S.; Yuan, W.; Zhu, W.; Zheng, Y.; et al. Dietary Administration of Bacillus Subtilis HAINUP40 Enhances Growth, Digestive Enzyme Activities, Innate Immune Responses and Disease Resistance of Tilapia, *Oreochromis niloticus*. *Fish Shellfish Immunol.* **2017**, *60*, 326–333. [CrossRef]
22. Santos, R.A.; Oliva-Teles, A.; Pousão-Ferreira, P.; Jerusik, R.; Saavedra, M.J.; Enes, P.; Serra, C.R. Isolation and Characterization of Fish-Gut *Bacillus* spp. as Source of Natural Antimicrobial Compounds to Fight Aquaculture Bacterial Diseases. *Mar. Biotechnol.* **2021**, *23*, 276–293. [CrossRef]
23. Sankar, H.; Philip, B.; Philip, R.; Singh, I.S.B. Effect of Probiotics on Digestive Enzyme Activities and Growth of Cichlids, *Etroplus suratensis* (Pearl Spot) and *Oreochromis mossambicus* (Tilapia). *Aquacult. Nutr.* **2017**, *23*, 852–864. [CrossRef]
24. Adorian, T.J.; Jamali, H.; Farsani, H.G.; Darvishi, P.; Hasanpour, S.; Bagheri, T.; Roozbehfar, R. Effects of Probiotic Bacteria *Bacillus* on Growth Performance, Digestive Enzyme Activity, and Hematological Parameters of Asian Sea Bass, *Lates calcarifer* (Bloch). *Probiotics Antimicrob. Proteins* **2019**, *11*, 248–255. [CrossRef]
25. Sumon, M.S.; Ahmmed, F.; Khushi, S.S.; Ahmmed, M.K.; Rouf, M.A.; Chisty, A.H.; Sarower, G. Growth Performance, Digestive Enzyme Activity and Immune Response of *Macrobrachium rosenbergii* Fed with Probiotic *Clostridium butyricum* Incorporated Diets. *J. King Saud Univ. Sci.* **2018**, *30*, 21–28. [CrossRef]
26. Afrilasari, W.; Widanarni; Meryandini, A. Effect of Probiotic *Bacillus megaterium* PTB 1.4 on the Population of Intestinal Microflora, Digestive Enzyme Activity and the Growth of Catfish (*Clarias* sp.). *HAYATI J. Biosci.* **2016**, *23*, 168–172. [CrossRef]
27. Ringø, E.; Zhou, Z.; Vecino, J.L.G.; Wadsworth, S.; Romero, J.; Krogdahl, A.; Olsen, R.E.; Dimitroglou, A.; Foey, A.; Davies, S.; et al. Effect of Dietary Components on the Gut Microbiota of Aquatic Animals. A Never-Ending Story? *Aquac. Nutr.* **2016**, *22*, 219–282. [CrossRef]

28. Martínez Cruz, P.; Ibáñez, A.L.; Monroy Hermosillo, O.A.; Ramírez Saad, H.C. Use of Probiotics in Aquaculture. *ISRN Microbiol.* **2012**, *2012*, 916845. [CrossRef]
29. Liu, X.-F.; Li, Y.; Li, J.-R.; Cai, L.-Y.; Li, X.-X.; Chen, J.-R.; Lyu, S.-X. Isolation and Characterisation of *Bacillus* spp. Antagonistic to *Vibrio parahaemolyticus* for Use as Probiotics in Aquaculture. *World J. Microbiol. Biotechnol.* **2015**, *31*, 795–803. [CrossRef]
30. Meidong, R.; Khotchanalekha, K.; Doolgindachbaporn, S.; Nagasawa, T.; Nakao, M.; Sakai, K.; Tongpim, S. Evaluation of Probiotic *Bacillus aerius* B81e Isolated from Healthy Hybrid Catfish on Growth, Disease Resistance and Innate Immunity of Pla-Mong *Pangasius bocourti*. *Fish Shellfish Immunol.* **2018**, *73*, 1–10. [CrossRef]
31. Lee, S.; Katya, K.; Park, Y.; Won, S.; Seong, M.; Hamidoghli, A.; Bai, S.C. Comparative Evaluation of Dietary Probiotics *Bacillus subtilis* WB60 and *Lactobacillus plantarum* KCTC3928 on the Growth Performance, Immunological Parameters, Gut Morphology and Disease Resistance in Japanese Eel, *Anguilla japonica*. *Fish Shellfish Immunol.* **2017**, *61*, 201–210. [CrossRef]
32. Niu, K.-M.; Kothari, D.; Lee, W.-D.; Lim, J.-M.; Khosravi, S.; Lee, S.-M.; Lee, B.-J.; Kim, K.-W.; Han, H.-S.; Kim, S.-K. Autochthonous *Bacillus licheniformis*: Probiotic Potential and Survival Ability in Low-Fishmeal Extruded Pellet Aquafeed. *Microbiologyopen* **2019**, *8*, e00767. [CrossRef]
33. Yaylacı, E.U. Isolation and Characterization of *Bacillus* spp. from Aquaculture Cage Water and Its Inhibitory Effect against Selected *Vibrio* spp. *Arch. Microbiol.* **2021**, *204*, 26. [CrossRef]
34. Nakharuthai, C.; Boonanuntanasarn, S.; Kaewda, J.; Manassila, P. Isolation of Potential Probiotic *Bacillus* spp. from the Intestine of Nile Tilapia to Construct Recombinant Probiotic Expressing CC Chemokine and Its Effectiveness on Innate Immune Responses in Nile Tilapia. *Animals* **2023**, *13*, 986. [CrossRef] [PubMed]
35. Elmnasser, N.; Hassen, W.; Zmantar, T.; Ashraf, S.A.; Hadj Lajimi, R.; Humaidi, J.R.; Alreshidi, M.; Hamadou, W.S.; Emira, N.; Snoussi, M. Antagonistic and Enzymatic Activities of Bacillus Species Isolated from the Fish Gastrointestinal Tract as Potential Probiotics Use in Artemia Culture. *Cell. Mol. Biol.* **2024**, *70*, 48–58. [CrossRef]
36. Tan, H.Y.; Chen, S.-W.; Hu, S.-Y. Improvements in the Growth Performance, Immunity, Disease Resistance, and Gut Microbiota by the Probiotic *Rummeliibacillus stabekisii* in Nile Tilapia (*Oreochromis niloticus*). *Fish Shellfish Immunol.* **2019**, *92*, 265–275. [CrossRef]
37. Saputra, F.; Shiu, Y.-L.; Chen, Y.-C.; Puspitasari, A.W.; Danata, R.H.; Liu, C.-H.; Hu, S.-Y. Dietary Supplementation with Xylanase-Expressing *B. Amyloliquefaciens* R8 Improves Growth Performance and Enhances Immunity against *Aeromonas hydrophila* in Nile Tilapia (*Oreochromis niloticus*). *Fish Shellfish Immunol.* **2016**, *58*, 397–405. [CrossRef] [PubMed]
38. Zuo, Z.-H.; Shang, B.-J.; Shao, Y.-C.; Li, W.-Y.; Sun, J.-S. Screening of Intestinal Probiotics and the Effects of Feeding Probiotics on the Growth, Immune, Digestive Enzyme Activity and Intestinal Flora of *Litopenaeus vannamei*. *Fish Shellfish Immunol.* **2019**, *86*, 160–168. [CrossRef]
39. Yi, C.-C.; Liu, C.-H.; Chuang, K.-P.; Chang, Y.-T.; Hu, S.-Y. A Potential Probiotic *Chromobacterium aquaticum* with Bacteriocin-like Activity Enhances the Expression of Indicator Genes Associated with Nutrient Metabolism, Growth Performance and Innate Immunity against Pathogen Infections in Zebrafish (*Danio rerio*). *Fish Shellfish Immunol.* **2019**, *93*, 124–134. [CrossRef]
40. Lin, Y.-S.; Saputra, F.; Chen, Y.-C.; Hu, S.-Y. Dietary Administration of Bacillus Amyloliquefaciens R8 Reduces Hepatic Oxidative Stress and Enhances Nutrient Metabolism and Immunity against *Aeromonas hydrophila* and *Streptococcus agalactiae* in Zebrafish (*Danio rerio*). *Fish Shellfish Immunol.* **2019**, *86*, 410–419. [CrossRef]
41. Hmani, H.; Daoud, L.; Jlidi, M.; Jalleli, K.; Ben Ali, M.; Hadj Brahim, A.; Bargui, M.; Dammak, A.; Ben Ali, M. A *Bacillus subtilis* Strain as Probiotic in Poultry: Selection Based on in Vitro Functional Properties and Enzymatic Potentialities. *J. Ind. Microbiol. Biotechnol.* **2017**, *44*, 1157–1166. [CrossRef]
42. Midhun, S.J.; Neethu, S.; Vysakh, A.; Sunil, M.A.; Radhakrishnan, E.K.; Jyothis, M. Antibacterial Activity of Autochthonous Bacteria Isolated from *Anabas testudineus* (Bloch, 1792) and It's in Vitro Probiotic Characterization. *Microb. Pathog.* **2017**, *113*, 312–320. [CrossRef]
43. Olmos Soto, J. Bacillus Probiotic Enzymes: External Auxiliary Apparatus to Avoid Digestive Deficiencies, Water Pollution, Diseases, and Economic Problems in Marine Cultivated Animals. *Adv. Food Nutr. Res.* **2017**, *80*, 15–35. [CrossRef] [PubMed]
44. Mohammadi, G.; Hafezieh, M.; Karimi, A.A.; Azra, M.N.; Van Doan, H.; Tapingkae, W.; Abdelrahman, H.A.; Dawood, M.A.O. The Synergistic Effects of Plant Polysaccharide and *Pediococcus acidilactici* as a Synbiotic Additive on Growth, Antioxidant Status, Immune Response, and Resistance of Nile Tilapia (*Oreochromis niloticus*) against *Aeromonas hydrophila*. *Fish Shellfish Immunol.* **2022**, *120*, 304–313. [CrossRef]
45. Torres-Maravilla, E.; Parra, M.; Maisey, K.; Vargas, R.A.; Cabezas-Cruz, A.; Gonzalez, A.; Tello, M.; Bermúdez-Humarán, L.G. Importance of Probiotics in Fish Aquaculture: Towards the Identification and Design of Novel Probiotics. *Microorganisms* **2024**, *12*, 626. [CrossRef] [PubMed]
46. Yang, S.; Du, J.; Luo, J.; Zhou, Y.; Long, Y.; Xu, G.; Zhao, L.; Du, Z.; Yan, T. Effects of Different Diets on the In-testinal Microbiota and Immunity of Common Carp (*Cyprinus carpio*). *J. Appl. Microbiol.* **2019**, *127*, 1327–1338. [CrossRef] [PubMed]

47. Setlow, P. Spores of *Bacillus subtilis*: Their Resistance to and Killing by Radiation, Heat and Chemicals. *J. Appl. Microbiol.* **2006**, *101*, 514–525. [CrossRef]
48. Li, X.; Kong, R.; Wang, J.; Wu, J.; He, K.; Wang, X. The Formation Mechanism of *Bacillus subtilis* Biofilm Surface Morphology under Competitive Environment. *Can. J. Microbiol.* **2023**, *69*, 251–261. [CrossRef]
49. Wang, J.; Li, X.; Kong, R.; Wu, J.; Wang, X. Fractal Morphology Facilitates *Bacillus subtilis* Biofilm Growth. *Environ. Sci. Pollut. Res.* **2022**, *29*, 56168–56177. [CrossRef]
50. Wang, Y.; Tyler, B.M.; Wang, Y. Defense and Counterdefense During Plant-Pathogenic Oomycete Infection. *Annu. Rev. Microbiol.* **2019**, *73*, 667–696. [CrossRef]
51. Mazanko, M.S.; Gorlov, I.F.; Prazdnova, E.V.; Makarenko, M.S.; Usatov, A.V.; Bren, A.B.; Chistyakov, V.A.; Tutelyan, A.V.; Komarova, Z.B.; Mosolova, N.I.; et al. Bacillus Probiotic Supplementations Improve Laying Performance, Egg Quality, Hatching of Laying Hens, and Sperm Quality of Roosters. *Probiotics Antimicrob. Proteins* **2018**, *10*, 367–373. [CrossRef]
52. Prazdnova, E.V.; Mazanko, M.S.; Chistyakov, V.A.; Denisenko, Y.V.; Makarenko, M.S.; Usatov, A.V.; Bren, A.B.; Tutelyan, A.V.; Komarova, Z.B.; Gorlov, I.F.; et al. Effect of *Bacillus subtilis* KATMIRA1933 and *Bacillus amyloliquefaciens* B-1895 on the Productivity, Reproductive Aging, and Physiological Characteristics of Hens and Roosters. *Benef. Microbes* **2019**, *10*, 395–412. [CrossRef]
53. Ponomareva, E.N.; Sorokina, M.N.; Grigoriev, V.A.; Mazanko, M.; Chistyakov, V.A.; Rudoy, D.V. Probiotic *Bacillus amyloliquefaciens* B-1895 Improved Growth of Juvenile Trout. *Food Sci. Anim. Resour.* **2024**, *44*, 805–816. [CrossRef] [PubMed]
54. Soltani, M.; Pakzad, K.; Taheri-Mirghaed, A.; Mirzargar, S.; Shekarabi, S.P.H.; Yosefi, P.; Soleymani, N. Dietary Application of the Probiotic *Lactobacillus plantarum* 426951 Enhances Immune Status and Growth of Rainbow Trout (*Oncorhynchus mykiss*) Vaccinated Against *Yersinia ruckeri*. *Probiotics Antimicrob. Proteins* **2019**, *11*, 207–219. [CrossRef]
55. Oyugi, D.O.; Cucherousset, J.; Baker, D.J.; Britton, J.R. Effects of Temperature on the Foraging and Growth Rate of Juvenile Common Carp, *Cyprinus carpio*. *J. Therm. Biol.* **2012**, *37*, 89–94. [CrossRef]
56. Heydarnejad, M. Survival and Growth of Common Carp (*Cyprinus carpio* L.) Exposed to Different Water pH Levels. *Turk. J. Vet. Anim. Sci.* **2012**, *36*, 245–249. [CrossRef]
57. Ghodrati, M.; Hosseini Shekarabi, S.P.; Rajabi Islami, H.; Shenavar Masouleh, A.; Shamsaie Mehrgan, M. Singular or Combined Dietary Administration of Multi-strain Probiotics and Multi-enzyme Influences Growth, Body Composition, Digestive Enzyme Activity, and Intestinal Morphology in Siberian Sturgeon (*Acipenser baerii*). *Aquacult. Nutr.* **2021**, *27*, 966–976. [CrossRef]
58. Pourgholam, M.A.; Khara, H.; Safari, R.; Sadati, M.A.Y.; Aramli, M.S. Dietary Administration of *Lactobacillus plantarum* Enhanced Growth Performance and Innate Immune Response of Siberian Sturgeon, *Acipenser baerii*. *Probiotics Antimicrob. Proteins* **2016**, *8*, 1–7. [CrossRef] [PubMed]
59. Gobi, N.; Vaseeharan, B.; Chen, J.-C.; Rekha, R.; Vijayakumar, S.; Anjugam, M.; Iswarya, A. Dietary Supplementation of Probiotic *Bacillus licheniformis* Dahb1 Improves Growth Performance, Mucus and Serum Immune Parameters, Antioxidant Enzyme Activity as Well as Resistance against *Aeromonas hydrophila* in Tilapia *Oreochromis mossambicus*. *Fish Shellfish Immunol.* **2018**, *74*, 501–508. [CrossRef]
60. Geraylou, Z.; Souffreau, C.; Rurangwa, E.; De Meester, L.; Courtin, C.M.; Delcour, J.A.; Buyse, J.; Ollevier, F. Effects of Dietary Arabinoxylan-Oligosaccharides (AXOS) and Endogenous Probiotics on the Growth Performance, Non-Specific Immunity and Gut Microbiota of Juvenile Siberian Sturgeon (*Acipenser baerii*). *Fish Shellfish Immunol.* **2013**, *35*, 766–775. [CrossRef]
61. Di, J.; Chu, Z.; Zhang, S.; Huang, J.; Du, H.; Wei, Q. Evaluation of the Potential Probiotic *Bacillus subtilis* Isolated from Two Ancient Sturgeons on Growth Performance, Serum Immunity and Disease Resistance of *Acipenser dabryanus*. *Fish Shellfish Immunol.* **2019**, *93*, 711–719. [CrossRef]
62. Darafsh, F.; Soltani, M.; Abdolhay, H.A.; Shamsaei Mehrejan, M. Improvement of Growth Performance, Digestive Enzymes and Body Composition of Persian Sturgeon (*Acipenser persicus*) Following Feeding on Probiotics: *Bacillus licheniformis*, *Bacillus subtilis* and *Saccharomyces cerevisiae*. *Aquac. Res.* **2020**, *51*, 957–964. [CrossRef]
63. Gautam, V.A. DNA and RNA Isolation Techniques for Non-Experts. In *Techniques in Life Science and Biomedicine for the Non-Expert*; Springer Nature: Berlin/Heidelberg, Germany, 2022.
64. Grant, J.R.; Enns, E.; Marinier, E.; Mandal, A.; Herman, E.K.; Chen, C.-Y.; Graham, M.; Van Domselaar, G.; Stothard, P. Proksee: In-Depth Characterization and Visualization of Bacterial Genomes. *Nucleic Acids Res.* **2023**, *51*, W484–W492. [CrossRef]
65. Schwengers, O.; Jelonek, L.; Dieckmann, M.A.; Beyvers, S.; Blom, J.; Goesmann, A. Bakta: Rapid and Standardized Annotation of Bacterial Genomes via Alignment-Free Sequence Identification. *Microb. Genom.* **2021**, *7*, 000685. [CrossRef] [PubMed]
66. Alcock, B.P.; Huynh, W.; Chalil, R.; Smith, K.W.; Raphenya, A.R.; Wlodarski, M.A.; Edalatmand, A.; Petkau, A.; Syed, S.A.; Tsang, K.K.; et al. CARD 2023: Expanded Curation, Support for Machine Learning, and Resistome Prediction at the Comprehensive Antibiotic Resistance Database. *Nucleic Acids Res.* **2023**, *51*, D690–D699. [CrossRef]
67. Kanehisa, M.; Sato, Y.; Morishima, K. BlastKOALA and GhostKOALA: KEGG Tools for Functional Characterization of Genome and Metagenome Sequences. *J. Mol. Biol.* **2016**, *428*, 726–731. [CrossRef]

68. Blin, K.; Shaw, S.; Augustijn, H.E.; Reitz, Z.L.; Biermann, F.; Alanjary, M.; Fetter, A.; Terlouw, B.R.; Metcalf, W.W.; Helfrich, E.J.N.; et al. antiSMASH 7.0: New and Improved Predictions for Detection, Regulation, Chemical Structures and Visualisation. *Nucleic Acids Res.* **2023**, *51*, W46–W50. [CrossRef]
69. Chaumeil, P.-A.; Mussig, A.J.; Hugenholtz, P.; Parks, D.H. GTDB-Tk v2: Memory Friendly Classification with the Genome Taxonomy Database. *Bioinformatics* **2022**, *38*, 5315–5316. [CrossRef] [PubMed]
70. Parks, D.H.; Chuvochina, M.; Rinke, C.; Mussig, A.J.; Chaumeil, P.-A.; Hugenholtz, P. GTDB: An Ongoing Census of Bacterial and Archaeal Diversity through a Phylogenetically Consistent, Rank Normalized and Complete Genome-Based Taxonomy. *Nucleic Acids Res.* **2022**, *50*, D785–D794. [CrossRef] [PubMed]
71. Letunic, I.; Bork, P. Interactive Tree Of Life (iTOL) v5: An Online Tool for Phylogenetic Tree Display and Annotation. *Nucleic Acids Res.* **2021**, *49*, W293–W296. [CrossRef]
72. Prasetyo, D.; Zubaidah, A.; Cahya Putra, R.D.; Anne, O.; Ariansyah, F. Growth Performance of Tilapia Fed Commercial Feed with Cellulolytic Bacteria from Ruminants. In Proceedings of the BIO Web of Conferences: The 3rd and 4th International Conference on Bioenergy and Environmentally Sustainable Agriculture Technology (ICoN BEAT 2022 and 2023), Les Ulis, France, 1 May 2024; Volume 104, p. 00009. [CrossRef]
73. Wang, Y.; Al Farraj, D.A.; Vijayaraghavan, P.; Hatamleh, A.A.; Biji, G.D.; Rady, A.M. Host Associated Mixed Probiotic Bacteria Induced Digestive Enzymes in the Gut of Tiger Shrimp Penaeus Monodon. *Saudi J. Biol. Sci.* **2020**, *27*, 2479–2484. [CrossRef]
74. Gänzle, M.G.; Follador, R. Metabolism of Oligosaccharides and Starch in Lactobacilli: A Review. *Front. Microbiol.* **2012**, *3*, 340. [CrossRef]
75. Blanco, P.; Hernando-Amado, S.; Reales-Calderon, J.A.; Corona, F.; Lira, F.; Alcalde-Rico, M.; Bernardini, A.; Sanchez, M.B.; Martinez, J.L. Bacterial Multidrug Efflux Pumps: Much More Than Antibiotic Resistance Determinants. *Microorganisms* **2016**, *4*, 14. [CrossRef] [PubMed]
76. Abdelsamad, A.E.M.; Said, R.E.M.; Assas, M.; Gaafar, A.Y.; Hamouda, A.H.; Mahdy, A. Effects of Dietary Supplementation with *Bacillus velezensis* on the Growth Performance, Body Composition, Antioxidant, Immune-Related Gene Expression, and Histology of Pacific White Shrimp, *Litopenaeus vannamei*. *BMC Vet. Res.* **2024**, *20*, 368. [CrossRef] [PubMed]
77. Barale, S.S.; Ghane, S.G.; Sonawane, K.D. Purification and Characterization of Antibacterial Surfactin Isoforms Produced by *Bacillus velezensis* SK. *AMB Express* **2022**, *12*, 7. [CrossRef] [PubMed]
78. Jeon, H.J.; Song, J.W.; Lee, C.; Kim, B.; Park, S.Y.; Kim, J.H.; Han, J.E.; Park, J.H. Antibacterial Activity of Bacillus Strains against Acute Hepatopancreatic Necrosis Disease-Causing *Vibrio campbellii* in Pacific White Leg Shrimp. *Fishes* **2022**, *7*, 287. [CrossRef]
79. Sam-On, M.F.S.; Mustafa, S.; Mohd Hashim, A.; Yusof, M.T.; Zulkifly, S.; Malek, A.Z.A.; Roslan, M.A.H.; Mohd Asrore, M.S. Mining the Genome of *Bacillus velezensis* FS26 for Probiotic Markers and Secondary Metabolites with Antimicrobial Properties against Aquaculture Pathogens. *Microb. Pathog.* **2023**, *181*, 106161. [CrossRef]
80. Yang, Q.; Zhang, H.; You, J.; Yang, J.; Zhang, Q.; Zhao, J.; Aimaier, R.; Zhang, J.; Han, S.; Zhao, H.; et al. Transcriptome and Metabolome Analyses Reveal That *Bacillus subtilis* BS-Z15 Lipopeptides Mycosubtilin Homologue Mediates Plant Defense Responses. *Front. Plant Sci.* **2022**, *13*, 1088220. [CrossRef]
81. Grady, E.N.; MacDonald, J.; Ho, M.T.; Weselowski, B.; McDowell, T.; Solomon, O.; Renaud, J.; Yuan, Z.-C. Characterization and Complete Genome Analysis of the Surfactin-Producing, Plant-Protecting Bacterium *Bacillus velezensis* 9D-6. *BMC Microbiol.* **2019**, *19*, 5. [CrossRef]
82. Gu, Q.; Yang, Y.; Yuan, Q.; Shi, G.; Wu, L.; Lou, Z.; Huo, R.; Wu, H.; Borriss, R.; Gao, X. Bacillomycin D Produced by *Bacillus amyloliquefaciens* Is Involved in the Antagonistic Interaction with the Plant-Pathogenic Fungus *Fusarium graminearum*. *Appl. Environ. Microbiol.* **2017**, *83*, e01075-17. [CrossRef]
83. Vibe, V.; Kulikov, M.; Prazdnova, E.; Mazanko, M.; Chistyakov, V.; Rudoy, D.; Shevchenko, V.; Kulikova, N. Effect of Growth Medium Composition on the Efficiency of Non-Ribosomal Synthesis in Bacteria of the Genus Bacillus. *BIO Web Conf.* **2024**, *113*, 02020. [CrossRef]
84. Sword, T.T.; Abbas, G.S.K.; Bailey, C.B. Cell-Free Protein Synthesis for Nonribosomal Peptide Synthetic Biology. *Front. Nat. Prod.* **2024**, *3*, 1353362. [CrossRef]
85. Yan, H.; Xin, Z.; Sang, Z.; Li, X.; Xie, J.; Wu, J.; Pang, S.; Wen, Y.; Wang, W. A Rational Multi-Target Combination Strategy for Synergistic Improvement of Non-Ribosomal Peptide Production. *Nat. Commun.* **2025**, *16*, 1883. [CrossRef]
86. Skripnichenko, R.V.; Chelombitskaya, D.S.; Prazdnova, E.V.; Kulikov, M.P.; Neurov, A.M.; Zaikina, A.A.; Grigoryev, V.A.; Sorokina, M.N.; Chistyakov, V.A.; Chikindas, M.L.; et al. Potential Probiotic Bacillus Strains with Antioxidant and Antimutagenic Activity Increased Weight Gain and Altered Hsp70, Cxc, Tnfα, Il1β, and lysC Gene Expression in *Clarias gariepinus*. *Fishes* **2024**, *9*, 476. [CrossRef]
87. Krysiak, K.; Konkol, D.; Korczyński, M. Overview of the Use of Probiotics in Poultry Production. *Animals* **2021**, *11*, 1620. [CrossRef] [PubMed]

88. Barba-Vidal, E.; Martín-Orúe, S.M.; Castillejos, L. Practical Aspects of the Use of Probiotics in Pig Production: A Review. *Livest. Sci.* **2019**, *223*, 84–96. [CrossRef]
89. Ma, L.; Wang, L.; Zhang, Z.; Xiao, D. Research Progress of Biological Feed in Beef Cattle. *Animals* **2023**, *13*, 2662. [CrossRef] [PubMed]
90. Van Doan, H.; Hoseinifar, S.H.; Ringø, E.; Ángeles Esteban, M.; Dadar, M.; Dawood, M.A.; Faggio, C. Host-associated probiotics: A key factor in sustainable aquaculture. *Rev. Fish. Sci. Aquac.* **2020**, *28*, 16–42. [CrossRef]
91. Zhang, Z.; Zhang, H.L.; Yang, D.H.; Hao, Q.; Yang, H.W.; Meng, D.L.; Zhou, Z.G. *Lactobacillus rhamnosus* GG triggers intestinal epithelium injury in zebrafish revealing host dependent beneficial effects. *IMeta* **2024**, *3*, e181. [CrossRef]
92. Rahayu, S.; Amoah, K.; Huang, Y.; Cai, J.; Wang, B.; Shija, V.M.; Jiang, M. Probiotics application in aquaculture: Its potential effects, current status in China and future prospects. *Front. Mar. Sci.* **2024**, *11*, 1455905. [CrossRef]
93. Sugiura, S.H. Evolutionary Loss of Acid-Secreting Stomach and Endoskeletal Ossification: A Phosphorus Perspective. *Fishes* **2025**, *10*, 48. [CrossRef]
94. Radchikov, V.F.; Astrenkov, A.V.; Gadlevskaya, N.N.; Prodennia, V.I.; Djulić, E.L.; Rohalskaia, S.U. Improvement of carp production efficiency through reduction of the cost of feed combinations. *Mech. Electrif. Agric.* **2022**, *55*, 52–56.
95. Huang, Z.; Hou, D.; Zhou, R.; Zeng, S.; Xing, C.; Wei, D.; He, J. Environmental water and sediment microbial communities shape intestine microbiota for host health: The central dogma in an anthropogenic aquaculture ecosystem. *Front. Microbiol.* **2021**, *12*, 772149. [CrossRef] [PubMed]
96. Zhang, B.; Xiao, J.; Liu, H.; Zhai, D.; Wang, Y.; Liu, S.; Xia, M. Vertical habitat preferences shape the fish gut microbiota in a shallow lake. *Front. Microbiol.* **2024**, *15*, 1341303. [CrossRef] [PubMed]
97. Khalid, F.; Khalid, A.; Fu, Y.; Hu, Q.; Zheng, Y.; Khan, S.; Wang, Z. Potential of *Bacillus velezensis* as a Probiotic in Animal Feed: A Review. *J. Microbiol.* **2021**, *59*, 627–633. [CrossRef] [PubMed]
98. Wang, X.; Deng, Z.; Gao, J. Exploring the Antibiotic Potential of Cultured 'Unculturable' Bacteria. *Trends Microbiol.* **2024**, *32*, 124–127. [CrossRef]
99. Li, J.; Wu, Z.-B.; Zhang, Z.; Zha, J.-W.; Qu, S.-Y.; Qi, X.-Z.; Wang, G.-X.; Ling, F. Effects of Potential Probiotic *Bacillus velezensis* K2 on Growth, Immunity and Resistance to *Vibrio harveyi* Infection of Hybrid Grouper (*Epinephelus lanceolatus*♂ × *E. fuscoguttatus*♀). *Fish Shellfish. Immunol.* **2019**, *93*, 1047–1055. [CrossRef]
100. Liu, X.; Zeng, S.; Liu, S.; Wang, G.; Lai, H.; Zhao, X.; Bi, S.; Guo, D.; Chen, X.; Yi, H.; et al. Identifying the Related Genes of Muscle Growth and Exploring the Functions by Compensatory Growth in Mandarin Fish (*Siniperca chuatsi*). *Front. Physiol.* **2020**, *11*, 553563. [CrossRef]
101. Yin, Y.; Zhang, Y.; Hua, Z.; Wu, A.; Pan, X.; Yang, J.; Wang, X. Muscle Transcriptome Analysis Provides New Insights into the Growth Gap between Fast- and Slow-Growing *Sinocyclocheilus grahami*. *Front. Genet.* **2023**, *14*, 1217952. [CrossRef]
102. Anee, I.J.; Alam, S.; Begum, R.A.; Shahjahan, R.M.; Khandaker, A.M. The Role of Probiotics on Animal Health and Nutrition. *J. Basic Appl. Zool.* **2021**, *82*, 52. [CrossRef]
103. Jlidi, M.; Akremi, I.; Ibrahim, A.H.; Brabra, W.; Ali, M.B.; Ali, M.B. Probiotic Properties of Bacillus Strains Isolated from the Gastrointestinal Tract against Pathogenic Vibriosis. *Front. Mar. Sci.* **2022**, *9*, 884244. [CrossRef]
104. Wu, P.-S.; Liu, C.-H.; Hu, S.-Y. Probiotic *Bacillus safensis* NPUST1 Administration Improves Growth Performance, Gut Microbiota, and Innate Immunity against *Streptococcus iniae* in Nile Tilapia (*Oreochromis niloticus*). *Microorganisms* **2021**, *9*, 2494. [CrossRef]
105. Assan, D.; Kuebutornye, F.K.A.; Hlordzi, V.; Chen, H.; Mraz, J.; Mustapha, U.F.; Abarike, E.D. Effects of Probiotics on Digestive Enzymes of Fish (Finfish and Shellfish); Status and Prospects: A Mini Review. *Comp. Biochem. Physiol. B Biochem. Mol. Biol.* **2022**, *257*, 110653. [CrossRef] [PubMed]
106. Kobayashi, K. Diverse LXG Toxin and Antitoxin Systems Specifically Mediate Intraspecies Competition in *Bacillus subtilis* Biofilms. *PLoS Genet* **2021**, *17*, e1009682. [CrossRef] [PubMed]
107. Yu, T.; Kong, J.; Zhang, L.; Gu, X.; Wang, M.; Guo, T. New Crosstalk between Probiotics *Lactobacillus plantarum* and *Bacillus subtilis*. *Sci. Rep.* **2019**, *9*, 13151. [CrossRef] [PubMed]
108. Payne, J.; Bellmer, D.; Jadeja, R.; Muriana, P. The Potential of Bacillus Species as Probiotics in the Food Industry: A Review. *Foods* **2024**, *13*, 2444. [CrossRef]
109. Mazanko, M.S.; Prazdnova, E.V.; Kulikov, M.P.; Maltseva, T.A.; Rudoy, D.V.; Chikindas, M.L. Antioxidant and Antimutagenic Properties of Probiotic Lactobacilli Determined Using LUX-Biosensors. *Enzym. Microb. Technol.* **2022**, *155*, 109980. [CrossRef]
110. Pang, X.; Fu, S.-J.; Zhang, Y.-G. Acclimation Temperature Alters the Relationship between Growth and Swimming Performance among Juvenile Common Carp (*Cyprinus carpio*). *Comp. Biochem. Physiol. A Mol. Integr. Physiol.* **2016**, *199*, 111–119. [CrossRef]
111. Yanbo, W.; Zirong, X. Effect of Probiotics for Common Carp (*Cyprinus carpio*) Based on Growth Performance and Digestive Enzyme Activities. *Anim. Feed. Sci. Technol.* **2006**, *127*, 283–292. [CrossRef]

112. Ji, Z.; Lu, X.; Xue, M.; Fan, Y.; Tian, J.; Dong, L.; Zhu, C.; Wen, H.; Jiang, M. The Probiotic Effects of Host-Associated *Bacillus velezensis* in Diets for Hybrid Yellow Catfish (*Pelteobagrus fulvidraco* ♀× *Pelteo-bagrus vachelli* ♂). *Anim. Nutr.* **2023**, *15*, 114–125. [CrossRef]
113. Chen, L.; Lv, C.; Li, B.; Zhang, H.; Ren, L.; Zhang, Q.; Zhang, X.; Gao, J.; Sun, C.; Hu, S. Effects of *Bacillus velezensis* Supplementation on the Growth Performance, Immune Responses, and Intestine Microbiota of *Litopenaeus vannamei*. *Front. Mar. Sci.* **2021**, *8*, 744281. [CrossRef]
114. Zhang, D.-X.; Kang, Y.-H.; Zhan, S.; Zhao, Z.-L.; Jin, S.-N.; Chen, C.; Zhang, L.; Shen, J.-Y.; Wang, C.-F.; Wang, G.-Q.; et al. Effect of *Bacillus velezensis* on *Aeromonas veronii*-Induced Intestinal Mucosal Barrier Function Damage and Inflammation in Crucian Carp (*Carassius auratus*). *Front. Microbiol.* **2019**, *10*, 2663. [CrossRef]
115. Amoah, K.; Tan, B.; Zhang, S.; Chi, S.; Yang, Q.; Liu, H.; Yang, Y.; Zhang, H.; Dong, X. Host Gut-Derived Bacillus Probiotics Supplementation Improves Growth Performance, Serum and Liver Immunity, Gut Health, and Resistive Capacity against *Vibrio harveyi* Infection in Hybrid Grouper (♀*Epinephelus fuscoguttatus* × ♂*Epinephelus lanceolatus*). *Anim. Nutr.* **2023**, *14*, 163–184. [CrossRef] [PubMed]
116. Ariyanto, Y.S.; Anika, M. Probiotics on Commercial Fish Growth: A Meta-Analysis. *J. Sumberd. Hayati* **2024**, *10*, 205–216. [CrossRef]
117. Chen, X.; Yi, H.; Liu, S.; Zhang, Y.; Su, Y.; Liu, X.; Bi, S.; Lai, H.; Zeng, Z.; Li, G. Probiotics Improve Eating Disorders in Mandarin Fish (*Siniperca chuatsi*) Induced by a Pellet Feed Diet via Stimulating Immunity and Regulating Gut Microbiota. *Microorganisms* **2021**, *9*, 1288. [CrossRef]
118. Lee, Y.; Nguyen, T.L.; Roh, H.; Kim, A.; Park, J.; Lee, J.-Y.; Kang, Y.-R.; Kang, H.; Sohn, M.-Y.; Park, C.-I.; et al. Mechanisms Underlying Probiotic Effects on Neurotransmission and Stress Resilience in Fish via Transcriptomic Profiling. *Fish Shellfish Immunol.* **2023**, *141*, 109063. [CrossRef] [PubMed]
119. Zhang, J.; Huang, M.; Feng, J.; Chen, Y.; Li, M.; Meng, X.; Chang, X. Effect of Beneficial Colonization of *Bacillus coagulans* NRS 609 on Growth Performance, Intestinal Health, Antioxidant Capacity, and Immune Response of Common Carp (*Cyprinus carpio* L.). *Aquac. Nutr.* **2023**, *2023*, 1451394. [CrossRef]
120. Agrawal, S.; Acharya, D.; Adholeya, A.; Barrow, C.J.; Deshmukh, S.K. Nonribosomal Peptides from Marine Microbes and Their Antimicrobial and Anticancer Potential. *Front. Pharmacol.* **2017**, *8*, 828. [CrossRef]
121. Hoste, A.C.R.; Smeralda, W.; Cugnet, A.; Brostaux, Y.; Deleu, M.; Garigliany, M.; Jacques, P. The Structure of Lipopeptides Impacts Their Antiviral Activity and Mode of Action against SARS-CoV-2 in Vitro. *Appl. Environ. Microbiol.* **2024**, *90*, e0103624. [CrossRef]
122. Leistikow, K.R.; May, D.S.; Suh, W.S.; Vargas Asensio, G.; Schaenzer, A.J.; Currie, C.R.; Hristova, K.R. *Bacillus subtilis*-Derived Peptides Disrupt Quorum Sensing and Biofilm Assembly in Multidrug-Resistant *Staphylococcus aureus*. *mSystems* **2024**, *9*, e00712-24. [CrossRef]
123. Horng, Y.-B.; Yu, Y.-H.; Dybus, A.; Hsiao, F.S.-H.; Cheng, Y.-H. Antibacterial Activity of Bacillus Species-Derived Surfactin on *Brachyspira hyodysenteriae* and *Clostridium perfringens*. *AMB Express* **2019**, *9*, 188. [CrossRef]
124. Wu, L.; Wu, H.; Chen, L.; Yu, X.; Borriss, R.; Gao, X. Difficidin and Bacilysin from *Bacillus amyloliquefaciens* FZB42 Have Antibacterial Activity against *Xanthomonas oryzae* Rice Pathogens. *Sci. Rep.* **2015**, *5*, 12975. [CrossRef]
125. Nannan, C.; Vu, H.Q.; Gillis, A.; Caulier, S.; Nguyen, T.T.T.; Mahillon, J. Bacilysin within the *Bacillus subtilis* Group: Gene Prevalence versus Antagonistic Activity against Gram-Negative Foodborne Pathogens. *J. Biotechnol.* **2021**, *327*, 28–35. [CrossRef]
126. Li, H.; Han, X.; Dong, Y.; Xu, S.; Chen, C.; Feng, Y.; Cui, Q.; Li, W. Bacillaenes: Decomposition Trigger Point and Biofilm Enhancement in Bacillus. *ACS Omega* **2021**, *6*, 1093–1098. [CrossRef]
127. Chakraborty, K.; Kizhakkekalam, V.K.; Joy, M.; Dhara, S. Difficidin Class of Polyketide Antibiotics from Marine Macroalga-Associated Bacillus as Promising Antibacterial Agents. *Appl. Microbiol. Biotechnol.* **2021**, *105*, 6395–6408. [CrossRef]
128. Yuan, J.; Zhao, M.; Li, R.; Huang, Q.; Rensing, C.; Raza, W.; Shen, Q. Antibacterial Compounds-Macrolactin Alters the Soil Bacterial Community and Abundance of the Gene Encoding PKS. *Front. Microbiol.* **2016**, *7*, 1904. [CrossRef] [PubMed]
129. Yu, C.; Liu, X.; Zhang, X.; Zhang, M.; Gu, Y.; Ali, Q.; Mohamed, M.S.R.; Xu, J.; Shi, J.; Gao, X.; et al. Mycosubtilin Produced by *Bacillus subtilis* ATCC6633 Inhibits Growth and Mycotoxin Biosynthesis of *Fusarium graminearum* and *Fusarium verticillioides*. *Toxins* **2021**, *13*, 791. [CrossRef] [PubMed]
130. Xu, Z.; Shao, J.; Li, B.; Yan, X.; Shen, Q.; Zhang, R. Contribution of Bacillomycin D in *Bacillus amyloliquefaciens* SQR9 to Antifungal Activity and Biofilm Formation. *Appl. Environ. Microbiol.* **2013**, *79*, 808–815. [CrossRef] [PubMed]
131. Nam, J.; Alam, S.T.; Kang, K.; Choi, J.; Seo, M.-H. Anti-Staphylococcal Activity of a Cyclic Lipopeptide, C15 -Bacillomycin D, Produced by *Bacillus velezensis* NST6. *J. Appl. Microbiol.* **2021**, *131*, 93–104. [CrossRef]
132. Lee, D.; Oh, T.; Kang, B.; Ahn, J.S.; Cho, Y. Throughput Screening of *Bacillus subtilis* Strains That Abundantly Secrete Surfactin in Vitro Identifies Effective Probiotic Candidates. *PLoS ONE* **2022**, *17*, e0277412. [CrossRef]

133. Neurov, A.M.; Zaikina, A.A.; Prazdnova, E.V.; Anuj, R.; Rudoy, D.V. Modulation of Stress-Related Protein in the African Catfish (*Clarias gariepinus*) Using Bacillus-Based Non-Ribosomal Peptides. *Microbiol. Res.* **2024**, *15*, 2743–2763. [CrossRef]
134. Yang, H.; Wu, J.; Du, H.; Zhang, H.; Li, J.; Wei, Q. Quantifying the Colonization of Environmental Microbes in the Fish Gut: A Case Study of Wild Fish Populations in the Yangtze River. *Front. Microbiol.* **2022**, *12*, 828409. [CrossRef]
135. Chao, R.; Abel, C.; Malachi, A.W.; Nancy, C.; Bui, C.T.D.; Joseph, C.N.; Mark, R.L. Identification of Bacillus strains for biological control of catfish pathogens. *PLoS ONE* **2012**, *7*, 1–9.
136. Giatsis, C.; Sipkema, D.; Ramiro-Garcia, J.; Bacanu, G.M.; Abernathy, J.; Verreth, J.; Verdegem, M. Probiotic legacy effects on gut microbial assembly in tilapia larvae. *Sci. Rep.* **2016**, *6*, 33965. [CrossRef]
137. Asaduzzaman, M.; Sofia, E.; Shakil, A.; Haque, N.F.; Khan, M.N.A.; Ikeda, D.; Kinoshita, S.; Abol-Munafi, A.B. Host Gut-Derived Probiotic Bacteria Promote Hypertrophic Muscle Progression and Upregulate Growth-Related Gene Expression of Slow-Growing Malaysian Mahseer *Tor tambroides*. *Aquac. Rep.* **2018**, *9*, 37–45. [CrossRef]
138. Sadeghi, J.; Chaganti, S.R.; Heath, D.D. Regulation of Host Gene Expression by Gastrointestinal Tract Microbiota in Chinook Salmon (*Oncorhynchus tshawytscha*). *Mol. Ecol.* **2023**, *32*, 4427–4446. [CrossRef] [PubMed]
139. Hemarajata, P.; Versalovic, J. Effects of Probiotics on Gut Microbiota: Mechanisms of Intestinal Immunomodulation and Neuromodulation. *Ther. Adv. Gastroenterol.* **2013**, *6*, 39–51. [CrossRef]
140. Hines, I.S.; Santiago-Morales, K.D.; Ferguson, C.S.; Clarington, J.; Thompson, M.; Rauschenbach, M.; Levine, U.; Drahos, D.; Aylward, F.O.; Smith, S.A.; et al. Steelhead Trout (*Oncorhynchus mykiss*) Fed Probiotic during the Earliest Developmental Stages Have Enhanced Growth Rates and Intestinal Microbiome Bacterial Diversity. *Front. Mar. Sci.* **2022**, *9*, 1021647. [CrossRef]
141. Ul Hassan, H.; Mohammad Ali, Q.; Ahmad, N.; Masood, Z.; Hossain, Y.; Gabol, K.; Khan, W.; Hussain, M.; Ali, A.; Attaullah, M.; et al. Assessment of Growth Characteristics, the Survival Rate and Body Composition of Asian Sea Bass *Lates calcarifer* (Bloch, 1790) under Different Feeding Rates in Closed Aquaculture System. *Saudi J. Biol. Sci.* **2021**, *28*, 1324–1330. [CrossRef]
142. Shadrack, R.S.; Manabu, I.; Yokoyama, S. Efficacy of Single and Mix Probiotic Bacteria Strain on Growth Indices, Physiological Condition and Bio-Chemical Composition of Juvenile Amberjack (*Seriola dumerili*). *Aquac. Rep.* **2021**, *20*, 100753. [CrossRef]
143. Speare, D.J.; MacNair, N.; Hammell, K.L. Demonstration of Tank Effect on Growth Indices of Juvenile Rainbow Trout (*Oncorhynchus mykiss*) during an Ad Libitum Feeding Trial. *Am. J. Vet. Res.* **1995**, *56*, 1372–1379. [CrossRef]
144. World Aquaculture Society. Meeting Abstract. Experimental Design and Statistical Power of Fish Growth Studies. Available online: https://www.was.org/MeetingAbstracts/ShowAbstract/42356 (accessed on 22 June 2025).
145. Rwezawula, P.; Waiswa Mwanja, W.; Vereecke, N.; Bossier, P.; Vanrompay, D. Advancing Aquaculture Probiotic Discovery via an Innovative Protocol for Isolation of Indigenous, Heat and Salt Tolerant, Quorum Quenching Probiotic Candidates. *Front. Microbiol.* **2025**, *16*, 1558238. [CrossRef]
146. Zhao, C.; Men, X.; Dang, Y.; Zhou, Y.; Ren, Y. Probiotics Mediate Intestinal Microbiome and Microbiota-Derived Metabolites Regulating the Growth and Immunity of Rainbow Trout (*Oncorhynchus mykiss*). *Microbiol. Spectr.* **2023**, *11*, e0398022. [CrossRef] [PubMed]
147. Serra, C.R.; Almeida, E.M.; Guerreiro, I.; Santos, R.; Merrifield, D.L.; Tavares, F.; Oliva-Teles, A.; Enes, P. Selection of Carbohydrate-Active Probiotics from the Gut of Carnivorous Fish Fed Plant-Based Diets. *Sci. Rep.* **2019**, *9*, 6384. [CrossRef] [PubMed]

MDPI AG
Grosspeteranlage 5
4052 Basel
Switzerland
Tel.: +41 61 683 77 34

Animals Editorial Office
E-mail: animals@mdpi.com
www.mdpi.com/journal/animals

www.ingramcontent.com/pod-product-compliance
Lightning Source LLC
Chambersburg PA
CBHW061621210926
51850CB00009B/2400
* 9 7 8 3 7 2 5 8 6 2 8 4 9 *